Y0-CAN-058

The Maunder Minimum and the Variable Sun-Earth Connection

The Maunder Minimum and the Variable Sun-Earth Connection

WILLIE WEI-HOCK SOON
Harvard-Smithsonian Center for Astrophysics, USA

STEVEN H. YASKELL

NEW JERSEY • LONDON • SINGAPORE • SHANGHAI • HONG KONG • TAIPEI • BANGALORE

Published by

World Scientific Publishing Co. Pte. Ltd.
5 Toh Tuck Link, Singapore 596224
USA office: Suite 202, 1060 Main Street, River Edge, NJ 07661
UK office: 57 Shelton Street, Covent Garden, London WC2H 9HE

British Library Cataloguing-in-Publication Data
A catalogue record for this book is available from the British Library.

THE MAUNDER MINIMUM AND THE VARIABLE SUN-EARTH CONNECTION

ISBN 981-238-274-7
ISBN 981-238-275-5 (pbk)

Typeset by Stallion Press.

Printed in Singapore by World Scientific Printers (S) Pte Ltd

To Soon Gim-Chuan, Chua Chiew-See, and
Pham Than (Lien & Van's mother) and Ulla and Anna

In Memory of Miriam Fuchs (baba Gil's mother) — W.S.
In Memory of Andrew Hoff — S.H.Y.

To interrupt His Yellow Plan
The Sun does not allow
Caprices of the Atmosphere —
And even when the Snow

Heaves Balls of Specks, like Vicious Boy
Directly in His Eye —
Does not so much as turn His Head
Busy with Majesty —

'Tis His to stimulate the Earth
And magnetize the Sea —
And bind Astronomy, in place,
Yet Any passing by

Would deem Ourselves — the busier
As the Minutest Bee
That rides — emits a Thunder —
A Bomb — to justify

Emily Dickinson (*poem 224. c. 1862*)

Since people are by nature poorly equipped to register any but short-term changes, it is not surprising that we fail to notice slower changes in either climate or the sun.

John A. Eddy, The New Solar Physics (1977–78)

Foreword

E. N. Parker

In this time of global warming we are impelled by both the anticipated dire consequences and by scientific curiosity to investigate the factors that drive the climate. Climate has fluctuated strongly and abruptly in the past, with ice ages and interglacial warming as the long term extremes. Historical research in the last decades has shown short term climatic transients to be a frequent occurrence, often imposing disastrous hardship on the afflicted human populations. The 17th century in North America, Europe and China provides an outstanding example of the onset of cold. The principal impact is on agriculture, i.e. the food supply, particularly near the northern limits. Famine begets social and political turmoil, as well as pestilence and death. It is as important to appreciate the social impact of the transient cold as it is to understand the cause of the cold. The onset of warm periods has a comparable destructive impact, but in other ways in other parts of the globe, e.g. the prolonged drought in what is now the Southwestern United States during the unusually warm 12th century.

Recent studies of ice cores from drilling through the Greenland ice cap show occasional strong transients, with the mean temperature dropping a couple of degrees or more over a decade, followed closely by a substantial reduction of atmospheric carbon dioxide. Is such a cold snap initiated by a temporary disruption of the Gulf Stream, for instance? And if so, then what disrupted the Gulf Stream? Then did a worldwide reduction in surface sea water temperatures consume the missing atmospheric carbon dioxide?

We may expect that there is no single cause driving a downturn in temperature, or driving the present upturn. The measured rapid increase of carbon dioxide in the atmosphere through the second half of the 20th century is largely a consequence of profligate burning of fossil fuels. We can only conclude that the greenhouse effect of the accumulating carbon dioxide (and other anthropogenic emissions) has a warming effect, while anthropogenic aerosols may push a little in the opposite direction. However, that leaves us wondering about the comparable warming through the first half of the century, when the accumulation of carbon dioxide was slow and slight. There we turn to the inconstant Sun.

The history of the scientific detective work that has led to the present partial understanding of the brightening Sun (since about 1880) is fascinating. The work really got underway with the invention of the telescope around 1610 and extends through the ongoing contemporary studies of atmospheric physics and of solar and stellar variability.

It is fair to say that the scientific inquiry thus far has established the enormous complexity of the 20th century warming problem. The two major drivers appear to be the varying brightness of the Sun and, of course, the accumulating greenhouse gases. These are imposed on the diverse dynamical modes of circulation of the terrestrial atmosphere and oceans. Indeed, it has been proposed by some that the decadal—and even century long—swings in climate might be nothing more than the natural consequence of nonstable modes of atmospheric circulation. That is to say, the climate cannot settle down because there is no truly stable mode into which to settle.

This view is perhaps too simple, but it makes the important point that the wandering and never-stable climate is hypersensitive to both external and internal stimuli. The slight deflection by a small tweak may divert the atmosphere into a diverging dynamical path. We would not go so far as to apply the popular saw from chaos theory that the fluttering of a butterfly in the rain forest of the Amazon basin may ultimately strongly influence the rainfall in Tennessee. We reject the applicability of this vivid overstatement for the simple reason that the atmosphere is buffeted so much more strongly by other effects. For instance, the eruption of Mt. Pinatubo represented a "butterfly" of much greater vigor than any Amazonian lepidopteron.

However, we cannot escape the fact that seemingly small effects may, by their special nature, ultimately lead to large consequences. Thus, in our search to understand the climatic changes, past and present, we should not immediately reject an exotic idea as absurd until we can be sure that we fully understand the sensitive atmosphere. And that full understanding of the dynamical atmosphere, with its 3D hydrodynamics, cloud formation, coupling to the dynamic oceans, and coupling to the mountainous topography, lies many years in the future. Thus we must give consideration to a variety of unsubstantiated, and even dubious, ideas. We must keep in mind that absence of understanding of an effect is not the same thing as understanding that the effect cannot be.

Many current ideas will ultimately prove to be false or inaccurate, just as many old ideas have already fallen by the way. One of the most frustrating aspects of the origins of climate variation is the intermittent nature of some of the effects. An example is the sometime solar correlation of the formation of troughs at the 100 mbar level over the north Pacific Ocean. The effect appears to come and go with the varying surface sea water temperatures through El Niño and La Niña. Or are we just kidding ourselves and there really is no connection with solar activity? Then what can we say about the once well established, but now long vanished, correlation of the level of Lake Rudolph (now Lake Turkana) with the solar sunspot cycle? Drought in the high prairie in the Western United States in association with the deep sunspot minimum that occurs every 22 years seems now to be a statistically robust effect, but the physical connection remains mysterious. So what are we to think?

H. Svensmark has recently pointed out the remarkably close correlation between cloud cover and the cosmic ray intensity, closer than with any other index of solar activity. The connection here would presumably be through the nucleation of aerosols, water drops, and ice crystals by the atmospheric ions created by the passage of cosmic ray particles. If the effect is real, it represents a highly leveraged control on climate. Unfortunately we do not know enough physics and chemistry of cloud formation to pass judgement at the present time.

Then to what extent does the temperature of the upper stratosphere influence the dynamics of the jet stream and the troposphere? The temperature of the upper stratosphere varies widely between solar maximum and minimum. Some atmospheric models have shown a substantial coupling to the dynamical behavior of the lower atmosphere, while others find little or no effect.

Progress in understanding cloud formation, the global dynamics of the atmosphere and oceans, the quantitative ocean-atmosphere carbon cycle, nitrogen cycle, etc, will eventually clear away the rubbish and leave the gems. However there are a lot of laboratory work and observations in the field before that will be possible. The worst mistake a scientist can make is to assert prematurely that some exotic new effect "cannot be" because our present limited knowledge does not cover the effect. Certainly some ideas are without a known physical basis at the present time. They should be catalogued under some such heading as "Curious idea, presently dubious, to be kept clearly in mind as the science progresses."

The history of ideas on terrestrial climate and weather change and the history of ideas on the magnetic activity of the Sun can be traced back to ancient times. Some of the historical interpretations of the observations bring smiles to our faces today. But it must be remembered that interpretation is the first essential step in setting up the possibility of negation, the crucial step in any scientific investigation.

The effort to establish a coherent body of knowledge of climate variation and solar activity variation has been fascinating and protracted. A connection between the two has been speculated off and on for at least a couple of centuries, but only with the concept of quantitative statistical correlations has it been possible to make serious progress, and only with the advance of physics and technology has it been possible to pursue the actual mechanisms. There is no single point in time when the science of the connection got underway, but the modern phase, connecting global climate with the activity of the Sun, was initiated by E. Walter Maunder just a little more than a century ago.

Maunder's principal contribution was to emphasize the varying level of solar activity over the centuries since the advent of the telescope, with particular attention to the scarcity of sunspots from about 1645 to 1715, beginning only 35 years after the start of telescopic observations. Maunder's point was conveniently ignored or even denied, because no one knew what it meant. Fortunately Jack Eddy took up the historical investigation about 30 years ago and turned up enough old records that the reality of the "Maunder Minimum" was established beyond any reasonable doubt. Subsequent historical research has unearthed detailed systematic records of sunspot numbers which show how peculiar the behavior of the Sun was during that time. Then with modern data on the atmospheric production rate of carbon 14 by cosmic rays, Eddy went on to show that such prolonged periods of solar inactivity have occurred ten times in the last 7000 years. So we may anticipate that there will be yet another Maunder Minima in the future. Finally Eddy showed that the mean annual temperature in the Northern Temperate Zone exhibits a remarkable tendency to track the general level of solar activity. And that set the stage for contemporary research into the Sun-climate connection.

Global warming has turned attention to the Sun-climate connection as part of the general warming process. At the same time, the superposition of the two effects, of accumulating greenhouse gases and the brighter more active Sun, has rendered the scientific problem of studying either aspect much more difficult. The consequences of greenhouse gases and solar variability are not readily separated. As already noted, an immense amount of laboratory and field work will have to be done if we are to disentangle and understand the complex response of the atmosphere.

Meanwhile, we are in the exciting position—perhaps a little too exciting if we stop to think about it—of living in the midst of the warming Earth. The social and economic impact has not yet touched us, the first warning tremors being the international political and economic efforts to reduce the emission of carbon dioxide, thereby slowing the rate of increase of carbon dioxide in the atmosphere. The warming is expected to be strongest at extreme high latitudes, melting the polar ice caps and raising the sea level. Thus we may anticipate desiccation of certain geographical regions in association with the warming, at the same time that the sea begins to encroach on some of the best crop lands. The enhanced carbon dioxide promises enhanced crop yields, but that seems a small gain when compared to the uncontrolled risks with which it is associated. Human society has embarked on an adventure through the 21st century and beyond, and we can only wonder where it will lead. The best preparation is to understand the history of climate changes and the physics of global warming up to the present time, so that we can, hopefully, make intelligent decisions in the face of the forthcoming social, economic, and political problems. We anticipate that the political problems will be the most difficult to manage, of course.

Acknowledgements

The authors wish to thank Mary T. Bruck, astronomer and expert on women in astronomy who read the entire manuscript, giving us valuable insights not only into the Maunders' private and scientific lives, but also good historical perspectives of England in Maunders' time. To have done so while her husband was very ill, and to have continued to do so even after his death was simply remarkable. In a similar way, we honor the memory of Dr. Jean Grove, who assisted us greatly in dismantling and re-explaining concepts on the "Little Ice Age" that are sure to widen all our understandings of past and future climate change. In addition, she commented on our manuscript at critical junctures in our modest attempts to help broaden the scope of the Sun-Earth connection study. Her sudden death saddened us.

We thank Eugene N. Parker, Emeritus Professor at the University of Chicago and delineator of the solar wind who read the entire manuscript with pleasure, and wrote the foreword to this book. He encouraged us with his words and tempered us with his sage advice and positive criticism—a gracious act on his behalf which gained our deepest gratitude and respect. Additionally, Kirill Ya. Kondratyev, an authority on climate and environmental sciences also read our book in draft form, strongly praising and supporting our effort.

We also want to thank Richard B. Marks, who assisted in straightening out certain details on the Ming Dynasty. Dr. Marks also gave us insights into the China of the Maunder Minimum period. Our best wishes extend to Victoria Holtby (Kings College, UK) who provided nearly-vanished information on E. Walter Maunder's education record. Then there is John M. McFarland (Armagh Observatory), who provided us with pictures of A.S.D. Maunder that could be redeveloped and scanned, since very few portrait photographs of this regrettably forgotten scientist exist. Also, we thank Alan Maunder, a direct descendant of Maunder's who read and approved parts of the manuscript relevant to his relative and filled in some dark spots—especially about Maunder's first wife, Edith.

With gratitude, we mention Dimitry Sokoloff and Douglas Hoyt, the former for his support and encouragement in reading our work, and the latter for his direct contributions to our efforts in the form of data. Bill (W.C.) Livingston and Dave Hathaway generously supplied us with superb images of the Sun used in this book. Tom Bogdan shared remarkable stories about the fate of the Maunders original art-works (the Butterfly diagram and Annie's 1898 eclipse photo of the structured solar corona) that are now under the safe-keeping at the High Altitude Observatory. Additionally, there were contributions from Melissa Hilbert, Maria McEachern, Barbara Palmer, William Graves, Donna Coletti (John Wolbach Library) and Ewa Basinska (M.I.T. Library) for invaluable assistance in researching the Harvard libraries for material that helped form the backbone of this work.

Last but not the least, S. Yaskell thanks Larry DiThomas and John Yaskell. W. Soon further thanks his colleague of many years, Sallie Baliunas, and his elder sisters (Diana Guk-Hua, Yoke-Thim, Yoke-Lay) and brothers (Wei-Lean and Wei-Sin) and Kamil Abdul Aziz, Gene Avrett, Shaun Cheok, Bob Ferguson, Peter Frick, Gil Fuchs, Richard Goh, Phil Gozalez, Lucy Hancock, Sharil Ibrahim, Robert Jastrow, Joe Kunc, Thu and Duyen Le, Dave Legates, Dick Lindzen, Jane Orient, Eric Posmentier and Art Robinson for their encouragement and friendship.

Willie Wei-Hock Soon (wsoon@cfa.harvard.edu)
Steven H. Yaskell (starthrower1@msn.com)
Cambridge, Massachusetts, Summer, 2003

Contents

1

A Sun Most Pure and Most Lucid

Solar Blemishes and Imputed Effects on Climate: Scientific Solar Study Begins

Our Sun, *Solis*, shines upon us from roughly 149 million km away. It is the only star like it known to harbor life on any of its nearby planets.[1] People have used solar cycles for telling time, determining seasons, and measuring financial periods since antiquity.[2] The notion of cycles implies "circles" or repetitive phenomena. This is similar to the tree rings Leonardo da Vinci examined and thought reflected age and weather conditions over long periods of time.[3] (Centuries later, scientists would prove this connection.) But did da Vinci know about more distant cycles? Solar cycles—either as a means of practical measurement or as an abstraction—were used and reflected upon.

However, other solar phenomena were also known long ago. So long ago, in fact, that perhaps even prehistoric people were familiar with them. Sunspots were seen and discussed in China even in antiquity and probably a thousand years

---

[1] That is, stars with a similar age and mass to our Sun. Since September 1995, the number of extra-solar planets detected around other stars has been growing steadily. Those planetary bodies around rho 55 Cancri, 14 Herculis, 70 Virginis, 47 Ursae Majoris, tau Bootis, rho Coronae Borealis, upsilon Andromedae, 16 Cygni B, 51 Pegasi, Gliese 86, Gliese 876, HD 187123, HD 75289, HD 217107, HD 195019, HD 168443, HD 114762, HD 210277 have masses ranging from about 0.4 to 11 times of Jupiter and orbital positions from the host stars of about 0.04 a.u. to 3 a.u. Much progress and many breakthroughs have yet to be made before any capability to search for life is obtained, including perhaps even those life forms unfamiliar to what we can currently define.

[2] Prehistoric Native American Cahokia mounds, Japanese Edo-era (1603–1864) sun clocks, and the four-quarter business year are some. See Krupp, E.C., *Echoes of the Ancient Skies* (Oxford University Press, 1983), pp. 29–33, Renshaw, S., Ihara, S. "Marking the noon hour: Sun clock at Kochi Castle, Japan" (March, 1997, Internet URL), and Aveni, Anthony *Conversing With the Planets* (Times Books, 1992), p. 94, on business years.

[3] Webb, G.E., *Tree Rings and Telescopes: The Scientific Career of A.E. Douglass* (University of Arizona Press), p. 101.

before any were referred to in the west.[4] A Chinese oracle bone translation from about 3,200 years ago states: "will the Sun have marks? It really has marks."[5] Many believed that there were connections between the weather and such "marks" on the Sun. In Classical Greece, Theophrastes observed spots on the Sun approximately 2,400 years ago,[6] and perhaps Pythagoras saw them earlier in some kind of an astronomical context as well.[7] In his intricate notations on weather, Theophrastes forecast rain if there were "black spots on the sun . . . ," or, wind if "red spots."[8] Fair weather will be witnessed if "the sun rises brilliant but without scorching heat and without showing any special sign in his orb."[9] Much later a Chinese astrologer/astronomer's log entry for January 10, 357 A.D. stated: "within the Sun there was a black spot as large as a hen's egg."[10] The scholarly pursuits of cloistered theologians in Medieval Europe—perhaps prompted by the observations of their parishioners—could have included drawings of these marks in diaries and books (see Fig. 1).

We assume from these ancient non-Christian and early Christian European notes that people throughout this part of the world were already associating the Sun with having marks or blotches on it that challenged the notion of the Sun being "perfect." That is, the Sun that is pure and perfect should be without any marks or blemishes.[11] We can also see from this that such observed "impurities" may have been thought to have had effects on Earth's weather.

---

[4] Needham, J., Ling, W., *Science and Civilisation in China. Vol. 3, Mathematics and the Sciences of the Heavens and the Earth* (CUP, 1959), p. 435.

[5] Schove, D.J., *Sunspot Cycles* (Hutchinson Ross, 1983), p. 26.

[6] Noyes , R.W. *The Sun: Our Star* (Harvard University Press, 1982), p. 83. Douglas Hoyt argues that Theophrastes's observation record was lost in the burning of the Library at Alexandria in 300 A.D. Noyes claimed that Chinese astronomers "have left a rich series of sunspot records dating back at least as far as 1 A.D."

[7] Letter to the B.A.A., 101, 5.,1991, by Ronald Hardy. His references to Pythagoras come from the Dictionary of Scientific Biography. References to Theophrastes are from his *De Signis Tempestatum* (*Enquiry into Plants and Minor Works on Odours and Weather Signs*, Volume II, Trans. Arthur Hort, [Heinemann, New York, 1916]).

[8] Ibid, Hort, p. 409, paraphrased from: "Also black spots on the sun or moon indicate rain, red spots wind."

[9] Ibid, Hort, p. 427.

[10] Witmann, A.D., Xu, Z.T., *Astronomy and Astrophysics Supplement Series*, 70 (1987), pp. 83–94.

[11] Baumgartner, F.J., "Sunspots or Sun's planets: Jean Tarde and the sunspot controversy of the early seventeenth century," *JHA*, xviii (Science Publications, Ltd., 1987). The Sun was held to be a tabernacle of God. The Jesuit Jean Tarde wrote, "The Sun is the father of light, and so how can it be diminished by spots? It is the seat of God, His house. His tabernacle. It is impious to attribute to God's house the filth, corruption, and blemishes of earth." (p. 46)

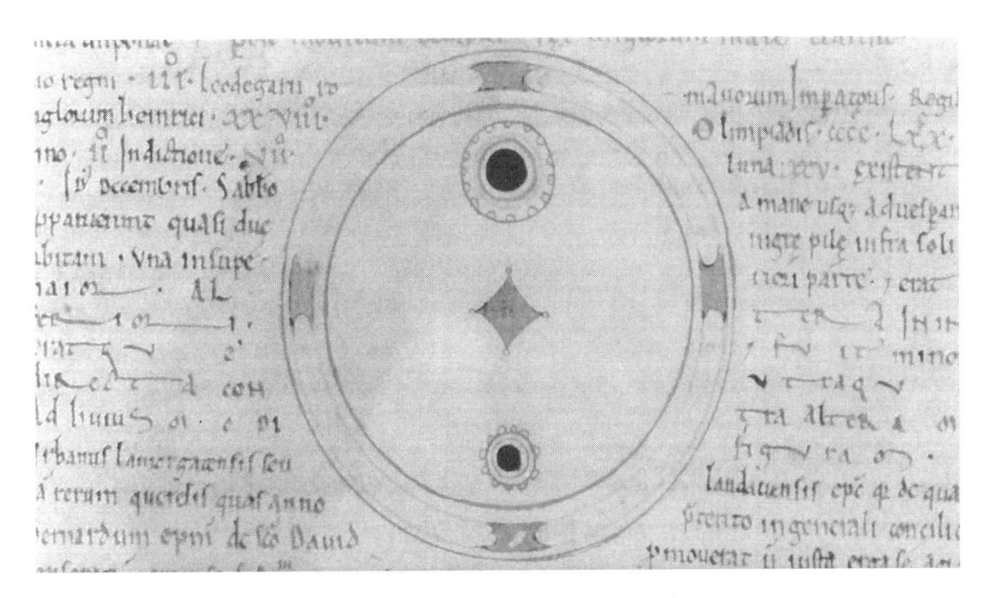

Fig. 1. Drawing possibly of sunspot by John of Worcester, 8 December, 1128 (after Stephenson and Willis).[12]

By 1610, Galileo and English telescopist Thomas Harriot were studying and recording spots with primitive spyglasses in Europe.[13] Johannes and David Fabricius[14] and the Jesuit scholar Christoph Scheiner did the same in 1611, which led to Johannes Fabricius's June, 1611 work,[15] *On the Spots Observed in the Sun, and Their Apparent Rotation with the Sun.*[16, 17]

---

[12] Stephenson, R.F., Willis, D.M., The earliest drawing of sunspots (*Astronomy & Geophysics,* Vol. 40 December, 1999), p. 621.

[13] At the time, still probably called by their Dutch appellation, "tubus"—or even perspicillum batavicum. Casanovas, J. "Early observations of sunspots: Scheiner and Galileo" 1st Advances in Solar Physics Euroconference/Advances in the Physics of Sunspots (*ASP Conference Series*, Vol 118, 1997), p. 3. In *Seeing and Believing* by Richard Panek (Fourth Estate, London, 2000, p. 55), Panek asserts that the name "telescopio" was given the instrument by Greek poet and theologian John Demisiani, during a party given in Galileo's name on April 14, 1611, in Rome.

[14] Johannes, who "rightly thought that the spots belonged to the Sun." For more on telescopes before Galileo, read Engel Sluiter's "The Telescope Before Galileo," *JHA*, xxviii (Science History Publications, LTD, 1997). The telescope's spread through the middle countries and England was driven in part by military use.

[15] Ibid, Casanovas.

[16] *De Maculis in Sole Observatis, et Apparente earum cum Sole Conversaione Narratio.* This book had virtually no circulation, according to the Internet source (Rice University's Galileo Project) and was not seen for years, but its contents paralleled what Galileo noted in print in 1613.

[17] This book remained obscure for many years. There is little literature on David or Johannes Fabricius in English, although their names do of course appear in the standard accounts of the discovery of

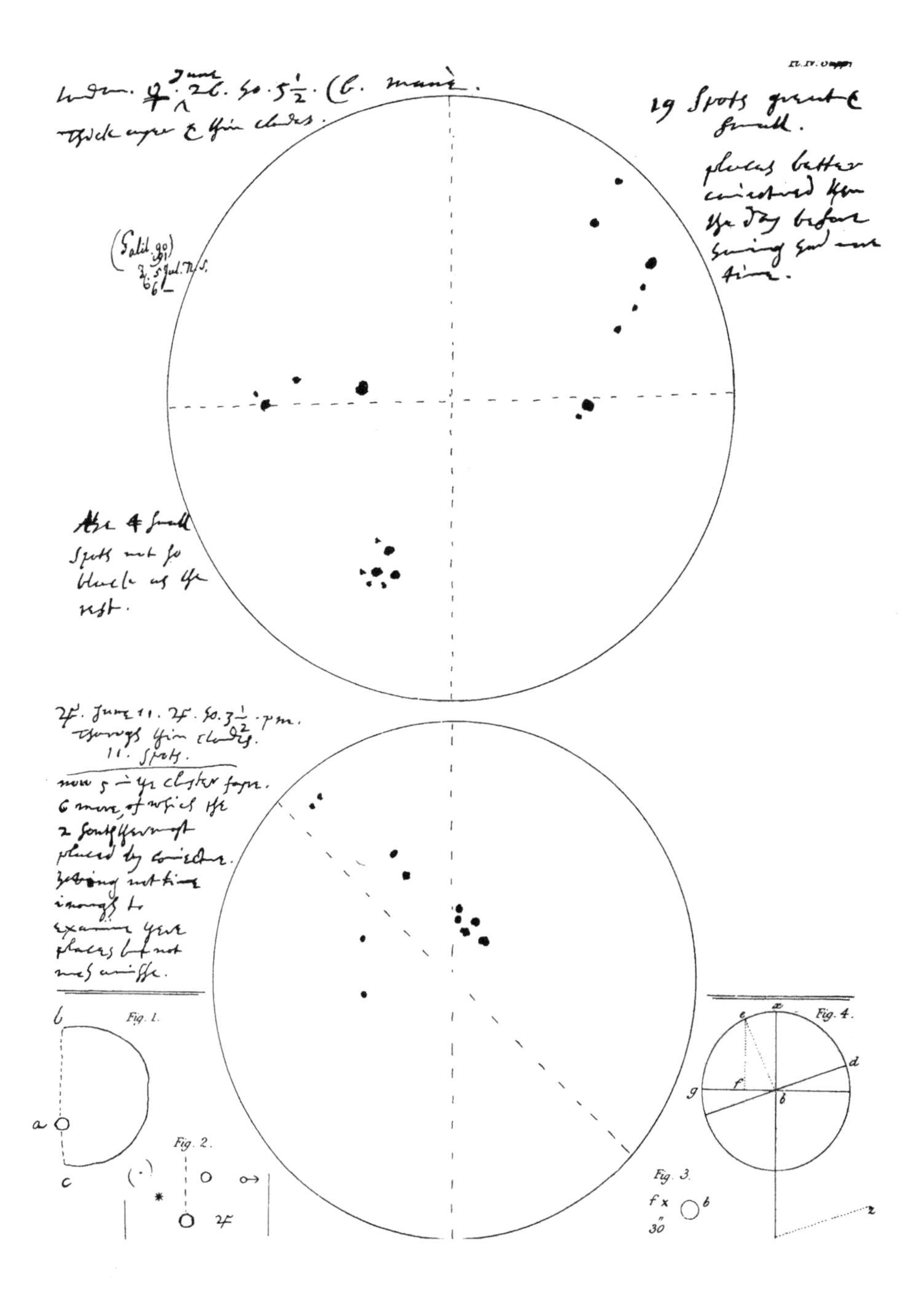

Fig. 2. One of Thomas Harriot's sketches of sunspots, drawn in 1610 (after Mitchell).[18, 19]

Though Scheiner issued to his correspondent, the science patron Mark Welser, *Three Letters on Solar Spots Written to Marc Welser,*[20] Galileo later exchanged letters with Welser as well. Galileo's are termed *History and Demonstrations Concerning Sunspots and Their Phenomena.* In these letter exchanges, Welser holds up to Galileo's examination the arguments of one "Apelles"[21] (Scheiner's pseudonym) who argued for the Sun's physical perfection, claiming that the spots were the Sun's satellites.[22]

When impartial examination of the Sun began in the seventeenth century, it did so in earnest. Contention arose in the West between scholars on this matter due to the paradigm shift away from Aristotelian physics at this time. There is biographical evidence[23] showing that contention arose between Scheiner and Galileo concerning who discovered what first,[24] about the Sun. The "who was first" contention between Scheiner and Galileo included: inventing methods of observing sunspots; the first spot observation through a telescope, and seeing spots on the Sun's surface or not. Additionally, there was a lack of clarity regarding who had made the first faculae observation, or the first determination of the tilt of the solar rotation axis, among others.

(*continued*)

sunspots. Edward Rosen, *Kepler's Somnium* (Madison, University of Wisconsin Press, 1967) has a brief appendix on David Fabricius (pp. 226–232). See also the introduction of Mario Biagioli and Albert Van Helden, *Galileo, Scheiner, and the Sunspot Controversy: Scientific Practice in the Patronage Context* (Note from the Rice University's Galileo Project).

[18] Shirley, J. W., *Thomas Harriot: A Biography* (Oxford, Clarendon Press, 1983).

[19] Mitchell, W. M., "The history of the discovery of the solar spots," *Popular Astronomy*, Vol. 24, 1916, pp. 149–162 (Plate VI, p. 151.)

[20] *Tres Epistolae de Maculis Solaribus Scriptae ad Marcum Welserum.*

[21] Legend has it that the famous Greek painter Apelles once hid behind one of his paintings to hear what people said about it. When a shoemaker praised the way Apelles had rendered shoes in the painting, Apelles revealed himself and thanked the shoemaker for the compliment, but this man now proceeded to give his not so complimentary opinions about other aspects of the painting. Apelles answered "Let the shoemaker stick to his last." Scheiner also referred to his posing under the cloak of this name as "Ajax did beneath his shield." (from *Discoveries and Opinions of Galileo*, Trans. Stillman Drake [Doubleday/Archer, 1957]).

[22] Ibid, Third Letter From Galileo Galilei to Mark Welser, p. 141 (Drake). But Scheiner in later years (as in his 1626–1630 publication of the *Rosa Ursina*) reversed his claim.

[23] See Sobel, D., *Galileo's Daughter: A Historical Memoir of Science, Faith, and Love* (Walker & Co., N.Y.) 1999. The book gives good insights into Galileo's life and controversies.

[24] There is a careful and detailed analysis of this contention shown by Walter M. Mitchell in "The history of the discovery of the solar spots," *Popular Astronomy*, Vol. 24, pp. 22, 82, 562, 1916. The source is old, and somewhat anti-clerical.

Scheiner could have used a pseudonym in his letter exchanges due to his

> "deference to the wishes of his Jesuit superiors, some of whom feared that the Society [that is, the Society of Jesus] would be embarrassed if the astronomer's theories about a maculate sun later proved false (and)...because [Scheiner] conceived of his artistic proficiency as the correlate of his observational skill."[25]

Scheiner's desire for anonymity could have been based more on fear of artistic/scientific failure than on defying the vows he took for his order. However, as a Jesuit Scheiner could not engage in damaging controversies. Jesuits such as Scheiner were imbued with the military-style regimen all members of the order were expected to go through. Taking vows or making pledges were therefore serious matters. Scheiner initially used a pseudonym in these controversies as the Jesuit Order's wish was that its members refrain from open participation in controversies.[26]

Galileo, a member of the Order of the Lyncei, followed a main dictate of the Order of the Lyncei's constitution. This was a vow to "not neglect the ornaments of elegant literature and philology, which like graceful garments, adorn the whole body of science."[27] While doing so, however, Galileo may have neglected one other dictate of the Lyncei—one that Scheiner as a Jesuit did not neglect: "to pass over in silence all political controversies and every kind of quarrels and wordy disputes, especially gratuitous ones which give occasion to deceit, unfriendliness, and hatred."[28]

In addition to Scheiner and Galileo, Johannes Kepler[29] was perhaps the first to comment on all three letters from Welser, convinced that the sunspots were physically

---

[25] Reeves, E., *Painting The Heavens: Art and Science in the Age of Galileo* (Princeton University Press, 1997), p. 6. In this book, one gets a good view of the scientist of the day, seeing his work as art, and the unity in these endeavours that are sadly missing in today's sometimes-lab coat and test-tube besotted sterility.

[26] Ignatius Loyola was a former soldier, and molded the Society of Jesus along the lines of a military command (see Harrison, J.B., and Sullivan, R.E., *A Short History of Western Civilization, Volume One: To 1776* [Knopf, 1975], p. 352.)

[27] Moss, J.D., "The significance of the sunspot quarrel," in *Novelties in the Heavens: Rhetoric and Science in the Copernican Controversy* (University of Chicago Press, 1993), p. 99.

[28] Ibid, Moss, p. 97.

[29] Ibid, Casanovas, p. 11.

on the Sun. He used a metaphor[30] to describe sunspots: "stains we observe on red hot iron, or like slag or dross on the surface of molten metal."[31–33]

We can assume that the debates and letter exchanges of men like Galileo, Kepler, and Scheiner had a lasting impression on all who came into contact with them. For instance, Mark Welser's dissemination of those debate letters helped motivate open inquiry regarding at least the sunspot controversy. What Galileo—for one—saw on the Sun was a tendency for it to darken and spot over at times, when the standard gospel was that it was perfect and unchanging, or as Galileo quotes others' descriptions of it: "most pure and most lucid."[34]

One of Galileo's pupils, Benedetto Castelli,[35] invented an easy way to observe sunspots by pointing a spyglass at the Sun and letting the disc's image fall on paper.[36] This way, the spots could easily be seen rolling across the Sun's face while seen on a white-painted wall, for instance.[37]

---

[30] In the words of Susan Haack, "metaphor can be the training wheels of inquiry." Haack, S., "Dry truth and real knowledge: Epistemologies of metaphor and metaphors of epistemology," in *Manifesto of a Passionate Moderate: Unfashionable Essays* (University of Chicago Press: 1998), p. 87. Here, she develops on the work by Donald Davidson. Perhaps more radical is Julian Jaynes's idea that metaphor is a fundamental structure of advanced human thought, itself (*The Origin of Consciousness in the Breakdown of the Bicameral Mind*, Princeton, 1990).

[31] Ibid, Casanovas, p. 11.

[32] This was apparently in advance of either Scheiner ("Apelles")—who responded at the end of 1611 to Welser. From Caspar, M., *Kepler* (Trans. C. Doris Hellman) with revised footnotes by Owen Gingerich and Alain Segoinds (Dover, 1993).

[33] This was before Galileo, who responded to the third letter from Welser in December, 1612. Ibid, Drake. Galileo responded to Welser on 1 December, 1612, regarding the third letter.

[34] Ibid, Drake, p. 92.

[35] Castelli was not the only one to invent a way of viewing spots. For example, C. Scheiner admitted in *Rosa Ursina* that his use of coloured glasses for observing the Sun was suggested to him by his student, J.B. Cysat (Walter M. Mitchell, *Popular Astronomy*, Vol. 24, p. 293, 1916). Jean Tarde developed a system called *caverne obscure*, which projected spots against a wall in a darkened room through a telescope. Fabricius also had a method. In fact, each astronomer had probably their own method, with variations.

[36] Read "Observing the Sun by projection," by Jeffery Sandel, *Sky & Telescope*, Vol. 94, No. 4, October, 1997, pp. 98–100.

[37] One can easily reproduce this technique today, and is a favorite pursuit of schoolroom science teachers, lecturing in astronomy. It is a way of studying sunspots still in use. (Details of the spots show up, in terms of penumbra and shape, and other characteristics.)

Galileo, either taking a cue from Kepler or of his own accord likened the sunspots alternately to "clouds or smokes" and said they were "produced and dissolved"[38] on the Sun's face.[39] He illustrated this by stating:

> Surely if anybody wants to imitate them [the spots] by means of earthly materials, no better model could be found than to put some drops of incombustible bitumen [coal] on a red hot iron plate. From the black spot thus impressed on the iron, there will arise a black smoke that will disperse in strange and changing shapes.[40, 41]

Galileo challenged the idea that the spots were other planets going by it—as Johannes Kepler supposedly first thought in November, 1607, when he confused a sunspot with a transit of Venus.[42] Galileo doubted that they were the Sun's satellites, as the Jesuit Scheiner and Father Jean Tarde originally thought.[43] It was held that sunspots trailed their shadows or "phases" across the Sun's disk. Galileo argued against the spots being stars, or even that the brownish areas were holes through a fiery shell concealing a dark surface that hid "the real" Sun. Former Jesuit colleagues of Galileo and fellow members of the Order of the Lincei (the Lynxes[44])

---

[38] Ibid, p. 143, "Third letter from Galileo Galilei to Mark Welser."

[39] Always called "him," or "his," until the 1900s.

[40] Ibid, p. 140, "Third letter from Galileo Galilei to Mark Welser," 1 December, 1612.

[41] Galileo had even noted brighter areas independent of sunspots on the Sun's surface that he referred to as faculae, though Scheiner may just as well have been the discoverer of faculae, as mentioned, earlier. Mitchell, W.M., *Popular Astronomy*, Vol. 24, p. 292, 1916). The discoverer of this fact is also debatable. Scheiner is the main candidate, who gave this phenomenon the name "torches" since they looked like these to him (faculare). Much later, an apt metaphor was given for these by William Herschel, who said they resembled "dried apple skin" that moved.

[42] Ibid, Casanovas.

[43] Tarde and Scheiner both apparently revealed a deeply-rooted Catholic belief in the purity of this heavenly body; see Baumgartner, F.J., "Sunspots or Sun's planets: Jean Tarde and the sunspot controversy of the early seventeenth century," *JHA*, xviii (Science Publications, Ltd., 1987). Jean Tarde, taking Galileo's lead in trying to gain royal favour—in his case, from the French King—hazarded that sunspots were circling planets, and called them the "Bourbon Planets," much as Galileo called Jupiter's satellites the "Medician Planets." Like Galileo, he was hoping for money and royal favour. (His deepest feelings were also one of purity of the Sun: "The Sun is the father of light, and so how can it be diminished by spots? It is the seat of God, His house. His tabernacle. It is impious to attribute to God's house the filth, corruption, and blemishes of earth." p. 46).

[44] A society of intellectuals founded by Federigo Cesi in Rome, and of which Galileo became a member.

such as the Jesuit John Schreck,[45] visiting China on invitation, described spyglass-viewed sunspots in 1628:

> On the sun there are spots of various sizes: one, two, or three or four, but no more. They are found always above a line running east and west across the sun. They constantly follow the same trajectory ... when the first spots are finished, others take their place. The largest spots can cause the light of the sun to be dimmed. When the spots were first discovered they were thought to be Venus, perhaps, or Mercury; but the trajectories did not agree. Observed recently with the telescope, they have, on the one hand, been seen not to be a part of the body of the sun, but, on the other hand, not distant from the sun, like red clouds; but to be exactly in front of it. What they are is unknown.[46]

Finally, René Descartes[47] somewhat belatedly developed a "foam theory" about sunspots. What is to be noted in all these matters is the interpretation of what these people observed, rather than denying that anything impure could be seen at all upon the Sun. Denial of seeing anything at all on the solar surface is what many a churchman and natural philosopher did at the time.[48] From Christianity's beginnings through the Middle Ages, changelessness in the Sun was tied to the Sun's spiritual perfection as reflected in Christ.[49] And although the Bible states instances when the Sun did change, it was a bidden spiritual belief that profundity brought on, for instance, by witnessing that something such as solar change could be

---

[45] D'Elia, P.M., *Galileo in China*, Trans. Rufus Suter and Matthew Sciascia (Harvard University Press: Cambridge, Mass, 1960), pp. 24–25.

[46] Ibid, D'Elia. Quote was taken from Schreck's work, *The Sphere*. (Note well Schreck's remark that the sunspots seemed to *dim* the light—as well as his reluctance to indicate that the spots were actually *on* the Sun—which approximates the reluctance of his fellow Jesuit, Scheiner.) For an insight into the accuracy of solar observations made by Jesuit astronomers in China, see F. R. Stephenson and L. J. Fatoohi's *Accuracy of Solar Eclipse Observations made by Jesuit Astronomers in China* (*JHA*, xxvi, 1995).

[47] In 1644. See Rizzo, P.V., Schove, D.J., "Early new world aurorae," *JBAA Papers Communicated to the Association*, 1962, 72, p. 8.

[48] When "Apelles" offered to show his newly seen solar spots to a colleague (Theodore Busaeus, the provincial of the Order), he was told that he "must be seeing things," and to obtain a better pair of glasses: "I have read my Aristotle from end to end many times and I can assure you that I have never found in it anything similar to what you mention. Go, my son, calm yourself, and be assured that what you take for spots on the sun are the faults of your glasses or your eyes. (Montucla, *Histoire des Mathematiques*, Vol. 2, 312)." From Walter M. Mitchell, *Popular Astronomy*, Vol. 24, p. 209, 1916.

[49] McCluskey, S., *Astronomies and Cultures in Early Medieval Europe* (CUP, 1998), p. 25.

beyond common understanding.[50] That is, behind the physical phenomena of the Sun was a deeper spiritual significance, and this spiritual significance was of paramount importance. All else could easily be ignored. Among the learned in the seventeenth and eighteenth centuries, the thought that the Sun was beyond one's understanding in a physical sense was decidedly on the wane. Additionally, the accepted understanding from Medieval times and earlier that the Earth was a hopelessly corrupt and defiled place was also being rethought.

Seen in this way, it is clear that at least a core of people in the Northern Hemisphere understood and took pains to record sunspots moving in groups and disappearing around the Sun's edge only to reappear. This and other recorded phenomena like faculae sightings found its way into written records. By the seventeenth century, select people in Europe and Asia knew that sunspots favored creation at certain locations on the Sun, and this knowledge was perhaps spread elsewhere.[51] Many among them began passing on the "light"[52] in trying to explain the Sun's more stubborn secrets, distinguishing between the spiritual understanding of the Sun and its purely corporeal substance and behaviour. This quest is ongoing whether the Sun is studied by itself, or is considered in tandem with Earth.

---

[50] Ibid, McCluskey, p. 31. The instance used here is from 4 Kings (20:1–11) and Isaiah (38:6–8) that the sundial would be turned back to add 15 years to the life of King Hezekiah. An explanation as to why this was possible was stated that "the created order was unchanging, although human minds might not be able to comprehend it fully." McCluskey draws this from Augustine.

[51] Sharma, V.N., *"Astronomical Efforts of Sawai Jai Singh—A Review"* (University of Wisconsin), pp. 235–236. Quotation taken from Zij Jadid Muhammad. Singh had sent a fact-finding embassy to Europe to learn more of European science and had observed sunspots. Perhaps he even used a telescope for this pursuit, and in any case, noted that "it [the Sun] rotates once on its axis within a period of one year."

[52] We used this phrase as a metaphor for increasing physical knowledge and understanding about the Sun mainly through Galileo's scientific views and approaches. Another reason for the choice of the phrase is simply to note that Galileo loved wine, referring to it as "light held together by moisture" (ibid, Drake). See also Michael Sharratt's *Galileo: Decisive Innovator* (Cup, 1994).

2

Background of the Maunder Minimum

This book centers on the time the redoubtable Galileo fades from the picture in the 1620s and ends before the finish of the first quarter of the eighteenth century.[1] This is one way to date the approximate length of the Maunder Minimum extended low-solar activity phase. This cycle was named by solar scientist John A. Eddy in the 1970s, Eddy attributing the popularization of the phenomenon to E. Walter Maunder (1851–1928). Maunder and his wife A.S.D. Russell Maunder were solar scientists who painstakingly helped delineate this reduced activity cycle.

Off center from this point of departure, however, is the not-so-straightforward history of how many of the searchers—Maunder included—worked either in tandem or off each others' findings on the route to discovery. Astronomer, physicist, mathematician, or climate scientist; historian, correspondent, or dedicated amateur worked either alone or together somewhat in the manner of Galileo, Welser, Kepler, and Scheiner to help establish not only a fundamental understanding of phenomena like extended solar minima such as the Maunder (which Maunder himself of course never completely knew) and its effects on Earth. Attempted here is at least an outline of how this jagged process has led to revealing how our Sun also can—at times—seep into a period of extended solar maxima, for example—also to influence Earth. As such, the prolongated activity minima and maxima have a marked influence on Earth's weather for longer and shorter terms but the exact relationship between them and Earth is little known. Much was to be learned in order to obtain just a hint of this odd varying habit our Sun maintains. The climate effects from such processes are more anomalous, still. We diverge from this well-trodden path ultimately to help open up another one. This leads to further understanding the basic functioning of such processes in the Sun, since more precise knowledge of what drives mechanisms so vital to the well being of life, itself, continues to elude science.

---

[1] These vary, some between 1645–1710. Others earlier, as this book's position assumes.

To begin putting the Maunder Minimum in perspective, we remain for a while longer with Galileo. He was one of the natural philosophers living during at least one part of the Maunder Minimum. However, he was also more than this. Galileo and his kind represent a revolution in thought. He is a touchstone in our appreciation of this revolution. A mathematician-physicist with interest in the cosmos such as himself saw a need to assume the role of a philosopher and become a question-asker as well as an explainer[2] of how things operated in nature. For where many asserted and speculated

> Galileo insisted on differentiating between speculation and assertion, between hypothesis and theory. His idea of the investigative process did not end with sense evidence, nor did it end with philosophizing. Instead it began with both, infused them with reason, and ended only after examining the alternatives.[3]

Furthermore, Galileo understood that even if mathematics could help in explaining phenomena, it was not the only or even the hardest "proof." Rather, there was always a middle ground where thought, observation, and mathematical explanation joined. It was in this middle ground where the truth could often be found.[4] Moving from "classificatory and philosophical arguments" to "mathematical terms" was Galileo's way of moving the frame of reference, and was emulated by others.[5] For Galileo, it was never the ultimate proof—though measurement, apart from all else, could stand out at least as partial knowledge that would not need constant revision.[6] Nor would pure mathematical reasoning be the hindering element for E. Walter Maunder in his solar research two and a half centuries later.

---

[2] Panek, R., *Seeing and Believing*, pp. 52–56. Galileo's "explanation method" described "earthshine," and helped remove the idea that Earth was "a pit where the universe's filth and ephemera collect" (*Sidereus Nuncius*).

[3] Ibid, Panek, p. 59.

[4] Drake, S., *Galileo at Work: His Scientific Biography* (Dover, 1995). In an revealing aside, Galileo's father had had to assert—in the face of much opposition—that what, for instance, was mathematically exact, was not necessarily to be immediately superimposed on other subjects—especially music. For instance, the fact that an "irrational twelfth root of two governed harmonic modulation," according to one of his father's defenders—not integers. But the surprising and natural phenomenon of harmonics is not only how modern pianos are tuned. Without this knowledge, Bach's "Well Tempered Clavichord" would never have been written (had Galileo's father lost the argument) (p. 16) Galileo's father—Vincenzio—is also purported to be the inventor of opera (see Sharratt).

[5] Ibid, Drake, *Galileo at Work,* pp. 29–20. One user of this was Christiaan Huygens and his work on relative motion.

[6] Ibid, Drake, p. 108.

Coincidental to Galileo's time, the prevailing understanding of angels in their crystal spheres[7] steering stars appeared less in the minds of the exalted—even the common prelates. John Donne[8] spoke of Galileo in a satire of 1611, the notice of whose discoveries came to him probably by way of Donne's friend, the English ambassador to the Doges Court, Henry Wotton[9]:

> I thinke it an honester part as yet to be silent, than to do Galileo wrong ... who of late hath summoned the other worlds, the Stars to come Nearer to him and give an account of themselves.[10]

So, figuratively if not intellectually, the stars had come closer. Nor was angelic statuary to be seen only in churches. Nobility throughout Europe erected new edifices geared more to secular pursuits, such as theatres and opera houses, and placed angelic statuary in these as well.[11]

There were other buildings and institutions dedicated to less artistic pursuits. One of these was the Paris Observatory[12] and the Royal (Greenwich) Observatory[13] in England. These and others were opened across Europe, most of which began to record methodical astronomical observations in the west. Both of these scientific institutions play important roles throughout this book.

* * *

It was often implied that the Sun was a pure body up to Galileo's time. That is, something without stain and so, spotless. Thus, spotlessness implies purity. But "spotless" can also mark a paucity or a total absence of things. That the Sun was

---

[7] Ibid, Drake, *Galileo at Work*, p. 25.

[8] Donne actually rose to become the Dean of St. Paul's.

[9] Wotton brought news of how Galileo used the telescope to the king. Bronowski, J., *The Ascent of Man* (Futura, 1981). See also quote in Sobel, *Galileo's Daughter*, p. 35.

[10] Donne, J., "Ignatius his Conclave," from *The Pseudo-Martyr. Complete Poetry and Selected Prose of John Donne*, ed. Charles Coffin (Modern Library, New York, 1952), p. 319.

[11] In letters to an associate, read by Cardinal Bellarmine, Galileo had printed crucifixes and pictures of saints on the title page.

[12] It was founded, intriguingly enough, by the self-declared "Sun King"—Louis XIV—an institution in which much solar work was—and still is—done. The Paris Observatory was to log in 8,000 days of observation on the Sun for the next 70 years ... a fitting tribute to the Sun King's largess (from "The Stellar Dynamo," by Elizabeth Nesme-Ribes, Sallie L. Baliunas, and Dmitry Sokoloff, *Scientific American*, Vol. 275, Number 2, August, 1996, pp. 31–36).

[13] For purposes pertinent to the history contained in this book, the earlier name for this now defunct institution—the Royal Observatory, alternately Greenwich Observatory—will be used. It was not called the "RGO" until quite recently.

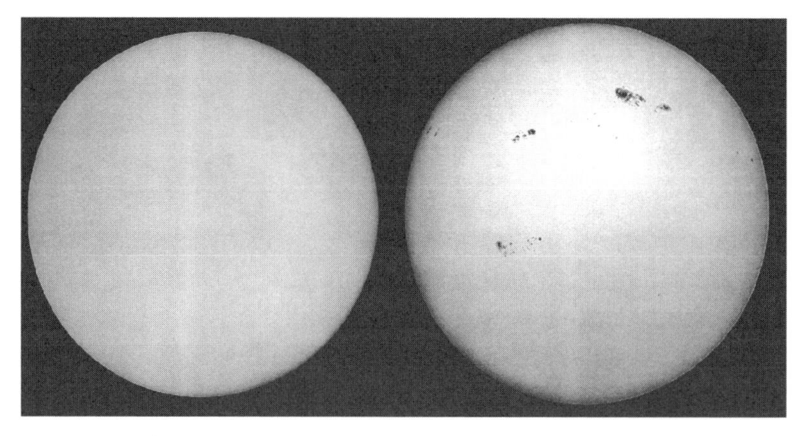

April 7, 1994 *February 12, 1989*

Fig. 3. The spotless Sun of April 7, 1994 (left). Occasional spotless days during solar activity minima are not uncommon. However, extended periods without sunspots appearing on the surface of the Sun during the 1645–1715 interval of the Maunder Minimum were truly unusual. In contrast to the spotless Sun, an active Sun with a moderately spotted surface on February 12, 1989 is shown (right).(Images courtesy of William C. Livingston, Kitt Peak Solar Observatory)

completely without spots or spot groups from about 1620 to 1720 is false. That their number was few (especially after 1640) is in little doubt.[14]

As if to commiserate with Kepler's, Galileo's and others' thoughts on this troublesome subject, such as what they were and how they appeared, the Sun's spots disappeared altogether at times after 1620. The weather was often unusual and did not match the memories of people then alive. Nor did the weather and storm intensity of occurrence mesh with the oral and in some cases, written records of these peoples' forebears—whether in Europe or in Asia. In fact, cooling weather in the Northern Hemisphere, at least, had been a reoccurring event since the mid-1500s and earlier times.

Before the Maunder Minimum—right after another, less-well-known solar minimum called the Spörer Minimum (c. 1460–c. 1540[15])—Pieter Brueghel

---

[14] Extensive works by Hoyt and Schatten (1996, *Solar Physics*, Vol. 165, pp. 181–192) had recently re-affirmed that the Sun was well observed during the Maunder Minimum. The best estimate of the fraction of time the Sun was observed is $68 \pm 7\%$. A "lower estimate" of the days which have recorded observations is 52.7%. Therefore, one can conclude that the lull in the sunspot numbers was real rather than due to a lack of observations.

[15] Kippenhahn, R., *Discovering the Secrets of the Sun*, "Sunspots and Radioactive Carbon," (1994), p. 31.

Fig. 4. Hunters In The Snow (Brueghel) in pre-Maunder times. H.H. Lamb claimed this picture was painted "during the first of the great winters of the next two hundred years"—1565.[17] (Delevoy)

showed from circa 1565 to 1568 nature's vibrancy on canvas. Depicted here are mainly farm scenes recording steely winter skies and skaters and workers (such as *Hunters in the Snow, The Numbering at Bethlehem*, and *Landscape With Skaters and Bird-Trap*) and tepid, summer alabaster ones. Yet the years following this would have an English preacher, John King, lament in 1595, "our years are turned upside down, our summers are no summers; our harvests, no harvests."[16]

The distinguished climate scientist H.H. Lamb pointed out that Brueghel the Elder, to accentuate the suffering of the infant Jesus in the manger, repainted a picture from 1563 to show snow chilling all the visitors in a revamped 1567 version. In Lamb's words, Brueghel did this "to emphasise the poverty and exposure of the accommodations."[18] We are sure the average person of Brueghel's time was not suffering any less. Was the sunspot number low or absent at times in the Spörer Minimum? The body of evidence in these pre-telescopic times is sketchy.

---

[16] Daley, A., "The Little Ice Age; was it big enough to be global?" (Internet topic, 5/8/98, daleyac@miavx1.muohio.edu)

[17] Lamb, H.H., *Climate, History and the Modern World* (Methuen, London, 1982), p. 223.

[18] Lamb, H.H., *Climate, History and the Modern World*, p. 225.

Since the development of perspective painting in Europe's Renaissance approximately 200 years before the Maunder Minimum, "real life," truth-to-detail painting and drawing among artists had progressed in parallel with the scientific use of art. In other words, the methods developed by Leonardo da Vinci, Giotto, and Titian became handmaidens to visual scientific description if not to practical and theoretical science itself.[19] If three dimensional painting and drawing served faithfully as a tool for deliberately helping to spread the reach and understanding of science, then this kind of skill spreading or describing scientific evidence unintentionally from at least the fifteenth century in Europe is also true. Hence we can appreciate the cold, snow, and dim Sun depicted in paintings by Brueghel (and others) as a kind of factual record. In other words, paintings such as these reveal empirical evidence, albeit non-quantitative, and can serve as a database for the pursuit of past climatic changes.

Such rather indirect archival evidence for recording the weather does not end there. Semi-fictional accounts of weather, a popular genre in the eighteenth and nineteenth centuries, are often quite revealing in this regard. Especially as they "describe the time of their forefathers,"—an almost mythical time when weather was simply "harder." Other human evidence left by the cold of these times is found in architecture and manufactured goods available at the time of the Spörer, and indeed, the Maunder, minima. As such, they are good subjects for archaeologists to study. Even items so apparently mundane as sixteenth and seventeenth century furniture and appliances (if this is the correct term) show practical attempts towards attempting to ward off the cold. Dining tables were often built with footrests that kept the feet suspended in the somewhat-warmer air hovering over cold floors, and the very buildings these kinds of tables were located in were often built low to avoid heat rising very high into the air as one of many practical benefits of architectural design. That less energy was expended in constructing smaller dwellings was a distinct benefit and the end products required less energy to heat. A very popular and presumably very useful item was the portable metal coal grate, which could be locked and transported to carriage, and thence to church, for example for the purposes of keeping feet or hands warm. Similar grates were used in heating beds. Furthermore, although the argument can be forwarded that the age was one without the benefit of central heating, why, then, were such items and architecture less prevalent in mid-nineteenth century Europe and

[19] Goldstein, Thomas, *Dawn of Modern Science*; (Chapter 6) "Art and Science in the Renaissance" (Houghton-Mifflin, 1988).

elsewhere when the technological improvement of central heating still lay some 70 years in the future? Tenuous as such justifications are, it should be kept in mind that invention for comfort often precedes the explanation as to why the inventions/adaptations were made in the first place.

* * *

These cooling trends identified as the Spörer and Maunder and others (as well as the warming trends we will review) have been identified and grouped into rough cycles. The cycles of the Sun around these periods are given names as they can denote solar minima (in very general terms, cooling) as well as solar maxima (warming, equally as general). Giving names to these solar cycles was a convention borne of convenience only lately and they were given names to accommodate research. The Maunder Minimum was a low-ebb solar activity cycle bearing more implications than merely effects inside and upon the Sun, itself, in its early manifestation in the early-to-mid 1600s. What was also noticeable were peculiar weather effects on Earth at the same time.

The scientific literature has a logical place—albeit non-trivially—for the Maunder Minimum. The Maunder Minimum is noted as a systematic and large deviation away from a smooth, background sine curve that represents the modulation of ^{14}C production by Earth's magnetic field, together with three other minima that are alike: the Wolf Minimum (after R. Wolf) of the thirteenth and fourteenth centuries; the Spörer Minimum (after F.W. Gustav Spörer) which preceded the Maunder in the 1500s, and the shorter Dalton Minimum (after John Dalton) which followed (see Fig. 5 on the following page). Lying further behind at least two of these minina was an already ongoing "Little Ice Age" (which will be discussed later, and analysed thoroughly in the concluding chapters.)

Prior to these minima, however, was a cycle of *increased* solar activity (the Great Maximum, or "Medieval" Maximum, in J.A. Eddy's terms) and what came after all these was the current state of increased solar activity combined with other elements, such as what we are facing today. Some would call what we are now witnessing with the Sun the "Modern (solar) Maximum."[20] (The significance of ^{14}C and other isotopes in helping record Earth's climate will be discussed more thoroughly in later chapters.)

* * *

[20] Note however that at least a large part of the ^{14}C dilution in modern times (today) is caused by human burning of buried hydrocarbons like fossil fuels.

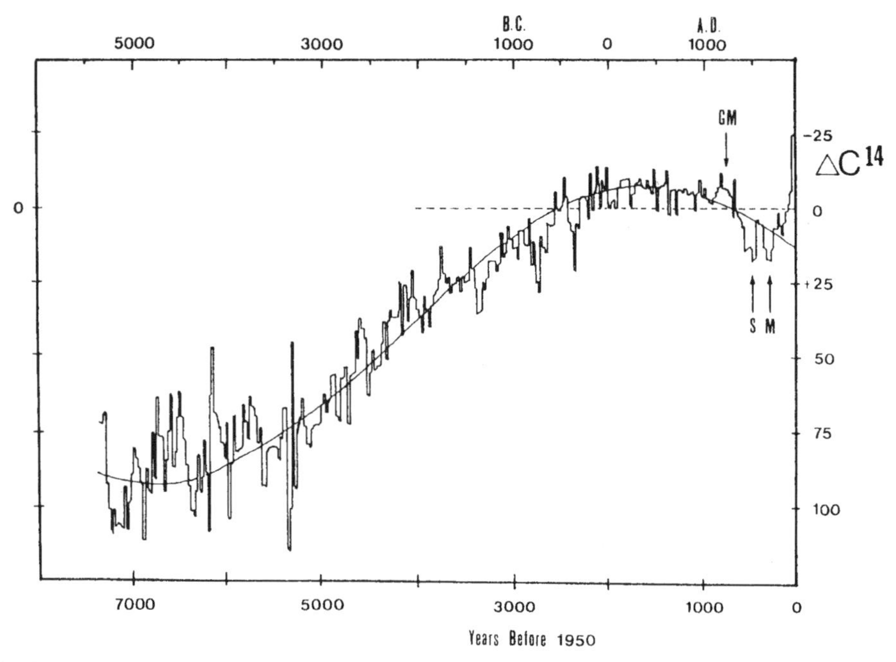

Fig. 5. History of relative radiocarbon concentration from tree ring analysis. S is for Spörer, and M is for Maunder, and both these dips mean more ^{14}C (relative to ^{12}C, in units of per mille or parts per thousand in concentration) to what was manufactured in the atmosphere, showing up in trees sensitive to moisture. Note that Eddy displayed more ^{14}C downward. In contrast GM (Grand or Great Maximum) denotes times of reduced ^{14}C production occurred during the Medieval Maximum. Note that the highly diluted ^{14}C level in modern times, say since the 1900s, is a result of fossil-fuel and coal burnings (the so-called Suess effect). The smooth sinusoidal curve ("making the ski-slope") overlaid on the ^{14}C data could be tied to the effects caused by a gradual variation of the Earth's magnetic dipole field intensity. (After John A. Eddy, 1980)

The Maunder Minimum—our "cold-weather" subject—occurred in a time also marked by many wars across the Earth's Northern Hemisphere. North America became a war zone of opposing native tribes backed up by warring European colonists. In Europe, for example, there was the Thirty Years' War on the continent, and civil war in England—as well as inter-nation strife throughout Northern Europe. There was also Counter-Reformation internecine strife and warfare in Asia—particularly in China. Protestant divided against Catholic in Europe (such as the plight of the Huguenots there) and perhaps elsewhere. Factions and cliques of either of these religions divided against each other. Other religions, fueled by their proponents' sharply polarized beliefs faced off against each other, as well (for example, in the Mediterranean basin). In the far south of Europe near the crumbling

Venetian Mediterranean sphere of influence the Ottoman Empire moved ever closer to Western Europe. Muslim and Catholic and Orthodox Christians fought like their Catholic and Protestant cousins to the north. These combatants left brushfires that smolder to this day.

It is difficult to say if the weather lay behind these tumults and similar dissent and wars, sometimes occurring simultaneously, in other parts of the world. This is not an unreasonable assumption. Neither is it absurd to connect the various plagues and illnesses—and their severities—to the weather's twists and turns. Colder than expected weather, such as crop-killing sudden droughts, damaging heavy rains and snows, and other phenomena certainly exacerbated effects on already-roughened politico-religious climates.

It is very ironic that, in this blossoming period of intellectual warmth and light in tiny corners of Europe, China, and America, the times should have been so politically and militarily murderous. These times were unevenly cold and often dry, windy, and hazardous to geographically-locked agricultural belts that could not deliver the quantities of food populations required. Wandering armies spread disease and famine in their wakes.

The age's misery is well acknowledged throughout Europe and Asia. Cold, drought, and famine were common. Ignorance, superstition, and fear exacerbated the effects of all of the above.[21] Philosopher of science Francis Bacon, a victim of sickness and death due to cold within the Maunder Minimum[22] in these years began his *Instauration* with a prayer that asked:

> Remembering the sorrows of mankind and the pilgrimage of this life wherein we wear out days few and evil, They [God and the Holy Trinity Bacon believed in] will ... endow the human family with new mercies ... I humbly pray that ... opening of the ways of sense and the increase of natural light there may arise in our minds no incredulity or darkness.[23]

The Maunder bore witness to the end of mathematician/astronomer Johannes Kepler's life in 1630, Kepler having just managed to get the *Rudolphine*

---

[21] Ibid, Welser, in Drake, and Bacon, F., "The Great Instauration," in *The English Philosophers from Bacon to Mill*, ed. Edwin A. Burtt (Modern Library, 1939).

[22] Ibid, Wilson, p. 23. On p. 25, Wilson notes diarist John Aubrey indicates that Bacon caught a chill during an experiment on seeing if cold snow could preserve flesh (using a chicken). Bacon died on April 9, 1626, probably—like Descartes—of pneumonia.

[23] Ibid, Burtt, p. 12.

Tables[24, 25] published. Sickened (his life had been marred by such episodes—as Galileo's apparently had) Kepler died en route to see the final land map for this publication. He was buried on November 17, 1630 in a Protestant graveyard which soon vanished as the land battles between differing religious factions raged.[26] Suffering for years[27] from the cold weather in these times like his daughter Virginia (Maria Celeste), Galileo died eight years after her on January 8, 1642.[28, 29] Fleeing Holland, where he spent his life more-or-less in exile from his native France, the Jesuit-educated René Descartes died of pneumonia on February 11, 1650 in Stockholm, Sweden: a land to which he had turned for continued patronage as a teacher of Queen Kristina. His was a condition contracted due to severe weather.[30] The interesting point, coincidental though it may be, is that so many of these now well-known contemporary scientists died from sicknesses contracted due to the influence of the cold during the Maunder Minimum. Among these, only Newton escaped a sickly end, dying, it might be added, after the Maunder Minimum could be construed as being over, late in the second decade of the eighteenth century.

* * *

That much observation of sunspots was made in the time these people lived is no mystery. Telescopes were being purchased in Europe and used for astronomical purposes after 1610 and in China by the 1620s—if not earlier, and somewhat later in America, albeit by a tiny fraction of those societies' inhabitants. Additionally, this fraction of observers making such observations were pioneers in the field and path breakers known for their acute attention to detail, skill, and honesty. Considering the period they lived in, and the simplicity of the instruments

---

[24] Named after their patron, Rudolph II.

[25] This work—the combined genius of the pre-telescopic observer Tycho Brahe interpreted by the ingenious theoretician Kepler, was probably one of the "shoulders of giants" upon which Newton later said he "stood."

[26] Caspar, M., *Kepler* (Dover Reprint, 1993), p. 385. No one knows exactly where he was buried, according to this version of Caspar.

[27] Ibid, Sharratt. See also Sobel, *Galileo's Daughter.*

[28] With physicist and fellow Lucasian Professor of Mathematics as Cambridge University, Stephen Hawking being born on January 8, 1942 (300 years to the day).

[29] At the beginning of the year Newton was born, on December 25, 1642. Hall, A.R., *Issac Newton* (Cambridge University Press, 1992), p. 3.

[30] Probably pneumonia. Read the description of his death and his attempts at remedies in Gaukroger, S., *Descartes: An Intellectual Biography* (Oxford University Press, 1995), p. 416.

they used compared with today—especially telescopes—much was accomplished, and much of it very high in quality.[31]

The surprising element to the observations by such Maunder Minimum-living astronomers as Johannes Hevelius, Jean Picard, John Flamsteed, and others was not the behaviour of sunspots. What was most perplexing to them in some cases was the sunspots' absence. These individuals had access to information produced by earlier observers (Galileo, Scheiner, and so on). The question was, "where were the spots?" or "where were the faculae?" and so on. Some among these natural philosophers—such as Royal Society member Robert Boyle—observed peculiar sizes to the spots in the Maunder Minimum and noted, in passing, just how long they could be seen on the Sun's face. So what was known, was the "what," and perhaps the "how," regarding basic solar activity, be it sunspot motion, the discovery and observation of faculae, and even "white flashes"—the latter having been observed near the Maunder Minimum's end for the first time. What was not known amongst these men and in some cases, their assistant wives (and which is still often the case), is the "why" behind such activity. Furthermore, weather activity was not generally connected by these individuals to solar activity, albeit intuition abounded. Nor was "aurora borealis," the absence of which was apparent, and the actual, physical witnessing of which existed only in forebears' memories—if even there. For aurora had been much more common in the centuries past, and handed down most commonly in legend, perhaps, than by any other means. Aurora were to resume—and to resume in abundance—early in the eighteenth century after a 50-year near total absence of sunspots as recorded by the best living observers.

In abundance, however, was extreme weather of all sorts, "instability," in the words of the climate researcher Jean Grove being "a leading characteristic of climate."[32] There were violent storms and cold weather that appeared not to have been part of people's forebears' memories. And not everywhere, at once, at all times, across the Northern Hemisphere. There were droughts that could desiccate nearly immobile agronomic peoples—which was what most of the world's population

---

[31] To note specifically here would be Douglas V. Hoyt and Kenneth H. Schatten's *The Role of the Sun in Climate Change* (Oxford University Press, 1997) where great pain is taken to point this out, amongst European observers.

[32] Grove, J., "The Century Time Scale" *Time-Scales and Environmental Change*, eds. Driver, T.S., Chapman, G.P., (Routledge, 1996), pp. 55–57. See also Zheng, S., Feng, L., *Scientia Sinica* (Series B), Vol. 24, (1986), p. 441, "Historical Evidence on Climate Instability Above Normal in Cold Periods." Instability during the cold period of 1620–1720 occurred at least for mid-latitudes of Eastern China. That is, cooler temperatures are associated with larger inter-annual rainfall variability.

then was—in the Northern Hemisphere, at least. There was a disappearance or diminution of the mild weather that had seen the start of Medieval Viking colonies across the North Atlantic—some colonies ceasing to exist.[33] Droughts, very unusual snows, weird cyclonic activity in the Northern and even Southern Hemispheres and unusually violent volcanic eruptions were often observed and suffered through.

The volcanic eruption dated to about 1600 of Huaynaputina, in Peru, is considered to be one of the most violent eruptions of the past 500 years.[34] Immediate cooling effects on climate[35] were due most likely to sulphur ash blocking the more beneficial rays of the Sun, and not to anything conversely "not emanating" from the Sun. The effects of massive volcanic eruption spread not only across a single year, but very often for years at a time while the ash clouds slowly disperse in the environment. Pointing out this particular geoclimatic effect, however, serves the purpose of showing year-long spreads in short-term climate change going on, independent of any lack of solar inputs into Earth's climate (though lower solar activity could contribute to Earth cooling, if the effects of reduced solar activity and reflecting atmospheric ash particles intertwine). The effects from Huaynaputina's ash clouds blocking sunlight could, in other words, have continued well into the decade of 1600–1610 if not longer. The same is true for the second major c. 25-year stretch of meticulously-observed sunspot rarity (and hence, reduced solar activity) when the massive volcano Krakatoa erupted (in 1680). This most likely added strongly to the global cooling events then being withstood across at least the Northern Hemisphere.

With the advent of the eighteenth century a gradual distancing from the severest parts of non-linearly-cold Northern Hemisphere weather occurred. Aurora borealis—a phenomenon linked to solar electromagnetism—was observed only

---

[33] Ibid, Grove. Lamb in some publications also discusses these at length, as well as many other sources, especially Scandinavian.

[34] Shanaka L., de Silva, Zielinski, G.A., "Global influence of the AD 1600 eruption of Huaynaputina, Peru," *Nature,* Vol. 393, June 4, 1998, p. 455. The fall deposit covered 300,000 square miles. The eruption itself lasted nearly a month. Core samples taken from ice for this period show global cooling through the distribution of sulphur aerosols in the atmosphere, probably due to this volcano.

[35] In fact, the temperature of summer 1601 was estimated from tree-ring climate proxy to be "the coldest of the past 600 years throughout much of the Northern Hemisphere and among the coldest of the past 1,500 years in Fennoscandia." (p. 457 of Shanaka L, de Silva and Zielinski, G.A., "Global influence of the AD 1600 eruption of Huaynaputina, Peru," *Nature*, Vol. 393, 4 June 1998).

a few times for most of the period from about 1620 to 1705. Yet from around 1700 on, the Maunder Minimum gradually faded away. Proof of increased magnetic activity was observed by the recorded return of sunspots, the witnessing of "white flashes" on the Sun, and a very robust renewal of aurora sightings—in the far north as well as at Earth's mid-latitudes. Naturally, no one at the time associated any of these phenomena with "increased magnetic activity" though room for doubt is always present. Nor did anyone have a clear understanding that there may have been a connection between the absence of any of these effects and poorer or stranger-than-usual weather on Earth.

* * *

What follows in the next two chapters is a region-by-region description of Northern Hemispheric meteorological and astronomical incidents (relevant ones such as aurora sightings, etc.) marked by their severities or rareness, or both, given the general knowledge of such phenomena. The events list is not complete—we do not pretend that it is and great holes are to be found. In fact, the reconstruction is mainly an example of what could be done, were all sources more tightly covered and complete, then woven into a larger picture. Thus with some data, we have constructed a large template into which the data we have neglected or overlooked might be added at a later time, or inserted by others, if found.

The period covered is not the usual timeframe ascribed the Maunder Minimum (one of the common dating frames is 1645–1710) but, rather, from 1620–1720. For although the matter of the Maunder occurring much before 1640 (or to reach, much before 1630) and much after 1710 is still debatable, the empirical evidence described builds a case for being able to explain the minimum as having spread over an 100 year time frame. Perhaps the strongest indicator of the Maunder's earlier impact is seen in the weather from quite early in the seventeenth century on, the recurring famines in, for instance, China, being one mirror reflection of the crop-crippling weather in the absence of clearer climate records (recurring famines in the Mediterranean and North Africa reflect in the same way). Additionally, what were the lingering effects of the Spörer, which ended c. 60 years before? The Maunder's later impact (after 1710) is also seen in lingering effects (heightened weather and storm intensity driving animal life from biomes, etc.) as well.

The events in the following two chapters are described in as chronological an order of occurrence as it was possible to make, given the disparity of incidence. The intention here is to inform the reader of what the average person saw and recorded; in some cases, even their opinions or impressions are obtained. At times, cumulative, modern reconstructed reports (of sunspots and weather, etc.) are given for added perspective, as well as commentaries and notations from modern commentators. The regions listed are for Western Europe and the Mediterranean (to include at times North Africa); China and the Far East, and North America (to include at times the North Atlantic–Western Europe).

If the narrative seems filled with *non sequitur* or perhaps even bombast, it is perhaps only because we have not divided the phenomena by type of climate incidence, type of solar observation, and so on, but adhering roughly to a chronological order of occurrence per region, as the people then alive would have lived through them. Later chapters (Chapter 5) will sort out these occurrences for more detailed scrutiny as per, for instance, type of phenomenon (storm, drought, etc.), how many, its relative severity, etc., with a count of the incidents listed in graphs and in tabular form. This should yield a rounded picture of the Maunder Minimum's phenomena and its effects. Standing on its own merits, it will exist in contrast to any reconstructions made of the same for, say, the Spörer or Wolf, or for any other minima or maxima that others might wish to construct.

3

The Maunder Minimum: Europe, Asia, North America (As Dated from c. A.D. 1620–A.D. 1650)

Western Europe and the Mediterranean Basin

Probably in the wake of Huaynaputina, as the general circulation spread the sulphur ash across at least the Northern Hemisphere, the summer of 1601 was, by all accounts, characterized by poor weather across England. For example, "the month of June was very cold, frosts every morning."[1] A similar story was related in Northern Italy that same summer of 1601, where freezing weather was reported into July, and the sky was overcast for much of the year. Iceland and Scandinavia also experienced the same sort of weather at this time.[2]

In 1608 the Thames River in England was frozen, and it snowed in Crete in the Mediterranean Sea. These two widely separated locations matched the cold effects witnessed in 1608. Northern Italy in the years 1614 (nonlinearly, to 1729) statistically reads like a study in winter severity,[3] though the average coldness of –3 Celsius (very cold) extends far into the late 1700s in some studies of this region. (As Jean Grove reports, record keeping between 1620 and 1648 shows famines, droughts, and multiple bad winters on Crete).[4] In 1614, "continual south winds" on Crete caused a drought of four months' duration. Furthermore, what

---

1 Pyle, D.M., "How Did the Summer Go?" *Nature,* Vol. 393, p. 415, 1998.

2 Ibid, Pyle, 1998

3 Camuffo, D., Enzi, S. in *Climate Since 1500 A.D.* (eds. Bradley, R.S., Jones, P.D., Routledge, New York, 1995). Information gleaned from Table 7.1, Great and severe winters in Northern Italy, compared with other works, p. 151.

4 Ibid, Grove. The Century Time-Scale. The record keeping shows from circa 1580–1598, cold winters and bad harvests, with attendant plague and famine. Significantly, so does the timescale between circa 1620 and 1648, when records stop. (Table 3.3. Climate and Crops in Crete, 1548–1648).

with famine and plague—not to mention drought—locusts invaded northern Italy over these years, either from the northeast or from the south, being transported mainly by wind and as the report notes, regardless of temperature. Prevailing winds assisted them in their flight during the Maunder Minimum, as these hordes of locusts were frequent visitors in Western and Central Europe in the seventeenth century.[5]

People were not much better off in the highly-civilized, deeply-cultured belt of land extending from Florence to Rome merely a few years later. Galileo, for example, had planned a winter visit to Rome about November, 1623. The journey was delayed until April, 1624 due to the severely cold weather.[6]

Since people in Europe had access to very primitive thermometers and barometers they could find out what was going on, to some degree. In Switzerland in 1628 the summer was cold and wet. Grapes were harvested on October 29th after a frost severe enough to disrupt agriculture.[7] Heavy snows covered mountains like the Engstligen Alp (at 1,964 meters in height) twenty three times within seven weeks, from May to the end of August. The culprit often cited is a "Little Ice Age" ("LIA") afflicting Earth at this time. The "LIA," which will be described, was a non-uniformly cold dip that spread out over a span of time in a non-linear manner for possibly more than 500 years. Naturally, this timespan includes an unusually deeper period of non-uniform cold: the Maunder Minimum.

Unseasonable rains in Crete in the summer devastated the olive and wine crop in July, 1632. As Galileo had suffered for years from unidentifiable ailments, sickness in 1634 took Virginia's life. The cold weather and sunspot inactivity was notable throughout the Northern Hemisphere. (This was to be a recurring theme for the next seventy or eighty years, on and off, to happen very noticeably again from c. 1795–1823, in the Dalton Minimum.)

As old age and political danger beset Galileo, he often retired to his daughter and loving confidante Virginia's convent on visits (where she was called Maria Celeste) the cold weather preventing him from visiting her in the early 1630s, though his prison/villa, Arcetri, lay close enough to her convent (San Matteo) for

---

[5] Ibid, Camuffo, D., Enzi, S. in *Climate Since 1500 A.D.*, p. 147.

[6] Letter from Maria Celeste to Galileo, November 21, 1623. (Rice University's Galileo Project)

[7] Pfister, C., "The Years without a Summer in Switzerland: 1628 and 1816" from *Klimageschichte der Schweiz,* 1525–1860 (Band I:140–141, 1984, Verlag Paul Haupt, Bern). Records were cross-compared with similar agricultural shortcomings recorded in 1816 (mid Dalton Minimum). The note on the Little Ice Age spanning these centuries, non-linearly, is from Jean Grove.

him to hear its bells.[8] In 1629–30 in fact, outside of the bubonic plague hitting Galileo's part of Italy, late in 1630, "tramontana winds" arrived to bring a deep cold to the region.[9] Towards the end of her life, Virginia also reported gales or tornadoes and hale storms near his residence.[10] In 1638 and 1639, severe snowstorms were reported on relatively-nearby Crete.

Additionally, it is interesting to note that in a detailed study done by D. Hoyt and K. Schatten,[11] "no sunspots observations exist" between the years 1636–1637, emanating from the likes of observatory astronomers such as Johan Hevelius, Jean Picard, Pierre Gassendi and others. To underscore the accuracy of their not seeing any, it is worthwhile to note that men such as these also kept in contact with one another to compare notes.

In England in March 1643, an English diarist made the following comment on what was truly a rare observation of "northern lights", or aurora borealis, later corroborated by a Dutchman for March 22 at about the same time (it was also seen in North America at this time):

> (From 11:00 p.m. to 1 a.m.): I must not forget what amazed us the night before ... a shining cloud in the air, in shape representing a sword, the point reaching to the North, it was bright as the moon, the rest of the sky being very serene ... being seen by all the South of England.[12]

In these years, Descartes proposed that these displays were the work of thundercloud-like activity exciting particles in the atmosphere to create these effects. It was an idea that was rejected in some quarters at the time, owing to the fact that northern lights were not usually seen in times of abundant clouds in the sky, taking into consideration the self-destructing effects of those particle-heating activities on clouds. But it was one of the few scientific theories proposed for these

---

[8] Rice University's Galileo Project. Maria Celeste (Virginia) was one of Galileo's daughters. The monastery they lived in reportedly was the poorest in Italy. Claims of Galileo's misogyny abound. Clearly, however, Virginia loved—and admired—him, evidence to wit being gleaned from her letters, and these letters show his ample love and near dependence, as well. In Sobel's *Galileo's Daughter*, the author reveals that the pair were subsequently buried together.

[9] Ibid, Sobel, *Galileo's Daughter,* p. 210.

[10] Ibid, Sobel.

[11] Hoyt, D., Schatten, K., "Group Sunspot Numbers: A New Solar Reconstruction" (*Solar Physics*, October, 1997).

[12] Ibid, Rizzo and Schove. The Dutch corroborating source is Acteboek van den Kerkenraad te Sluis van 1605–1648.

lights then.[13] Significantly, the period of 1645-to the early eighteenth century—or, most of the Maunder Minimum—is marked as a time typically *lacking* in these displays of light in Europe.[14]

As continental Europe struggled in religious wars, Anglican England was cut up by civil wars; an England to which fleeing Huguenots (Protestants from France) ran, escaping religious intolerance. A time of Commonwealth emerged, under the puritan Oliver Cromwell (after 1649, as it did in Puritan-British North America's northeast).[15] Civil-war inundated England reported a mortality rate between 1642 and 1646 that was more than double the 30% norm in death rate, at the time[16] in Berkshire. Additionally, there were reports of a total absence of sunspots: in 1648, according to Hoyt and Schatten.[17] In 1649 the Thames River was frozen hard, once again.

Long before Cromwell in Elizabethan England, Shakespeare wrote the play *Othello*. The backdrop of this play is a scene that described events that were contemporary for him—the Venetian struggles against a rising Ottoman presence on Crete—an area of "strategic and no doubt psychological importance" to Venice[18] at this time. The struggles continued throughout the 1600s. Weather had a part to play in the desperation of the conflict which resulted, by 1648, in the nearly total occupation of Crete by Ottoman Turks.

From January, 1645 to December, 1651 Johan Hevelius, at work in his observatory in what is current-day Poland, saw no sunspots (according to accounts made by Spörer, much later).[19] He not only knew Pierre Gassendi, but was perhaps even his friend, and along with Robert Boyle and Robert Hooke, was a fellow member of the English Royal Society: as such, these people kept in contact with one another on scientific matters.

---

[13] From *Meteora*, quoted in Brekke, A., Egeland, A., *The Northern Light* (Springer Verlag, New York, 1983), p. 55.

[14] Ibid, *The Northern Light,* p. 41.

[15] Massachusetts is still called the "commonwealth."

[16] Dils, J., "Epidemics, Mortality and the Civil War in Berkshire, 1642–6," in *The English Civil Wars*, ed. R.C. Richardson (Sutton, 1997), pp. 145–156.

[17] Ibid, Hoyt and Schatten (Solar Physics, 1995) (Guess is Picard or Gassendi, after a report by Kraft in 1745).

[18] Ibid, Grove, The Century Time-Scale, p. 72.

[19] Hoyt, D., Schatten, K., "How Well Was the Sun Observed During the Maunder Minimum?" (*Solar Physics,* September, 1995).

Fig. 6. Example of Hendrick Avercamp's painting from around the Maunder Minimum period.

Earlier, probably in what was the Spörer Minimum, the cultural component of increasing cold weather in Europe was shown to be a factor in proof—however subliminal—of an increasingly colder climate. The example used in that case was Brueghel's work in that period of time located between the end of the Spörer Minimum and before the start of the Maunder Minimum. Cited was the fact that three-dimensional artistic representation was driven by scientific innovation, and that such representation can serve to unintentionally (as well as intentionally) assist scientific appreciation of empirical evidence.

To appreciate the accumulating cooling effects in Europe for the Maunder Minimum one has only to turn to the paintings of Hendrick Avercamp (1585–1634), another Dutchman, to get a cultural record—and perhaps even the left-over accumulated cooling of the Spörer (cumulative cooling being a matter that will be discussed later)—in the early seventeenth century.

This painting is from European's Maunder Minimum period[20] and is titled "A Scene on the Ice near a Town" (town unidentified). It is one of many that

[20] This depends on how the Maunder Minimum is dated. Some date it from 1645 to 1710. Some date it earlier.

Avercamp painted on this theme in these times that reveal an ice-dominated world. Aert van der Neer and Gijsbrecht Lijtens also painted such scenes, as did the painter, van Beerstraaten. All were contemporaries.

Most of these artistic expressions may be emotional reflections on the then-ongoing activities of then-totally unknown solar cycles and periods as Earth was influenced by them. Also in this "Little Ice Age"—a misleading rubric for a rather complex phenomenon (as will be discussed)[21]—there is a familiarity with snow and ice, as the painted figures seem to indicate, there was little getting away from it, and that they indeed used it as an extended highway (workers, travellers, on the ice[22]). Yet, this familiarity with cold, snow, and ice is even recorded in the sixteenth century paintings of Brueghel.

It is not true to state that most painters of this period in Holland or elsewhere in the Northern Hemisphere went especially out of their way to record snow and ice. Additionally, not every painter, Dutch or otherwise, well-known or not, rushed to their easels to make a conscious record of the weather. As with a photo album, however, paintings that show climatic effects serve as a kind of archive, and Renaissance art was, indeed, originally a handmaiden of science, as outlined earlier. This archival tendency on behalf of the Dutch at this time happened to be more the case, since seventeenth century Netherlanders "had a passion for depiction of city and countryside, either real or imaginary."[23] Significantly, a local pride in their sea commerce prompted a drive for *technical accuracy* in the on-canvas reproductions of their merchant ships, for example. So it is not hard to see the tendency for punctilious realism slipping on over into their landscape painting, as well—whether the landscape contained ice, snow, or vales of grain—or dimly glowing suns.

The 1650s brought a dry time of anticyclone wind circulation to England with drought conditions over most of it, and on the continent as well. It is peculiar to

---

[21] The "Little Ice Age" ("LIA") was previously assigned to the timeframe 1550 A.D. to 1880 A.D. However, in personal communications with Jean Grove, 1550–1880 should no longer be linked to this term (i.e., LIA). "LIA should only be related to periods of glacial extensions, not climate, as such. I believe that considerable confusion has been caused by the idea that LIA was a long period of lower temperatures, and that the medieval warm period [was] one of unbroken benign conditions." (Jean Grove, personal communication, September, 2000.)

[22] Brueghel and Avercamp were not the only painters painting cold scenes in the 1600s: nor were only Dutch painters recording snow scenes.

[23] Dutch Landscapes and Seascapes of the 1600s. (National Gallery of Art)

note that Scotland[24] fared worse, weather wise, than the south of England even if these political entities are connected by the same landmass. The damage seemed worse on the Scottish coasts and coastal seas. Shifting sands, damaging gales on coastline properties, and bad crop yields characterized at least the summer of 1659 from one Scottish diarist's notation, aggravating the tenant farmers' collective plight. A four day long gale in September wiped out not only crops, but 16 mills. More on the conditions in Scotland will be shown later.

Coastal storms continued in the hemisphere, in the north of Great Britain and south of it, as well. Back in the coastline-dominated Mediterranean basin—which P. Gassendi had once diligently mapped—sea storms such as the one in 1658 all but stranded the defenders on their weather-torn island of Crete. Their opponents in the territory-grab, the Ottoman Turks—having a good supply base—alternately starved or invaded them. Prices of food skyrocketed so that even the defending soldiers had to pay high prices for what they ate. The Cretan peasant, upon whose back most toil fell, was forced to subsist mainly on wild plants due to the stress of war and bad harvests.[25]

A year after Descartes's death in 1651, Italian astronomer Giambattista Riccioli recalled his 1642 observation where he was thinking back even earlier (to 1632): "the warm and dry September of 1632 when there were no spots and the cold June of 1642 when there were many," thus already making a connection between sunspots and weather conditions. Perhaps Riccoli did not foresee the Sun's fluctuating condition between strong and weak.[26]

China and the Far East Asian Region

One of the first well-reported appearances of a cold wave to grip and strangle southern China was made in 1614–1615 (the same year Crete suffered a four-month long Spring drought). Guangdong (23rd parallel) is a northerly province of China in relation to extreme Southern China (that is, it is somewhat above Hong Kong) containing two lesser entities, Guangzhou to the south, and Huizhou to the

[24] Schove, D.J., Reynolds, D., "Weather in Scotland, 1659–1660: The Diary of Andrew Hay," *Annals of Science*, Vol. 30, No. 2, June, 1973, pp. 172, 177. –3 degrees Celsius is calculated in modern measurement from the original temperature measurements in the data.

[25] As Scots later did in Scotland.

[26] Citation traced back through Schove (1983) to Riccioli, G.P. (1651, 1653, Vols. I and II) *Almagestum Novum Astronomian Veterem Novamque Complectens, Observationibus Aliorum et Propis, Novisque Theorematibus Problematibus as Tabulis Promotam.* (Bologna, Italy)

north, which for these years was ravaged by a drought. (In 1613, China was extremely dry, as well.) In 1614–1615, snow fell in the extreme southwest of China, above what is today Vietnam (in Nanning): this is a region not known for Christmas-like snowstorms, then or now. At this same time, the years leading up to 1615 were noted as being "particularly cold" elsewhere in the Northern Hemisphere.[27]

The years 1618–1619 reprised the cold performance south Chinese endured in 1615 late in the Ming Dynasty, with snows as far south as the Pearl River delta and in Guangzhou. The snows were so heavy "elders said it had never been like this before."[28] This was in Northeast China: exactly one year before English pilgrims landed on the northeastern coast of America. Among extreme events noted, the cold period recorded in China between 1620 and the early eighteenth century, on the whole, was remarked upon as being "absolutely [the] coldest period during the past 500 years."[29]

China was routinely visited and lived in by Jesuit scholars active in astronomy from about 1620 on. Former Jesuit colleagues of Galileo and fellow members of the Lincei (the Lynxes[30]) such as the Jesuit John Schreck,[31] visiting China, described sunspots using a telescope around 1628. Curiously enough at this time, he reported seeing many of them. Chinese imperial bureaucrats and the Emperor were very keen on the advantages of calendar reform, if not on telescope use and solar study. On June 21, 1629 for example, three astronomical rivals (one representing the Chinese, another, a Muslim, and the third, European) were asked to predict the length of the expected eclipse the next day. Only the Europeans (represented by the missionaries) were correct in determining its length, according to one source.[32]

In China in the winter of 1634, cold episodes sprang up which lasted two more years, so that residents of "coastal Huilai county (northern Guangdong province) reported not only snow, frost and ice four to five inches deep on ponds, but also

---

[27] Hoyt, D., Schatten, K., *The Role of the Sun in Climate Change* (Oxford University Press, 1997) quoting Landsberg (from 1980), p. 197.

[28] Marks, R.B., *Tigers, Rice, Silk, and Silt* (Cambridge University Press, 1998), p. 138.

[29] Ibid, Domrös and Gongbing, "The Gazetteer Period 1400–1900 A.D.," p. 134.

[30] A society of intellectuals founded by Federigo Cesi in Rome, and of which Galileo became a member.

[31] D'Elia, P.M., *Galileo in China*, Trans. Rufus Suter and Matthew Sciascia (Harvard University Press, Cambridge, Mass, 1960), pp. 24–25.

[32] Ibid, D'Elia, p. 41.

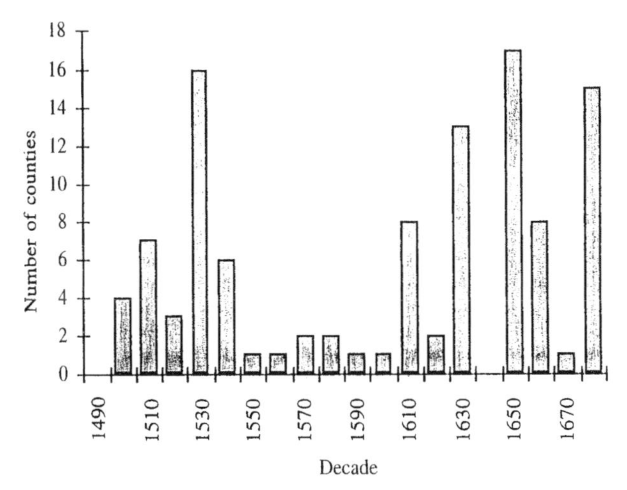

Fig. 7. Number of Chinese counties reporting frost or snow, 1490–1680. Notice the absence of reports from the 1640s. (Source compiled from Guangdong Sheng Ziran Zaihai Shilian [Guangzhou, Guangdong Sheng Wenshi Yanjiu Guan, 1961] in p. 139 of R. Marks 1998)

that they had never experienced anything like it before."[33] This was experienced through the years 1635 and 1636, and possibly even later.

In fact, the usually well-kept records of cold and precipitation in China between the 1630s and 1650s contain a curious absence in the 1640s. It is not believed that the weather suddenly experienced a ten-year normalization. What is believed to have occurred is civil and social strife so violent and turbulent that record keeping in at least southern China, so well maintained by the Ming Dynasty, effectively broke down—as perhaps all concerted civil bureaucracy (at least in the southern Lingnan provinces) did as the Ming Dynasty fell.

However, the end of the Ming Dynasty actually witnessed attempts to modernize China after Mongol rule had purportedly ravaged the empire.[34] Hence, the acts of the last Ming emperor, Chung Chen (1627–1644) who openly welcomed the missionaries from the west. For example, one of them, Johann Adam Schall von Bell, was eventually made a Mandarin and tutor to the child emperor Shin-chih.[35] The "Grand Design" appointees spreading western science saw the Jesuit missionaries

[33] Ibid, Marks, p. 139.

[34] Billington, M., "A Grand Design: Kepler and Renaissance Science in China," *21st Century*, Summer, 1996.

[35] Ibid, Billington.

sent to reform Chinese science assuming top positions in the Chi'ing hierarchy at this time. Von Bell became Director of the Chinese Bureau of Astronomy, for example.[36]

The acts of the "Grand Design"—as this was called by the German mathematician Leibnitz—were presumably noble on behalf of Jesuits. This "Grand Design" would spread science to the "darker" regions of the east—even to what is now South Africa and Brazil.[37] Yet it was not as if the Chinese of the Ming period came empty-handed to the bargaining tables of their Christian benefactors. Although lacking in modern scientific method, the Chinese had managed to preserve a wealth of detailed astronomical observation from before the thirteenth century.[38] This data may be providential for academic astronomy historians. And fortunately for those who reconstruct climate, the Chinese also generated numerous written reports by gazetteers—who were known for their comprehensiveness in recording weather.[39] It should also be kept in mind that the weather was faithfully recorded in dynastic China in more official and regular ways.

Recording cold weather on canvas was not merely a European phenomenon of the Maunder Mininum years. In the early Ch'ing Dynasty (1644–1912) are stark, icy winter landscapes by Wang Hui (which may have been made around the time of Galileo's death in the 1640s). Ice and snow is also a major motif in Nan Ling Chinsen's *Animals and Birds.*

A dynasty was broken and another one built while the Maunder occurred across China (Ming into the Ch'ing)—mainly due to the machinations of the Imperial Eunuchs. There was further political instability, for example, in Shaanxi Province because of famine. As the Ming Dynasty was effectively eliminated in 1642, and loyal elements were hunted down by Chi'ing insurgents, and wide scale killing and starvation grew, it is thought that weather measurement was stopped,

---

[36] Three paragraphs paraphrased from Marks and Billington, respectively.

[37] Evans, D.S., "Historical Note on Astronomy in South Africa," *Vista in Astronomy*, Vol. 9, pp. 265–282, 1967. Notes made on Father Guy Tachard stopping at Cape Town South Africa (on his way to Siam) in 1685 to make determinations of longitude. For the Brazil notation, see A. Udias and L.M. Barreto in *Exploring the Earth: Progress in Geophysics since the 17th Century,* eds. Wilfried Schröder, Michele Colacino, and Giovanni Gregori (Interdivisional commission on history of IAGA, 1992), pp. 128, 139, respectively.

[38] For more on the purposes of Chinese astrological observations and their copious data collection in antiquity, see "Early Chinese Observations and Modern Astronomy" by F. Richard Stephenson, *Sky and Telescope*, February, 1999, pp. 48–55.

[39] Domrös, M., Gongbing, P., *The Climate of China,* pp. 133–135.

Fig. 8. Early Chi'ing Dynasty (1640s) painting by Nan Ling Chinsen of animals and birds. Notice the ice on trees. Weather reporting could have broken down in parts of China during the 1640s. (Meridiane, 1981)

altogether, in at least the 1630s and 1640s. The weather was recorded regularly and diligently in at least the Ming period. The extent of the war was such that the north, having effected a Chi'ing dominance there, was forced to repeat a take-over in the south in the same year Descartes perished from pneumonia in Sweden—1650. Estimates of those who perished range from 80,000 to 700,000. (It is interesting to recall that 1650 marked a change of regime in England, as well [Commonwealth from monarchy] due to civil war.) The years 1653, 1654 (and again in 1660, 1665, 1670–71; 1683, and 1690–91) marked "years of severe winters in the Yangtse River basin in China (30 degrees N.), with freezing of some or all of the large lakes and tributary rivers."[40] This coolness continued there, off and on until the early decades of the eighteenth century.

The 1640s, according to records in China, showed declining harvests due to a cooler climate and this cooling is viewed as a main reagent behind political unrest. (Notably no sunspot observations exist, according to Hoyt and Schatten, for the

[40] Ibid, Lamb, *Climate Past, Present and Future,* p. 612.

year 1641.[41]) Other ingredients in this recipe for disaster were ongoing financial difficulties such as falling silver prices and the Japanese crushing any trade links to China. (Japan—like southern England and Siberia—escaped the brunt of the Maunder Minimum weather.) The fighting for resources on a now land-locked China was such that many soldiers and sailors were no longer loyal (or divided in loyalties) between Ming and Chi'ing ravaged the population—who themselves were hard strapped by bad weather to obtain enough to eat. One can easily imagine armies wandering far and wide to keep themselves fed. It is believed that these marauding armies spread disease in their wake (it was probably likewise in continental Europe and in England over this same time period).

Seen in China at this time were paid assassins feeding off of political discord and renegade armies whose members numbered in the thousands. For example, there appeared the "Associated Bandits"—along, presumably, with a surge of knowledge in martial arts: enforced relocations, and piles of bodies which could have been a mirror image of the continental European wars occurring at the same time, and perhaps worse. As the decade progressed, famine broke out in parts of China in 1647. To the west of Guangzhou countless people died of starvation, and many died from epidemics, as well, diseases probably having been spread by warring bands of soldiers and displaced civilians. Common people, forced to survive on a meagre diet in the unexpected cold in the wakes of passing armies, succumbed in their weakened state, in part, to dysentery.

A severe typhoon in 1652 off the Guangdong coast (perhaps as bad as the hurricanes off Cape Hatteras, in North America, near this time) renewed famine among a population scarcely recovered from the last one. By 1654 the Chinese cultivation of oranges and mandarins in Kiangsi province—a practice for centuries there—was halted. (It was not resumed until 1676.)[42]

Not far south of China, in a climatic note collected from the Indian subcontinent, the seventeenth century records indicate "more frequent interruptions and failures of the monsoon than in our times."[43] This was a probable sign of an extension of the polar cap: southerly winds swept Siberia, for a change—which, like Japan, was spared much of the Maunder-period cold and unsettled weather. Meanwhile, this extending polar ice cap washed Northern, Western and Central

---

[41] Ibid, Hoyt and Schatten (*Solar Physics*, 1997).

[42] Lamb, H.H., "An approach to the study of the development of climate and its impact in human affairs," in *Climate and History* (eds. Wigley, T.M.L., Ingram, M.J., Farmer, G., CUP, 1981), p. 304. Lamb quotes Chu Ko-Chen (1961).

[43] Lamb, H.H., *Climate, History and the Modern World* (Methuen, London, 1982), p. 227.

Europe with icy northerly winds, cooling the air as well as cooling the northern seas (for instance, the North Atlantic).

North American Region

The year 1615 marked the beginning of epidemics among American native tribes in what is now Massachusetts where, in some cases, an 80% mortality rate reduced native populations by thousands of individuals. Due to this, many contact-period bands and cultures were so seriously disrupted that no true record of how exactly these people existed before heavy European contact will ever be obtained. The archaeologist Dean R. Snow indicates that many became refugees to other tribes, thus muddling cultural identities. It was in this state that most Europeans (English and French in particular) came to know them tentatively. The culprit of the contact-period sickness interval is cited by Snow as being European-introduced sicknesses, though, given the weather across the Northern Hemisphere, the effects of cold may already have helped many malnourished natives further toward their graves.[44]

In North America for the landing of the Puritan pilgrims from England, it was later recorded that "the decision of the Pilgrims to land on the shores of Massachusetts was dictated solely by the weather." This found their vessel the *Mayflower* at daybreak on November 19, 1620 in Province Bay. The actual landing at Plymouth was at Christmas, the Pilgrims being greeted by an English-speaking native named Squanto. The *Mayflower* had been blown there by foul Massachusetts winter weather. Though the winter of 1620–1621 was described as mild by historians, the season began with harsh weather early in December just at the time the pilgrims arrived. William Bradford, Plymouth Plantation governor and historian, reported around Christmas, 1620, that

> Now the heart of winter, and unseasonable weather, was come upon us, so that we could not go upon coasting and discovery, without danger of losing men and boat, especially considering what variable winds and

[44] Snow, D.R., *The Archaeology of New England* (New World Archaeological Record series, ed. James B. Griffin; Academic Press, 1980). See chapter on 'historic baseline,' where Snow gives even incubation times for big killers like measles among native Americans. Around these years (1614–15) large numbers of Europeans presumably make their appearance in Massachusetts and environs. Particularly cold weather—if apparent in Massachusetts that year—could have exacerbated effects among a people already weakened by several kinds of illnesses—malaria perhaps being one of them. It should also be remarked that another reason given by Snow for the uprooting of different coastal tribes in Massachusetts was due to a dominating Mohawk nation to the west which assisted in a political/military weakening of mid-to-coastal agronomic tribes.

> sudden storms do there arise. Also cold and wet lodging has so tainted our people (for scarce any of us were free of vehement coughs) as if they could continue long in that state, it would endanger the lives of many, and breed disease and infection amongst us.

There is no detailed information as to the nature of subsequent winters during the first decade of settlement at Plymouth. Yet each winter was reportedly "a dreaded season" in this timeframe.[45]

Some years later, far after settlement in New England had taken root and spread in 1635, a storm was recalled on a Saturday in August in Rhode Island:

> This year (1635) on Saturday, the fifteenth day of August, was such a mighty storm of wind and rain, as none now living in these parts, either English or Indian had seen the like, being like unto those Hirracanes or Tussins that writers mention to be sometimes in the Indies.[46]

Further descriptions of the wind and waves—and their destructiveness—were also made. An August storm in 1638 caused double tides in Massachusetts.[47] Two months later in Maine, a hurricane wrecked ships and blew down trees.[48]

The severity of the cyclones is apparent in these early North American diary notations, in that people bothered to record them using some descriptive terms. Noticeable in all cases, however, be they from China, the Mediterranean basin, North America or elsewhere is that the recorders take care to point out the *severity*, and also, that they usually occurred along coastal or near-coastal locations accompanied by cloud cover, rain, hail, or other precipitation usually in long-lasting or in violently-concentrated amounts.

Carefully researched literature serves as a database. Outbreaks of war in the North American colonies, diligently collected by American historian Francis Parkman, were the norm throughout the 1640s–1660s, and were labeled "The Iroquois War." In deadly accompaniment, sickness and cold followed, and

---

[45] Sections on Plymouth and Bradford taken from Stormfax homepage. Internet: other sources used were: Ludlum, D., *Early American Winters 1604–1820,* Boston, American Meteorological Society, 1966, pp. 7–12. Bradford, W., *Of Plymouth Plantation 1620–1647*; ed. Samuel E. Morison, New York, Alfred A. Knopf, 1952, pp. 60–71. Mourt's Relation or Journal of the Plantation at Plymouth. ed. Henry M. Dexter, Boston, John Kimball Wiggin, 1865, p. 39.

[46] Nathaniel Morton's, New England's Memorial, 1699 edition, annotated with manuscript additions of Peter Easton, one of the first Newport (Rhode Island) settlers (Courtesy Maris S. Humphreys, Special Collections Librarian, Redwood Library and Athenaeum in Newport, RI).

[47] Ludlum, D., *New England Weather Book* (Houghton Mifflin, Boston, 1976).

[48] Ibid, Ludlum. Jocelyn quote.

Parkman attributes death from cold and drought in the 1620–1660 timeframe in lower (southern) Canada.[49]

In 1640s New England, a 1643 hurricane in July in Massachusetts, in the words of Governor Winthrop, "through God's mercy (did) no hurt, but only killed one Indian."[50] In November two years before this, he also witnessed "a great tempest of wind and rain from the S.E. all the night, as fierce as a hurricane ... and thereupon the highest tide we have seen since our arrival here."[51] He was referring to the arrival in 1620, most likely, and perhaps had not seen a hurricane so violent before.

Other interesting "meteorological" phenomena in the Maunder Minimum—and rare for this period—were northern lights, even those recorded in the British North American colonies.

Governor Winthrop—whose eldest son, John, Jr., was destined to become the first colonial American member of the Royal Society—recorded weather and celestial phenomena in British New England that was a relatively peaceful haven from the conflicts abounding overseas.[52]

Some of these "lights" were electrical storms, the composition of which Governor Winthrop probably had no idea but recorded, anyway. In January, 1644[53] Winthrop wrote:

> About midnight, three men, coming in a boat to Boston, saw two lights arise out of the water near the north point of the town cove, in form of a man, and went a small distance to the town, and so to the south point, and there vanished away. They saw them about a quarter of an hour.

---

[49] Morison, S.E. (ed.), *The Francis Parkman Reader* (Da Capo Press, 1998).

[50] Winthrop quote.

[51] Ibid, Ludlum. Assorted Winthrop quotes.

[52] Winthrop's son obtained one of Edmond Halley's telescopes and donated it to Harvard College—the 'scope vanishing in the great Harvard College fire of 1764. See Bedini, S.A., *Thinkers and Tinkers* (Scribners, New York, 1975), pp. 74–76. See also Jones, B.Z., Boyd, L.G., *The Harvard College Observatory, 1829–1919* (Harvard University Press, 1971), p. 14. Halley's quadrant was used by Thomas Brattle to observe the solar eclipse of June 12, 1694 where the longitude of Boston was deduced to be 4 hours 43 minutes or 70°45′ west of London (of course with help from observers in London).

[53] Or 1643, as the governor used a system in which January was regarded as the eleventh month of the previous year. Rizzo, P.V., Schove, D.J., "Early New World Aurorae, 1644, 1700, 1719," *JBAA, Papers Communicated to the Association*, Vol. 72, pp. 396–397, 1962.

A month later, Winthrop made another report on this type of phenomenon off the coast of Boston:

> The 18th of this month two lights were seen near Boston ... and a week after the like was seen again. A light like the moon rose about the N.E. point in Boston, and met the former at Nottles Island, and there they closed in one, and then parted, and closed and parted divers [that is, many] times, and so went over the hill in the island and vanished. Sometimes they shot out flames and sometimes sparkles. This was about eight of the clock in the evening, and was seen by many.

These were probably northern lights, termed "aurora borealis" by a European contemporary of the Winthrop's,[54] the shapes of which even assumed the proportions of terrific electrical flashes, wherein people often imagined flames (like Winthrop did); persons, armaments (swords, etc.) and portents (of wars, etc.). In some corners of Earth, people probably still do.

---

[54] Probably de Mairan (1678–1771) in *Traite Physique et Historique de l'Aurore Boreale.*

4

The Maunder Minimum: Europe, Asia, North America (As Dated from c. A.D. 1650–A.D. 1720)

North Atlantic/Western Europe and Mediterranean Basin/North African Region

From February, 1653 until January of 1660 the French astronomer and instrument maker Jean Picard—fresh from observing with Gassendi, in Paris—observed no sunspots.[1] Again, in 1654, the Thames was frozen solid in the Winter once again. This seemed to be part of a recurring pattern of ocean currents and cold air in Northern Europe. In 1657, the sea between the large Danish islands froze, allowing a Swedish invading force to march straight across—and hence to victory.[2] At the furthest point between these islands the distance is nearly fifty kilometres, and an entire army with horses and artillery marched almost this far. Also, from January 1675 through to 1688, the observer Siverus did not see a single sunspot, according to Hoyt and Schatten. Neither did the diligent Jean Picard at about the same time (1676–1683).[3]

Sunspots were then seldom seen, if at all. Two recorders, one named Fogel—who was one of Picard's associates—and the other, Weigel, saw no spots on the Sun from 1661 to 1671, and from 1662 to 1664, respectively, while the Thames routinely froze over in the winter.[4] When a spot *was* seen, it was nearly a *cause célèbre*.[5] For not only were the spots big or in groups, but they often hung onto the

---

[1] Ibid, Hoyt and Schatten (*Solar Physics*, 1995) Keill reported this from Picard's now-lost notebooks.

[2] Carl X defeated Denmark after crossing the "Stör Bält" (Big Belt), a larger inland sea between two Danish Islands, from the west (Jutland). At its narrowest point, the distance is perhaps 4–5 km. At its largest, closer to 50 km.

[3] Ibid, Hoyt and Schatten (*Solar Physics*, 1995).

[4] Ibid, Hoyt and Schatten (*Solar Physics*, 1995). Fogel is also seen as "Fogelius," or "Vogel."

[5] Or, in the words of John Eddy, enough to occasion the writing of a paper on it (which Boyle did).

sun for longer periods of time. This was in contrast to more active solar times, such as now, when the spots fairly speed past our eyes across the Sun's face. So the sight of spots in Maunder Minimum times often occasioned much curiosity and ease and length of study.

Robert Boyle, a Royal Society member (and later, Newton's valued friend) described a spot he and other "ingenious persons"—one being Robert Hooke—noted on April 27 (through to May 25th) 1660—sometimes, but perhaps not always, using a telescope:

> *Friday, April 27, 1660*
> About 8 of the clock in the Morning, there appear'd a [large] Spot in the lower limb of the Sun, a little towards the South of its Aequator ... it disappear'd upon Wednesday Morning (May 9th) though we saw it the day before, about 10 in the morning, to be near about the same distance from the West-ward limb a little South of its aequator, that it first appear'd to be from the East-ward-limb, a little South also of its aequator. It seem'd to move faster in the middle of the Sun than towards the limb. It was a very dark spot almost of a quadrangular form, and was enclosed round with a kind of duskish cloud ... we first observ'd this very same Spot, both for figure, colour and bulk [on May 25th], to be part of the same line it had formerly traced ... At the same time, there appear'd another [smaller] Spot.[6]

Despite careful and regular viewing, Picard saw only a few spots from 1661 to 1665[7] and was excited to observe what for him was their return after a six-year absence on August 11, 1671.[8] Even if some observers of note saw no spots, despite "careful observing", this did not mean they were correct. And, in these pre-email times, contact was not often well kept between the observers. Immediate contact was seldom kept between astronomers in these times except in rare cases.[9]

---

[6] Boyle, R., "An Observation of a Spot in the Sun." First printed in the *Philosophical Transactions*, Vol. 74, p. 2216, for April 27, 1671.

[7] Ibid, Hoyt and Schatten, *The Role of the Sun in Climate Change*, p. 22. "Between 1653 to 1665. Picard saw only one or two sunspots."

[8] Flamsteed, then-Royal Observator (and first of what became known as Astronomers Royal) as early as Jan 31, 1671 reported after numerous observations over the period, that he "could never see any Macula [that is, spots] upon him but his whole disck, perfectly cleare." (From *The Correspondence of John Flamsteed, the First Astronomer Royal*, Vol. I, 1666–1682, Compiled by E.G. Forbes, Institute of Physics Publishing, Philadelphia. Letter 84, Flamsteed to Collins.)

[9] Transits of Venus being one.

Nor were observations in some cases made so assiduously as claimed—especially by casual or half-hearted observers. To digress, Chinese observers in 1665, for instance, were either more fortunate or more accurate than some Europeans were between 1661 and 1671 and often reported on the solar dimming and occasional spot phenomena. One report for 15 February stated: "on the Sun there was a black shimmering light,"[10] and another for the 20th, stating "two black spots on the Sun, moving about for a long time."[11] A Korean report on May 22, 1660 described a spot in Galilean terms: "black vapor on the Sun", the same that was noticed by Robert Boyle almost simultaneously (on May 25th). Here we see an early—though uncoordinated—synergy of Eastern and Western astronomical observation.[12]

Again, the lengths of these spots' stay on the Sun's face were long compared with current observations, which state that only one sunspot group in about 250 lasts for four solar rotations.[13] In 1661, Polish brewer/astronomer Johan Hevelius (who, like Flamsteed, and long after him, Maunder, formed observing teams with their wives[14]) saw a spot group from February 22 to 26. This same group reappeared in March, from the 12th to the 22nd. Boyle and the Jesuit Jean Picard noted this, and the group appeared again through May and into June, through July, to early August.

So from 9 May to 7 August, the sunspot group had been prevalent 91 days—or, nearly *four* solar rotations (circa 27 days equals one solar rotation). D. Hoyt and K. Schatten indicate that, if Hevelius's February group was the same one seen in August—and this seems to be the case—then this sunspot group would have lasted a total of *seven* solar rotations. This is virtually unheard of. As Hoyt and Schatten conclude in one work, the spots' presence on the Sun's disk were few and longer lasting in these years.[15] And, as noted from the Boyle observation, much study time—at quite a leisurely pace—of sunspots was thus afforded.

Mention of sunspots' absence actually managed to find itself located in verse. For that is how apparently common the topic had been in some quarters. The

---

[10] Ibid, Witmann and Xu.

[11] Ibid, Witmann and Xu.

[12] *Philosophical Transactions of the Royal Society, London* 6, Number 74, 1671. Vol. 74, p. 2216, for Boyle. Korean report in Witmann and Xu.

[13] Ibid, Hoyt and Schatten, *The Role of the Sun in Climate Change*, p. 21.

[14] In Helevius's case, with his bride, Elizabeth. See also Cook, A., *Edmund Halley* (OUP, 1998).

[15] Ibid, Hoyt and Schatten, *The Role of the Sun in Climate Change*, p. 21.

English poet Andrew Marvell wrote a satirical poem, *The Last Instructions to a Painter* in 1667, which contains perhaps the first reference in English to the disappearance of sunspots in the late seventeenth century[16]:

> *... Man to the Sun apply'd*
> *And Spots unknown to the bright Star descry'd;*
> *Showed they obscure him, while too near they please ...*
> *Through Optick Trunk the Planet seem'd to hear,*
> *And hurls them off, e'er since, in his Career.*[17]

In music near this time, Thomas Campion, British poet and madrigal writer, wrote the madrigal *"Never, Weather Beaten Sail,"* the title alone suggesting rough seas, gales, and tempestuous weather. Or it was at least a paean to the rising might of the English navy; at least the fishing end of it perhaps getting a boost, as will be shown—ironically, by the weather.

England in the 1660s improved economically (as did China). A period of royal restoration had many Londoners breathing a collective sigh of relief after the demise of Cromwell's puritanical Commonwealth era. Charles II was the king at this time. The writers, the dramatic performances, circuses and other things that had been officially absent on the English scene since the late 1640s returned.[18] Edmund Halley was "a boy when *Paradise Lost* and *Pilgrim's Progress* appeared. John Dryden, the playwright, was in business. Defoe's *Robinson Crusoe* [was already written, and his] *Moll Flanders* came somewhat later. Addison, Swift, and Pope were all contemporaries of Halley and quite probably known to him."[19] In fact, Defoe would write a well-researched book on the "plague year of 1665" with notes outlining not only the weather in 1664–1665 London, but also the devastation created by the "great fire."[20] Defoe, in his *Journal of The Plague Year*, relates

---

[16] Weiss J.E., Weiss, N.O., *Quarterly Journal of the Royal Astronomical Society*, Vol. 20, pp. 115–118, 1979.

[17] Marvell, A. *"The Last Instructions to a Painter"* (September 4, 1667) (from Weiss and Weiss, 1979, quote.)

[18] See for example the diaries of Samuel Pepys (ague [malaria]-sufferer like John Flamsteed) chronicler of Restoration England. Cromwell—like Pepys and Flamsteed, also had this ague—termed tertian ague in Cromwell's case by Paul Reiter in "From Shakespeare to Defoe: Malaria in England in the Little Ice Age" in *Emerging Infectious Diseases,* Vol. 6, No. 1, January–February 2000, p. 5. Cromwell perished from this in September, 1658.

[19] Ibid, Cook on Halley.

[20] Cook, A., *Edmund Halley* (Oxford University Press, 1995). Though only six when it occurred, Daniel Defoe wrote an account based on factual occurrences called *A Journal of the Plague Year* (Penguin, 1986) later in life. Some of this book's highlights include weather descriptions.

the following description of weather in London from December 1664, to March of 1665:

> But in the beginning of the year a hard frost, which lasted from December to almost March, and after that, moderate weather, rather warm than hot, with refreshing winds, and, in short, very seasonable weather, and also several very great rains.[21]

One solar observer at this time, Weickman, reportedly saw no sunspots for the year that nearly coincided with the English plague year—the year of the "great London fire" (1666–1667).[22]

A city of international commerce, sometimes at odds with the royalty that usually needed it more than it needed royalty, London was a city alive with coffee houses,[23] a growing world-class navy (of which Samuel Pepys and Edmund Halley were the partial architects[24]); intellectual ferment, and "clubbs", some of which were formed overnight and vanished nearly as quickly. Robert Hooke,[25] astronomer and secretary of the Royal Society that Halley would one day become secretary of, was out and about in a social whirl that, "like Athenians did in classical times," Londoners hid amidst to "retreat from savage political conflicts."[26] An excerpt from Hooke's journal for 18 December, 1675 reveals the London weather when he makes a reference to "Wild Cold."[27] Meanwhile, Hooke insisted

---

[21] Defoe, D., *A Journal of The Plague Year* (Penguin, 1986).

[22] Ibid, Hoyt and Schatten (*Solar Physics*, 1995).

[23] Spreading rages for these things is not new: in Salem Village at this time, a "London Coffee House" was founded in 1698 off what is now Essex Street in Salem, Massachusetts (it is often forgotten that New England was a colony of England for over 150 years, and whatever was in style in England, was usually imitated, there.) To illustrate how close the colonial ties were to Britain then, George Downing (of 10 Downing Street fame)—a colonial resident of what is now Peabody, Massachusetts (known then as "Brooksby") was a prominent member of Oliver Cromwell's cabinet. Downing hired Pepys as a clerk under Commonwealth times (1658).

[24] Ibid, Cook on Halley.

[25] Espinasse, M., *Robert Hooke* (William Heinemann Ltd, Toronto, 1956) on "Hooke's Social Life."

[26] Ibid, Cook on Halley, p. 9.

[27] Ibid, Espinasse, "Hooke's Social Life." Hooke wrote, "Wild Cold" in reference to this night. Whether Hooke meant a person named Wild was complaining of the cold, or was being imperative on describing how cold it was that day, is up for debate. Obsolete capitalization conventions from that time confuse the meaning here.

that there was water in the Moon (there are indications that there are water ice caps under permanently-shaded regions at the lunar poles) and they "discussed aids for crossing bogs and walking on ice."[28]

Where ice had covered the Thames for periods of days in recent years—probably 1621, 1635, 1649, and 1655—not to mention a good many years in between,[29] the mid-1670s (particularly 1677) would record it freezing over hard for months on end,[30] perhaps necessitating skating as a mode of transport. H. H. Lamb quotes from E. Walford's 1887 book, *Frost Fairs on the Thames* that in the winter of 1663 and in the winter of 1666 (1666 had a hot summer) the Thames froze very hard. [31] Depending on whose side you were on it was a good idea, elsewhere that year, to walk the ice. In the "colonies," possibly while Hooke and the architect Christopher Wren spoke, Her Majesty's troops crossed the ice on a large swamp near Kingston, Rhode Island and managed to kill 600 Narragansett native Americans this same winter (1675).

The "Royal Observator" of the Greenwich Observatory (as the post of Astronomer Royal was then referred to) John Flamsteed had developed "ague," which could have been malaria[32] sometime around 1670. Having suffered from

---

[28] Ibid, Espinsasse, p. 109.

[29] Lamb, H.H., *Climate Present, Past and Future Vol. 2, Climatic History and the Future* (Methuen, London, 1977). See appendices to Part III in particular, pp. 568–569.

[30] Ibid, Hoyt and Schatten, *The Role of the Sun in Climate Change,* p. 182. Also, Lamb.

[31] Walford, E., *Frost Fairs on the Thames* (London, 1887).

[32] For an insightful discussion on Malaria (Italian for "bad air," that is, *mala aria*) in England during the non-uniformly cold period called the "Little Ice Age," read Reiter P., "From Shakespeare to Defoe: Malaria in England in the Little Ice Age," in "Perspectives," *Emerging Infectious Diseases,* Vol. 6, No. 1, January–February, 2000, pp. 1–11. [We are grateful to Sallie Baliunas for originally calling W. Soon to this paper]. Reiter commented that "During the Medieval Warm Period, mention of malaria-like illness was common in the European literature from Christian Russia to Caliphate Spain. ... The English word for malaria was *ague*, a term that remained in common usage until the 19th century. The Medieval Warm Period was already on the wane when Geoffrey Chaucer (1342–1400) wrote in the *Priest's Nun Tales*: *You are so very choleric of complexion/Beware the mounting sun and all dejection, /Nor get yourself with sudden humours hot; /For if you do, I dare well lay a groat/That you shall have the tertian fever's pain,/Or some ague that may well be your bane.* Such mention of agues did not disappear when the coldest years of the this non-uniform cold period began. Indeed, in 16th century England, many marshlands became notorious for their ague-stricken populations, and remained so well into the 19th century. William Shakespeare (1564–1616), who was born in the autumn of Brueghel's first fierce winter, used the word in eight of his plays.

rheumatism and other unexplained illnesses his entire life, he then began suffering from this ailment. Flamsteed, who had never left England in his lifetime, probably got the sickness as it was spread by mosquitoes, which some attribute to the marshy areas around Greenwich at this time (malarial and yellow fever episodes were common in northern mid-latitudes in the 1600 and 1700s[33]). That the prevalence of this and other recorded viral illnesses have all but vanished in the Northern Hemisphere is a point worthy of interest to climate research, relative to prevalence or non-prevalence of certain kinds of sicknesses during extended solar minima or maxima.

In 1676, Flamsteed[34]—whose wife Margaret was a valuable mathematical assistant[35]—noted a sunspot, "the firs [*sic.*] I ever saw" (that is, the first in some time) which he suspected might take a week to cross the disk.[36] Observations were probably made of this same spot in between July and August by Edmund Halley as well as by Cassini at Meudon, in France (the Paris Observatory). The spot's motions spawned the idea that this could have been caused by the "rotation of the sun about his Inclined axis."[37]

References to paucity—and in some cases, lack of movement—in sunspots by Flamsteed and his numerous collaborators continued in this vein. Flamsteed

---

(*continued*)

For example, in *The Tempest* (Act II, Scene II), the slave Caliban curses Prosper, his master: *All the infections that the sun sucks up/From bogs, fens, flats, on Prosper fall and make him/By inch-meal disease!* but is later terrified by the appearance of Stephano, who mistaking his trembling and apparent delirium for an attack of malaria, tries to cure the symptoms with alcohol: ... *(he) hath got, as I take it, an ague ... he's in his fit now and does not talk after the wisest. He shall taste of my bottle: if he have never drunk wine afore it will go near to remove his fit ... Open your mouth: this will shake your shaking ... if all the wine in my bottle will recover him, I will help this ague.*" It is also peculiar to note that the presumed location for *The Tempest* was Bermuda.

[33] See P. Reiter (2000) and Foster, K.R., Jenkins, M.E., Toogood, A.C., "The Philadelphia Yellow Fever Epidemic of 1793," *Scientific American*, August, 1998, pp. 69–73. Discussions on the complex factors controlling such diseases are given in Reiter. It was stressed that malaria is endemic and common in many temperate regions, and major epidemics have been found extended as far north as the Arctic Circle, so there is often the incorrect claim that the hypothetical future man-made global warming will cause more "tropical" disease like malaria.

[34] Murdin, L., *Under Newton's Shadow* (Adam Hilger, Boston, 1985).

[35] Ibid, Murdin.

[36] Ibid, Forbes, Letter 268, Flamsteed to Moore, July 27, 1676.

[37] Ibid, Forbes, Letter 271, Flamsteed to Oldenburg, 30 November, 1676.

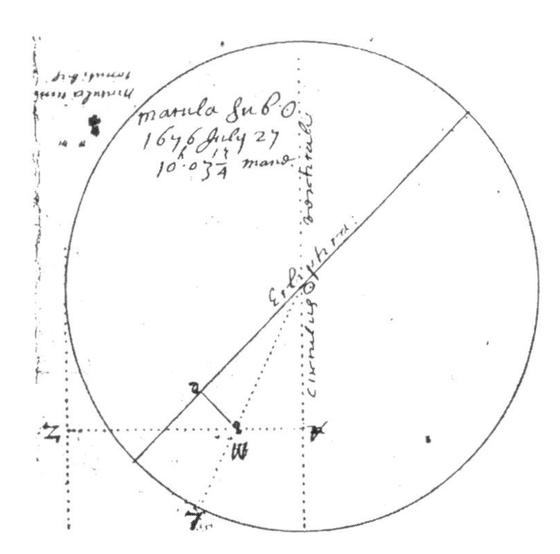

Fig. 9. Flamsteed's drawing of the July 27, 1676 sunspot. Flamsteed carefully noted that this image is inverted "as it appeared through the tube" (after Forbes, Letter 268 of John Flamsteed).[38]

reported a spot from July, 1679, in a Gresham College lecture as having "kept the same Constant place in his body."[39]

An interesting Flamsteed correspondent, the day-labourer but also methodical and careful solar astronomer Stephen Gray[40, 41] perhaps informed him of faculae; these are bright streaks or flashes seen on the Sun. He also related precise measurements and descriptions of spots to Flamsteed. By 1681 another of Flamsteed's correspondents, John Wallis, confessed that "As to the Faculae solares, which you seem to disbelieeve, [*sic.*] I can say nothing, having never seen any ... and scarce any of the maculae (sunspots)"—though he admitted that more spots were being seen lately.[42] In May 1684, Flamsteed commented to a friend on the absence of spots, followed by the

[38] Ibid, Forbes. For good insights into how accurate micrometers were in these times, or how the development of compound eyepieces went, read "The Development of Micrometers in the Seventeenth, Eighteenth, and Nineteenth Centuries," by Randall C. Brooks, (*JHA*, xxii, pp. 127–173, 1991) and "The Development of Compound Eyepieces, 1640–1670," by Albert Van Helden, (*JHA*, viii, pp. 26–37, 1977).

[39] Clark, D.H., Murdin, L., "The Enigma of Stephen Gray Astronomer and Scientist (1666–1736)," from *Vistas in Astronomy,* 1979, Vol. 23, pp. 351–404 (Pergamon Press, U.K.). Footnote, p. 378.

[40] Gray, whose name will probably become better known now that his contributions to solar astronomy are becoming familiar, eventually became an F.R.A.S. and a winner of the Copley Medal, after Newton's death. Murdin speculates that there may have been blockage of his letters to the Royal Astronomical Society (RAS) as Gray was perceived to be the ally of one whom Newton often disagreed with John Flamsteed (Ibid, Murdin, *Under Newton's Shadow*).

[41] Ibid, Clark and Murdin.

[42] Ibid, Forbes, Wallis to Flamsteed, August 13, 1681.

remark, "tis near 7½ yeares since I saw one before they have been of late so scarce how ever frequent in the days of Galileo and Scheiner."[43]

The reference to the "days of Galileo and Scheiner" is a curious one. Astronomers such as Flamsteed were well versed in Galileo's astronomy, and must have pored over books like *Rosa Ursina* by C. Scheiner, wondering why sunspots were neither so numerous, nor (possibly) so fast, as in these mens'

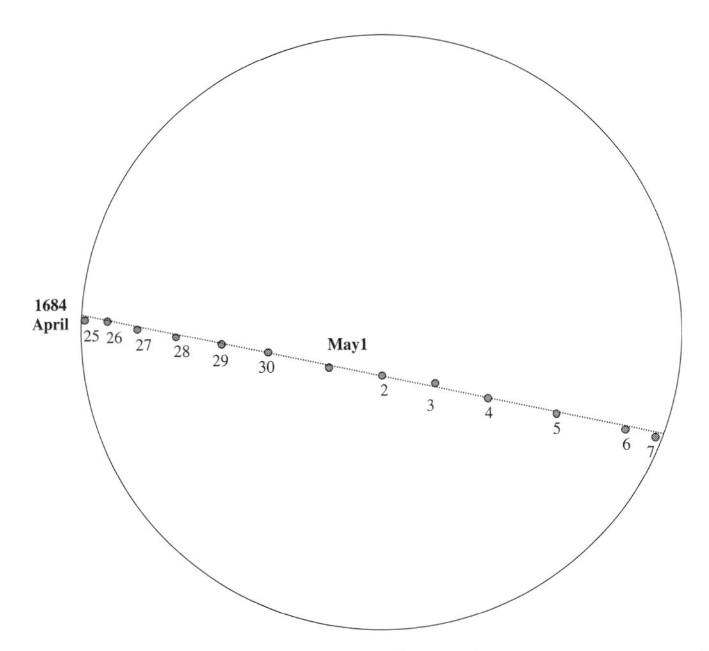

Fig. 10. Reproduction of Flamsteed's illustration of a passing sunspot, to support his theory of spots. He claimed the revolution of any point of the Sun to be "compleaste [*sic.*] in 25 dayes six hours praecise."[44] Flamsteed claimed to see this same sunspot in July of this same year, when activity witnessed by him in this group may have been a sign of the coming increased solar activity—or maybe not, since spots did not reappear for some years.[45, 46] (Sunspots moved over the Sun's disk slowly over many solar rotations in these years, and were not very common.) (Drawing adapted from Flamsteed's Letter 512)

---

[43] That is, a sunspot: he claimed that the last one he saw before April 25, 1684, was in December, 1676. Ibid, Forbes, Letter 521, Flamsteed to Molyneaux, 2 May, 1684. (*Nota bene:* he also described one in 1679 in a Gresham College lecture.)

[44] Another figure is, "25 daies 9½ hours sufficiently exact," done by Flamsteed and Halley (Ibid, Clark and Murdin).

[45] Ibid, Forbes, Letter 516, Flamsteed to Caswell.

[46] Ibid Forbes, Letter 745 (written 3 May, 1698) Flamsteed to Leigh. "As for Spots on the Sun there have been none since 1684." See also Hoyt and Schatten and their discussion of the significance of this.

day: or as to why they were so big, as well as so strangely rare. For from the May 1684 comment—like Boyle's view on his spot—Flamsteed noted: "When the Spot was near the middle of the Sun it appeared very broad and almost square, the nucleus of the same figure about 40" in diameter."[47]

Both Flamsteed and Gray[48] tied sunspot movement to solar rotation. This was remarkable work, considering it is hard to imagine the conditions these London astronomers worked under—Gray, as well as the immensely better off Flamsteed. For one thing, the air was continually smoke-filled.[49] This affected the ability to see, though perhaps it aided in causing a natural filter to be made so that absences or abundance of phenomena could be seen on the Sun's disk. It was not only difficult to see, in general, in London—but to breathe, as well, the air being filthy with sometimes-toxic smoke. The smoke "rose in a cloud from the furnaces of the soap-boilers, dyers [Gray's occupation], brewers, and limeburners, whose shops and factories were wedged in among the tenements of the old city."[50] Add to this the foul smell of uncollected horse dung and other faeces, and the olfactory picture becomes even more complete.

Gray's own dyer's establishment was located in this section of town, and we can imagine that conditions were hardly much better in the latter half of the 1600s than they were at mid-century, before the Great London Fire of 1666. Houses lay so close that people could touch hands across alleys filled with garbage that poor souls known as "scavengers" came to remove only when the stench became overwhelming. In these less-than-romantic conditions, Stephen Gray performed his science, in between rounds of back-breaking and tedious labour. It is almost incredible that anyone in his state would have had the energy to accomplish anything in this regard, then, at all—let alone the quality achieved. It was conditions such as these which fueled the writings of Maunder Minimum thinkers such as Francis Bacon who lamented man's "days, few and evil," spent amid such ignorance, misery, and filth.

---

[47] Flamsteed, J., Philosophical Transactions of the Royal Society of London, 14, 535, 1684 (as quoted in "Notes," *JBAA,* Vol. 51, No. 3, 1941, p. 101).

[48] Before Newton became RAS president, Sir John Hoskins asked Gray to make observations to determine the rate of the Sun's revolution on its axis. He did so, with letters on sunspots and relevant speculations. But the letters were never published. (Ibid, Murdin, *Under Newton's Shadow*)

[49] Gould, H., *Sir Christopher Wren: Renaissance Architect, Philosopher, and Scientist* (Franklin Watts, New York, 1970), Chapter 9, "London Aflame," pp. 103–114.

[50] Ibid, H. Gould, p. 104.

So, as it can be seen, much progress was made, not only in observing the recurring phenomena, but also measurements over the exact period of solar rotation. Known to many was the odd fact that spots moved slowly and were often big, as well as few. Over two hundred years later Walter Maunder and his wife, Annie, would write this about spots and the "twenty-seven day rotation":

> The sun appears to turn around in about twenty-seven days. In reality, he turns round in less time by a couple of days, for as the earth is traveling round the sun in the same direction as the spots do, we keep the spots longer in view than if we remained still; just as we might follow down the platform a train that is moving out of the station, to keep a departing friend a few seconds longer in sight.[51]

* * *

On the Thames River in England in the winter of 1683–1684 people could cross from Parliament to the other side on carriage and foot. There was even a winter, or frost, fair[52] held on this river that year—admittedly, something that was not regularly possible earlier than two decades before, or much later. Not far away on the mainland in 1684 the Lake of Bodensee also froze over, as it did in 1672, and would again in 1695.[53]

During the frost fair on the Thames, perhaps visiting colonists discussed the wicked weather experienced in New England just a few months before, for as they skated, the London Bridge had water frozen around and beneath it.[54] Though this 1684 winter in England was considered to be the country's harshest in the entire post glacial period—and was also perhaps quite terrible for Scandinavia[55] too (as the British Isles' anticyclone[56] weather system covered this region as well)—but it was *not* the coldest in Europe for the past 300 years.

---

[51] Maunder, A.S.D., and Maunder, E.W., *The Heavens and Their Story* (Dana Estes, 1908), p. 111.

[52] Soon, W., Yaskell, S.H., "Cycles of the Sun," *Astronomy Now*, October, 1998, p. 15. This is also pointed out by several other authors.

[53] Ibid, Lamb, *Climate Past, Present and Future*, p. 590.

[54] Ibid, Hoyt and Schatten, *The Role of the Sun in Climate Change*, p. 182.

[55] Close by the Scandinavian anticyclone, in 1700, Nova Zemlya (New Land, in Russian) was reported by Halley to have undergone a three week long fog and gale.

[56] Dragàn, J.C., Airinei, S., *Geoclimate and History* (NAGARD, 1989), p. 330.

Fig. 11. Contemporary drawing of the Thames' winter, perhaps called the Thames' "frost"[57] fair of 1684. Note pavilions and replica ships filled with people being drawn as if in taxis—a testament to the thickness of the ice. (After Lamb, 1982)

The previously-mentioned ice walking was perhaps common in Scotland, where a small lake to the north, upon which there "was always ice, even in the hottest summer",[58] was located. This "tarn" was about 750 meters above sea level at 57°22′ N, 5°02′ W, and today, versus the 2.5 degree Celsius (possibly more) lowering in temperature reported at that time, we can assume that the lake is a lot less icy in the winter today. In fact, English travellers from 1600 to the late eighteenth century routinely reported seeing permanent snow on the Cairngorm Mountains (1,200 to 1,300 metres) where only a few semi-permanent snow beds are seen, "in this or that hollow"[59] at that elevation now. Ice capping on other mountain tops is also noted in some flimsy sources from Africa. It is not confirmed for Maunder Minimum times, but Portuguese travellers in East Africa described greater snowlines on Mount Kilimanjaro and Mount Kenya.[60] More reliable Maunder Minimum activity shows up in the north central parts of Africa, where famines due to drought occurred frequently, somewhat like the Mediterranean

---

[57] Ibid, Ludlum, p. 35.

[58] Lamb, H.H., "Climatic Variation and Changes in the Wind and Ocean Circulation: The Little Ice Age in the North Atlantic," *Quaternary Research*, Vol. 11, 1–20 (1979), p. 15.

[59] Ibid, Lamb, p. 15.

[60] Ibid, Lamb, "An account ...," p. 304.

case, on Crete.[61] Weirdly—like the Scottish incidents where ancient villages were covered over with sand near this time—the traveller/diarist John Evelyn reported sand dunes overtaking estate lands in a North African context (Libya) in 1677—a scene repeated in parts of Turkey and other parts of central Europe at about this same time.[62]

H.H. Lamb[63] commenting on the year 1675 of Hooke's "Wild Cold," but from the context of the mid-Atlantic, had this to say based on research on polar and North Atlantic water-current convergence:

> Between 1675 and 1700, the water temperatures prevailing about the Faeroe Islands presumably were on overall average 4 degrees to 5 degrees Celsius below the average of the last 100 years.

Surges southward of cold water from the polar cap—forcing the northward-bound, warmer tropical Atlantic water backward—deadened the local cod industry off the Icelandic coast. This occurred even if deeper-ranging vessels in the warmer currents near Iceland had better fishing, overall. The fish's kidneys react badly to temperatures below 2 degrees Celsius,[64] which caused their diminution in previously populous areas off the south-western Icelandic coast. So this fishing industry diminished. But "total failures" in this industry were reported for 1625, 1629, *and no cod at all, from 1675*, onward.[65] Fish exports from here did not significantly increase until 1750.[66] But the fleeing fish gave rise to a growing southern North Sea fishing trade, where the fish now resided, forcing Norwegians and Icelanders out of this market. This occurred while a literal diaspora of Scots from the north of England, to Ireland, took place, where they joined the Irish poor. The Scots also fled to North America.

On bad terms with England, and burdened with climate-irritated growing seasons on land, coupled with poor fishing off the coast, many Scots turned to combat arms as mercenary soldiers. It was during the politically as well as

---

[61] Lamb, H.H., *Climate, History and the Modern World* (Methuen, London, 1982), p. 226.

[62] Ibid, Lamb, *Climate, History and the Modern World*, p. 232.

[63] Ibid, Lamb, H.H., *Quaternary Research*, Vol. 11, (1979), p. 15.

[64] Ibid, Lamb, H.H., *Quaternary Research,* Vol 11, 1–20 (1979), p. 11. For this survivability note, Lamb quotes Beverton and Lee (1965): "Hydrographic fluctuations in the North Atlantic Ocean and some biological consequences, in *The Biological Significance of Climatic Changes in Britain* (Institute of Biology and Academic Press, London).

[65] Ibid, Lamb, *Quaternary Research.*

[66] The primitive fishing equipment is in part blamed for Iceland's lack of being able to reach the cold-avoiding cod, as Lamb states in *Climate, History and the Modern World* (Metheun, London, 1982, p. 207). But, the hazard of a suddenly-changing climate in this lesson, shown, is quite clear.

ESTIMATES OF THE MEAN MONTHLY TEMPERATURE (°F) IN THE LONDON REGION, 1680–1706

These estimates are primarily based on Derham's Upminster record, 1699–1706, partially overlapped by Locke's incomplete record, 1692–1703. Earlier years are less reliable, as there is no satisfactory overlap with the later records. Estimates for this period, and for Locke's missing months (in brackets) have taken into account the daily observations of wind and weather in other contemporary journals, and are given to whole degrees only (1699 onward to 0·5°).

	Jan.	Feb.	Mar.	Apr.	May	June	July	Aug.	Sept.	Oct.	Nov.	Dec.	Year
1680	(41)	(39)	(43)	45	52	57	63	61	60	52	45	34	49·3
1681	34	36	40	48	53	59	61	63	59	54	45	38	49·2
1682	44	37	41	45	55	59	61	60	57	51	43	43	49·7
1683	39	37	43	51	55	63	62	58	57	45	41	33	48·7
1684	27	31	38	45	57	61	63	62	56	53	38	40	47·6
1685	34	39	42	49	56	60	59	60	55	54	46	44	49·8
1686	45	44	46	49	56	62	63	60	57	50	45	43	51·7
1687	39	41	41	45	54	57	62	61	53	53	44	43	49·4
1688	39	36	39	43	53	57	62	60	55	46	40	38	47·3
1689	34	41	42	48	53	56	62	61	57	48	42	41	48·7
1690	40	41	41	48	51	57	62	61	56	49	45	41	49·3
1691	35	35	42	45	52	58	61	62	54	(50)	(42)	39	47·9
1692	37	32	40	47	50	58	61	61	54	45	42	39	47·2
1693	38	42	38	45	50	60	61	61	55	51	44	38	48·6
1694	33	42	39	47	50	57	61	57	52	(47)	43	(37)	47·2
1695	31	33	39	43	50	(57)	58	(57)	(54)	50	(43)	40	46·2
1696	43	41	39	43	53	(57)	(62)	(62)	(55)	(50)	(43)	37	48·7
1697	35	34	43	46	55	57	(62)	(60)	(56)	(50)	(40)	37	47·8
1698	33	34	39	47	49	56	61	(61)	(57)	(50)	40	39	47·2
1699	38·5	39	40·5	45	51·5	60·5	65·5	61	58	50·5	42·5	39	49·3
1700	40	37·5	39·5	45	55·5	58·5	61	61	57·5	49·5	41·5	40	48·9
1701	38	37·5	38	41	53	59·5	67·5	63	60·5	47	44·5	39	49·0
1702	42	45	44	44·5	52·5	58	61	63	60	51·5	41	40·5	50·3
1703	36·5	40	43·5	49	55	59	63	63	52·5	47	46	42	49·7
1704	36·5	38·5	43	49	54	60	64	64·5	55·5	48·5	44·5	39	49·7
1705	37	39·5	41	47·5	53·5	56	62	65·5	55·5	49·5	40	40·5	49·0
1706	37·5	40·5	45	50	55·5	62	63	64	56·5	53·5	44	41	51·0
1681–90	37·5	38·3	41·3	47·3	54·3	59·3	61·7	60·6	56·2	50·3	42·8	40·4	49·1
1691–1700	36·3	37·0	39·9	45·3	51·5	57·8	61·3	60·4	55·2	49·3	42·1	38·5	47·9
1701–10	37·9	38·5	42·4	47·4	54·2	59·6	62·7	63·4	57·7	49·7	44·6	40·5	50·0

For the years 1707–10 approximations are based on Dutch monthly means, adjusted by reference to contemporary English accounts.

Probable values at Upminster today, based on *rural* stations around London:

	Jan.	Feb.	Mar.	Apr.	May	June	July	Aug.	Sept.	Oct.	Nov.	Dec.	Year
1921–50	39·9	40·3	43·8	48·3	54·1	59·8	63·4	62·7	58·5	51·3	44·4	40·4	50·6

Fig. 12. Estimated mean monthly temperature (Fahrenheit) in the London region, 1680–1706. (After Manley, 1961)

climatically-cruel mid-1600s that the Scottish mercenary soldier fled Scotland for Russia, Sweden, Denmark, and France, if not elsewhere, becoming a stock figure in the works of Sir Walter Scott.[67] To this day, for example, there are seventh or eighth-generation Swedes with Scottish last names. These trace their ancestry to mid-1600s Scotland. (Notably, the Russian—and even Swedish—yearly weather was milder, on average, during the Maunder Minimum.)

[67] The part of this paragraph dealing with the Scotland case paraphrases H.H. Lamb in *Climate, History and the Modern World* (Methuen, 1982), pp. 209–212.

The summer weather in England in these years was moderate and dry, with—importantly enough—an improving fishing industry. Coupled with the fishing and other wealth of the British North American colonies, at least urban England was faring well, weather aside. A comet in the summer of 1684 purportedly announced the coming of "the dry season."[68, 69] Concerns in any case had fallen to the weather, and its importance, most likely to navigation and trade. Thus it comes as no surprise that Robert Hooke and colleagues at the Royal Society were already keeping daily weather records[70] in London and its vicinity in these years (1660s/1680s-early eighteenth century).

The thermometers and barometers were understandably primitive. "Scale readings were liable to change with time, as the glass was not annealed, the spirit [no mercury yet] was not standardized and the bore of the tube might be irregular."[71] Nevertheless, from several weather observers—Locke, Hooke, Derham, Gadbury, and others—deductions were made regarding the meteorology of Southern England between 1680–1720. (One wonders what these pioneering weather observers would have said, were they shown that trees would act as thermometers one day![72]) In England, "this resulted in the high price of corn recorded in the phrase 'King William's dear year'" with Scotland noting—as previously implied without surprise—"seven ill years,"[73] In Scandinavia, where many Scots

---

[68] Ibid, Bernard to Flamsteed, translation of letter 517 from Latin to English. Bernard's knowledge of Theophrastes is notable here, as Theophrastes connected comets to dry weather, and a comet had been seen.

[69] Theophrastes. *Enquiry into Plants and Minor Works on Odours and Weather Signs*. Trans. Sir Arthur Hort (Heinemann: New York, MCMXVI), p. 415. It is interesting to see how long such beliefs by Pliny, Aristotle and Theophrastes persisted into near-modern times.

[70] Manley, G., "A Preliminary Note on Early Meteorological Observations in the London region, 1680–1717, with Estimates of the Monthly Mean temperatures, 1680–1706," *Meteorological Magazine,* Vol. 90, 1961, p. 304. Even John Locke was among these recorders.

[71] Ibid, Manley, p. 303.

[72] Libby, Marshall, Leona *et al.*, Isotopic tree thermometers, *Nature*, Vol. 261, May 27, 1976.

[73] Ibid, Lamb, *Climate, History and the Modern World,* p. 212. England was sending Scot Presbyterians to Ulster as early as 1608, to disrupt Irish nationals (see Paul Johnson's *A History of the American People* [Weidenfeld & Nicolson, 1997], p. 18.) "Early in the 1600s, Ulster was made a theater of the largest population transfer ever carried out under the crown—thousands of Scots Presbyterians being allocated parcels of confiscated Catholic land along a defensible military line running along the Ulster border: a line which is still demographically significant to this day and explains why the Ulster problem still exists, in part. This major Ulster planting took root because it was based on agriculture and centered around hard-working Scots lowlands farmers who were also ready to take up arms to defend their new possesions." By mid-century, when Scots were forced to flee Scotland in increasing numbers to places such as Ireland, due to weather, Catholic-Protestant animosity must only have been all the more inflamed.

had fled, poor Swedish harvests and late springs were common, but probably tolerable. Winter weather in this entire anticyclone (England to Scandinavia) could be described generally as blustery, cloudy, and cold for most of the season, heading well into spring.[74] This source also refers to Scandinavian (including Iceland, and primarily Norway and Finland) having a noticeable glacial advance in the first quarter of the eighteenth century (1715–1720).[75] From 1691–1706, "snow frequency was about as twice that we should expect today."[76] This same source corroborates the 1695 surrounding of Iceland by ice packs as far out as could be seen with the human eye. The pushing southward of cold water—and hence the fish—even produced—purportedly—an Eskimo in his kayak, found floating on the river Don in Aberdeen, Scotland in these years. If true, this event was certainly a cruel joke to people needing meat.[77]

Figure 12, showing monthly averages for the years between 1680 and 1706—using mainly Derham's measurements—gives an idea of the relative coolness of London's air at the time.

The diligent solar observer and star-cataloguer G. F. Maraldi saw no sunspots from the Paris Observatory, it is reported, from March 1689, to May 26, 1695. Siverus also saw no sunspots from 1689–1690.[78] To underscore this, John Flamsteed at Greenwich Observatory saw none, either: this in spite of the fact that he had bevies of correspondents keeping him abreast of the latest solar news. These observations—or perhaps the lack of them—are laid side by side with the evidence for polar ice cap recrudescence, cooling North Atlantic water and air temperatures; the re-dispersal of fish populations, and the absence of much aurora over these years.

The winter weather was now very cold in England—especially in 1695. It was worse than it had been in 1683–1684—but perhaps not enough to cause excessive misery in the South, with London's relative wealth pouring in, as well as being

---

[74] The year 1695 for England, Manley points out as being as bad as 1879 in England. Was the weather just as bad in Scandinavia, lying behind the poor harvests that sent most of southern Sweden to Denmark, and then to America in the late 1800s?

[75] Ibid, Manley, p. 305.

[76] Ibid, Manley, p. 309.

[77] Lamb, H.H., *Climate, History and the Modern World* (Methuen, London, 1982), p. 209. Related reports were not isolated incidents, apparently: there were six such reports of Eskimauxs (or Inuit) hunters in kayaks being sighted through these years in Scotland (Ibid, Reiter, "From Shakespeare to Defoe: Malaria in England in the Little Ice Age," p. 2).

[78] Ibid, Hoyt and Schatten (*Solar Physics*, 1995).

within reach of a more prosperous Europe.[79] On the continent, windmill-guarded canals froze enough to give Hans Brinker of the Silver Skates his fame in Dutch folklore and a preferred winter mode of locomotion previously mentioned by Englishmen—skating—becoming apparently widespread.

Deep in the south of Europe around at this time Venice's canals froze.[80] This was a sight which might have surprised Galileo. But even in his own lifetime the weather was often adverse, involving driving snowstorms. Exceptions to this global cooling showed up in three areas in particular during the Maunder Minimum. These are mid-central Europe (including Russian Siberia), parts of Scandinavia and Japan—as previously mentioned.[81] Bad harvests characterized Sweden, and droughts did continue to negatively influence Central Europe, where the surprise noted is that Japan was not so unfavourably impacted, climate wise, compared to its neighbour, China.

In England in October 1697, Edmund Halley—who was interested in the source of rain—gave an account of a "very large hailstorm with very large hailstones at Chester."[82] A month before this, or perhaps the same month, sand completely covered a settlement in the Outer Hebrides (Udal). (The same had happened to coastal land in Scotland in 1675[83] and as noted, in Libya in 1677.)

Stuck in the middle between these extremes of British world power was tiny Scandinavian Iceland, in the throes of a moribund fishing industry perhaps much worse than had ever affected them in the memories of their eldest living relatives. It may even have been the worst fishing witnessed since their Viking forebears arrived some centuries earlier. The year in question was not too far from 1697: that is, the year 1695. Sea ice virtually surrounded the island for months with only one point on the west coast being free.[84] Water (that is, the ocean) could not be seen from the highest mountains. This same year, the fishing industry along the entire coast of Norway failed, "south to Stavanger" except in isolated fjord pockets

---

[79] Ibid, Dragàn and Airinei.

[80] Difficulty in the severity of events presents itself, here. D. Camuffo and S. Enzi, documenting Northern Italy in the "Little Ice Age" in *Climate Since 1500 A.D.* (eds. R.S. Bradley and P.D. Jones, Routledge, 1995) make a case of the hydraulics work being done on the canal since the 14th century limiting the amount of sea water into the canal works.

[81] Borisenkov, Y.P., in *Climate Since 1500 A.D.*, pp. 173–175, and H.H. Lamb in *Climate, History and the Modern World,* pp. 214–224.

[82] Cook, Alan, *Edmond Halley* (Oxford University Press, 1998), p. 195.

[83] Ibid, Lamb, 1979, p. 16.

[84] Ibid, Lamb, 1979. p. 15.

where warmer North Atlantic drift water allowed fish to survive above where their anatomies failed. We can presume, as Lamb did, that polar water by 1695 had stretched across the whole surface of the Norwegian Sea and south of the Faeroes. (We presume this was the same case, from Iceland to Greenland.) The cod fishing industry fared poorly in the Shetland Islands region as well.

The Isle of Wight in Britain was covered with snow on October 30, 1698 when Edmund Halley sailed past on his way to Cape Verde.[85] (The author of this source—Alan Cook—made the statement that the [English] winters "were colder then.") The conditions in central Norway were no better: farming in the village of Hoset, for example, was abondoned for the second or perhaps third time in the 1690s in a period of marked climate deterioration.[86]

* * *

By the early 1700s, sunspot sightings by the earlier-noted solar astronomer Stephen Gray were compiled by him for publication in the Philosophical Transactions of the Royal Society, but never made it there. Gray was especially productive after 1700, making reports on spots to Flamsteed at Greenwich in June, 1703.[87] Intriguingly Gray, between breaks in his job as a cloth dyer, made reports of sunspots during the first very noticeable 11-year solar cycle[88] seen for quite some time between 1703 and 1705—though this cycle was not then known. (The cycle—called the Schwabe Sunspot Cycle—was only confirmed by examining data much later.) After 1700 solar activity was noted to be inexplicably picking up by careful observers.

Gray was also one of the first to see with his telescope a famous "white flash" (seen again by like-minded astronomers such as Richard Carrington years later). And Gray not only "belieeved in," but witnessed, at this time, the faculae that Scheiner and Galileo noted and other Flamsteed correspondents heard only rumours about[89]:

> The faculae doe not appear, as was said, but when near the limb of the sun [that is, toward the edge] and the nearer thereto the Brighter, the

---

[85] Ibid, Cook, p. 271.

[86] Ibid, Lamb, 1982, p. 192. Lamb further relates "... each time in periods of colder climate."

[87] Ibid, Clark and Murdin, p. 377.

[88] This was figured out in retrospect: Gray had no idea of the 11-year sunspot cycles—this was not known until Heinrich Schwabe's breakthrough in the 1840s.

[89] Proof of the veracity of this observation is found in the "limb darkening effect" on Gray's note that they appeared brighter "when most contrasted." There *is* greater contrast for faculae near the Sun's limb.

> Reason whereof I take to be that Property of flame to apear brighter when most contrasted. ...

and somewhat further on, this very skillful and pioneering observer noted that

> The spots being thrown up soe suddenly argueth that they are ejected with a most violent force, as those that are sensible of their Magnetude will easely belive. They seem to proceed from the Internal parts of the sun being Cast out from thence in the same or some such like manner as sand stones, ashes, etc. are thrown from Mount Etna. The vertical motions of the spots that appeard in March plainly show that parts of the sun are fluid and that these were disjointed from the rest of the solid parts of the sun and that either they floated in the liquid fore of the sun or moved at some distance from it in the Soler Atmosphear.[90] [*sic.*]

Even these rather Galileo/Kepler-like metaphors seem to betray increased solar activity as they, it will be remembered, observed the Sun in more sunspot-active, solar-active times. Curiously, in a much-earlier letter (1684) Flamsteed's correspondent, Bernard, also compared the Sun's antics to "a new Etna." These may have been the kinds of descriptions these witnesses were probably giving to each other when they tried to describe the phenomena on the Sun's face. Interestingly enough, a paucity of sunspots was still being reported. Ägerholm, probably reporting for the famous Danish scientist Ole Römer, saw no sunspots from June 1, 1695 to October 31, 1700: the post-October sightings coinciding perhaps with annual solar maximum. However, the observer Stancarius saw none from January 1, 1697 to December, 1702, either.[91]

That winter in England (1703) a storm associated with a cyclonic centre passed over southern England and became a "legend" in the popular New-England diary style across a cod-poor North Atlantic.[92] At least insofar as an account of it was written by Daniel Defoe called *The Storm.*[93] Much like the coastal North

---

[90] Ibid, Clark and Murdin, p. 378.

[91] Ibid, Hoyt and Schatten (*Solar Physics*, 1995).

[92] Cod fishing was good enough off Georgia's bank, so much so, that the cod became a colonial American symbol. It is not hard to believe that these fish—forced out of higher North Atlantic schools, could have then congregated in any warmer currents off the New England coast. Catch records from colonial New England times confirm that the fishing was plentiful and prosperous in the Maunder Minimum.

[93] Ibid, Lamb, 1979, p. 16. A good perspective of not only the sickness and despair of 1665 and 1666 is found in *A Journal of the Plague Year*, by Daniel Defoe.

American storms, this one left its telltale trail of damaged buildings and uprooted trees, no worse a storm being witnessed in Southern England until very recently.

Northern light events increased world wide around or after 1720. Understandably enough, many were seen in nautically powerful England and New England. A large number of these events were reported from 1710, on. Although the more realistic drawings of these phenomena emanate from the likes of reports by Halley and, significantly, the Danish astronomer and mathematician O. Römer—who witnessed these twice, in February and March, 1707 in Denmark—the overriding verbal descriptions of them up until this time were virtually unanimous as "portents of war" as shown in the examples drawn from *lauboks*[94] of around this time in continental Europe. Like their British and colonial American counterparts called the "chapbook," the laubok was a common press item of the time.

Being sold door to door in some cases,[95] these prints could merely have been the "yellow journalism" of the day. Whether the yellow journalism influenced the description of the aurora, or whether aurora actually influenced the way they were

Fig. 13. Imaginative drawing probably of an aurora during an entire night in Central Europe in the 1730s, when aurora started becoming more abundant. (Courtesy, J.-P. Verdet, 1987)

[94] See Verdet, J.-P., *The Sky: Order and Chaos* (Gallimard, Thames and Hudson, 1987), pp. 95–97.
[95] Ibid, Verdet.

Fig. 14. Another imaginative view of probably what was an aurora, witnessed in Cartagena, Spain in 1743. (Courtesy, J.-P. Verdet 1987)

described is a matter of debate. Verbal descriptions, quite apart from the imaginative artwork, oddly parallel these pictures. For example, this one from the sixteenth century near modern-day Poland:

> On the 12th day of January [in 1570] ... an unusual omen was seen in the sky between the clouds. It lasted for four hours. Firstly, a black cloud, almost like a great mountain splattered with many stars appeared in the sky. Above this cloud was shown a strong band burning with sulfur in the form of a ship. From this arose many burning torches, almost like candles, and among these stood big pillars, one towards the east and one towards the west. The flames were running down along the pillars like drops of blood, and the town was illuminated as if on fire.[96]

From these years and well into the mid 1800s, the sight of aurora could confuse, mystify, and awe superstitious people. Northern lights seen around the sky at lower latitudes supposedly have a more reddish tinge than those from the Northern polar zone. It is now known that the mid-latitude lights are not true northern lights. These reddish glows fooled (and can still fool) many into believing fires were burning, locally or at a distance.

The effects of trailing rivers of blood dripping from things were very real to some minds. To make matters worse for the fearful uneducated and the superstitious, these displays often weirdly fulfilled their imagined prophecies. To wit,

[96] Ibid, *The Northern Light,* p. 7.

"Lord Derwentwater's Lights" of 1716, the verbal descriptions of which persisted in Scotland and elsewhere for many years after they were witnessed.

Having seen all he could see about exotic phenomena regarding weather and astronomical mysteries, Edmond Halley got a chance late in life to witness a northern light display in London in March, 1716. Admittedly in these times it was a rare occurrence, especially at the latitude he was observing from. This display occasioned the famous "Lord Derwentwater's Lights" attributed the Scottish Jacobite rebel, killed by the British, with his blood flowing in a stream on his land: a direct inference being made between the appearance of reddish aurora lights and his blood.[97]

Were these Scottish jacobite emotions spurred on by bad luck due in part from the weather? Or to where the Scots had fled some years earlier—Ireland? One of the oldest surviving Irish diaries on weather to date was kept from 1711 to 1725 by Thomas Neve of County Derry. Though the instrumentation used to measure precipitation was unsurprisingly faulty, giving incorrect measurements, his verbal detail on auroral displays and severe snow and frost over these years managed to endure. Notably, Neve noted the frost of 1715–1716, and the freezing of Lough Neagh, both of which are outstanding for the detail of their description.[98]

After following a rather early Cartesian explanation concerning "vapors" that rose into the atmosphere from Earth, Halley, long a proponent of Earth magnetism,[99, 100] struck upon the idea that northern lights were due to a "magnetic evaporation" from the Earth's interior. By using a magnet and iron filings in a Galileo-like experiment (see Fig. 15), he suggested how aurora bands might describe a phenomenon in this shape in the sky.[101] As facts would bear out, magnetism did indeed have much to do in explaining northern lights. But not in quite the way he reasoned and envisioned it.

Sunspots, as stated earlier, also became more common around this time. Many never knew them to ever have been gone. Fairly recently, solar astronomer

---

[97] Fara, P., "Lord Derwentwater's Lights: Prediction and the Aurora Polaris," *JHA*, xxvii (Science History Publishers, 1996), p. 241.

[98] Dixon, F.E., "Some Irish Meteorologists," *Irish Astronomical Journal*, Vol. 9, No. 4, December, 1969, p. 113.

[99] Ibid, Cook.

[100] More article readings on this, for a good perspective on what went into Halley's and others' thinking (from William Gilbert to Kepler) can be read in Martha R. Baldwin's "Magnetism and the Anti-Copernican Polemic" (*JHA*, xvi, 1985) or J.A. Bennett's "Cosmology and the Magnetical Philosophy" (*JHA*, xii, 1981).

[101] Ibid, *The Northern Light*, p. 55.

Fig. 15. Edmund Halley's drawing of magnetic field lines around Earth. In addition to this, Halley had a theory for the earth's hollowness.[102] (*Philosophical Transactions*, Vol. 30, No. 363, 1099–100) (Cook, 1998)

J.A. Eddy noted that literary references to sunspots were made.[103] In *Gulliver's Travels*, published anonymously in October, 1726 by Jonathan Swift[104] after the Maunder Minimum's construed end, Gulliver hears of the fears of the Laputian "learned men," who see that

> The face of the sun will by degrees be encrusted with its own effluvia, and give no more light to the world.[105]

Whether Swift was familiar with Descartes's sunspot theory based on the analogy of foam is hard to know. The fear of sunspots dispatching our celestial light, on the other hand, was certainly believed.

[102] Kollerstrom, N., "The Hollow World of Edmond Halley," *JHA*, xxiii (Science History Publishers, 1992).

[103] As well as Nigel Weiss. These references are to Marvell (already mentioned), to John Milton; Defoe, and Jonathan Swift—all living and writing in or close to the Maunder Minimum.

[104] *The Concise Cambridge History of English Literature* (Cambridge University Press, 1947), p. 468.

[105] Swift, Jonathan, *Gulliver's Travels* (Riverside Press, 1960), p. 132.

China and the Far East Asian Region

The 1660s in China witnessed a stabilization in power of the new Chi'ing Dynasty, and this no doubt had a positive effect on such mundane matters as rice prices[106] and the normalcy of trade, though walled-in cities and warlords ruling protectorates were still the norm. But occasional severe droughts and frosts continued throughout Southern China in spite of the fact that years grew "warmer and wetter"[107] in Southern China at this time. Snow reporting in counties dipped at this time, but civil bureaucracy was being restored, there, nevertheless in the wake of the previous dynastic tumult. The Emperor Shin-chih died in this time, and many Papal agents of the previously described "Grand Design" were repressed.[108] But the new emperor effectively kept up a dialogue with the French missionaries then active in China to bring forward both Copernicus's and Kepler's universal views (versus Ptolemaic ones) into seventeenth century China.

In China in perhaps the same year of the "Great Snow" in Massachusetts (1717) the Kangxi Emperor declared at the end of his 61-year reign that "the climate has changed."[109] This he remarked, in spite of the fact that cultivation of oranges and mandarins in Kiangsi province had resumed in 1676 after a 22 year lapse. The emperor—on record not as a casual, but methodical and regular interpreter of the connection between climate and harvests—made the following observation (Southern China):

> I remember that before 1671, there was already new wheat [from the winter wheat crop] by the 8th day of the fourth month. When I was touring Jiangnan, by the 18th day of the third month new wheat was available to eat. Now, even by the middle of the fourth month, wheat has not been harvested ... I have heard that in Fujian, where it never used to snow, since the beginning of our dynasty, it has.[110]

---

[106] See Secular Trends of Rice Prices in the Yangsi Delta, 1638–1935 by Yeh-chien Wang for a good statistical view of these prices in parts of southern China in the Maunder Minimum (from *Chinese History in Economic Perspective*, eds. Thomas G. Rawski and Lilian M. Li, University of California Press, Berkeley, 1991).

[107] Ibid, Marks, p. 201 (Table 6.1, Climate changes in Linguan, 1650–1859).

[108] Ibid, Billington.

[109] Ibid, Marks, p. 195.

[110] Ibid, Marks, p. 195. Dating months and days exactly at the time they were reported is problematical, no matter where in the world you were. European's and American's dating systems were sometimes as primitive as their clocks and thermometers. China was no exception. "The most that [can be said] is the emperor experienced the availability of wheat around 1670 as being earlier than around 1717, and hence that it was warmer then. As for the place where this observation occurred: Jiangnan. This is actually in central China, and is basically the same as current Shanghai." (Qualified footnote, courtesy of Robert Marks). Note: the emperor was wrong about it never snowing in Fujian.

Decade	Zhu Kezhen	Zheng Sizhong	Marks
1470	cooler	warmer	0
1480	cooler	warmer	2
1490	cooler	warmer	0
1500	cooler	cooler	4
1510	cooler	cooler	1
1520		cooler	1
1530		cooler	4
1540		cooler	4
1550	warmer	cooler	1
1560	warmer	warmer	1
1570	warmer	warmer	2
1580	warmer	warmer	2
1590	warmer	warmer	1
1600	warmer	warmer	1
1610	warmer	warmer	4
1620	cooler	cooler	1
1630	cooler	cooler	3
1640	cooler	cooler	0
1650	cooler	cooler	2
1660	cooler	cooler	3
1670	cooler	cooler	1
1680	cooler	cooler	6
1690	cooler	cooler	3
1700	cooler	cooler	4
1710	cooler	cooler	4
1720	warmer	cooler	3
1730	warmer	cooler	3
1740	warmer	warmer	1
1750	warmer	warmer	4
1760	warmer	warmer	5
1770	warmer	warmer	1
1780	warmer	warmer	5
1790	warmer	warmer	1
1800	warmer	warmer	3
1810	warmer	warmer	3
1820	warmer	warmer	1
1830	cooler	cooler	7
1840	cooler	cooler	3
1850	cooler	cooler	

Fig. 16. Temperature fluctuations in China (1470–1850). Note well the period between 1620–1720 and some time afterward. (after R. Marks, 1998)

And although the emperor was wrong about it never snowing in the southern province of Fujian, it was rather surprising to see it fall there.

The link between cooler temperature and fluctuations between wet and cold, and crop yields, was well established. Paradoxically, the more control inhabitants of China had gained over the environment in the centuries preceding the sharp Maunder Minimum climate change, the "more impact climatic changes could have on a less diverse agro ecosystem."[111]

If the weather had already been worsened in an accumulative fashion for the last 40 or so years, a near-repeat of the Huaynaputina effect reoccurred, this time

[111] Ibid, Marks, p. 225.

in the Asian basin. The volcano Krakatoa erupted in 1680.[112] The gaseous emission with attendant sulphur aerosols could have contributed to the hyper-cooling in some parts of the Northern Hemisphere for the next decade, particularly Eastern North America, England, and near-Central Europe. This may well have added to the cold effects beneath an already activity-reduced Sun.

* * *

Deforestation of colonized areas in the Asian basin and in the Pacific and elsewhere (such as in the Mediterranean Sea, the South Atlantic, and the Indian Ocean) was often stopped by colonial legal action in these years. Patrolled by agents of the Dutch East India Company from the 1660s on, islands like St. Helena were actively protected from rival traders who degraded land and resources to deny competitors a leg up on cross-oceanic trade. Introduced animals such as goats had ravaged the native flora. Dutch colonists later on in the eighteenth century were already establishing conservation movements. The United Provinces took "highly interventionist" roles in places like St. Helena (in the South Atlantic Ocean) and elsewhere, such as Mauritius. Of paramount concern here was the re-planting of trees.[113]

This conservation effort may have been due in part to the after-effects of the Maunder Minimum on the mid latitudes and the Southern Hemisphere. Actions like the prescription of cheap vegetable foods that could be used to relieve famines, either for colonial workers/slaves or for people in general in the Southern Hemisphere were carried out well after the Maunder Minimum's close.[114] (Captain Bligh's factual breadfruit mission on *H.M.S. Bounty* comes immediately to mind.) These programs were not purely mercantile in nature. Some colonial-inspired natural restorative ideas afoot were "Edenic" in intention and pursuit, at the time. That is, based on an idealism (Utopian) of "restoring paradise" to some colonial areas where it had been shattered.[115]

---

[112] See Simkin, T. *et al.*, *Volcanoes of the World* (Hutchinson Ross: Stoudsburg Pennsylvania, 1981), p. 52. This 1680 Krakatoa eruption has a volcanic explosivity index rating of 3 compared to a much stronger 6 during the 1883 eruption.

[113] Grove, R., *Green Imperialism* (1995), pp. 95–141.

[114] Roxburgh, W., "Suggestions on the Introduction of Such Useful Trees, Shrubs and Other Plants as are Deemed the Most Likely to Yield Sustenance to the Poorer Classes of Natives of These Provinces During Times of Scarcity," (Report to the President's Council, Tamil Nadu State, Vol. CLXXXL, 8 February, 1793). (From Grove, 1998)

[115] Ibid, Grove, *Green Imperialism*. "Stephen Hales and some Newtonian antecedents of climatic environmentalism, 1700–1763." The question of "what happened to Easter Island" in regards to vanished "utopias" is related to this whole discussion—most incongruously yet strikingly in the renowned military historian, John Keegan's *A History of Warfare* (Pimlico, 1994), pp. 24–28. Fascinating in this

North American and North Atlantic Region

Since it was settled most successfully first, perhaps the residents of colonial Massachusetts had reasons to witness and record the questionable weather most. Or more likely the most dramatic weather could be seen in those latitudes at that time. The majority of these recorders were able diarists: educated, religious dissenters whose plights as such perhaps found more pity amongst their sometime native American-saviours than malice.

Colonies in British North America continued to have their hard autumns and winters, exacerbated by drought and its terrible cousin: famine. They seemed to have been getting harder still. In June 1662, for instance, prayers were given to ask release from a killing drought in Salem, Massachusetts.[116]

Along the entire British colonial North American coast hurricane season was as active then as it is now. Or, was cyclonic activity more common or even more violent? (The numbers of Spanish, Portuguese and British ships sunk along this coast from Cape Hatteras to the Florida Keys still being dredged up from these years would seem to say so.) Three recorded hurricanes affected North Carolina during the seventeenth century, all occurring within a four-year period (late 1660s–early 1670s) some—like that in September, 1667—which covered Virginia as well as North Carolina with rains lasting for two weeks. Another one was in August 1669 and another in 1670, the Outer Banks being targets. South Carolina witnessed one in 1699. All of these could be considered unusually strong, since people bothered to report them—though collateral damage on meagre human populations in America at the time would have been almost negligible.

The southerly storms of the 1670s had their Gulf Stream-assisted lift to the north, as well—as one long-lived Newport, Rhode Island resident remembered:

> On Saterday night forty year after [a 1635 storm] came much the like storme blew doune our windmill and did much harme the 28 of August 1675.[117]

(*continued*)

regard is the theory that "forest clearance reduced rainfall, and fields yielded less; it also reduced the yield of timber from which canoes were built, thus reducing the harvest of the sea." A Dutch voyager, Roggeveen, by 1722, noticed that "anarchy was already far advanced." The end of the sixteenth century presumably saw the last raisings of the great stone statues there. "Then something went wrong," Keegan states, citing overpopulation as the unravelling thread that saw the culture vanish. As the culture started to vanish, during the Maunder Minimum, one wonders if the weather had more to do with aggravating warring groups by denying them access to food sources?

[116] Ludlum, D., *New England Weather Book* (Houghton Mifflin, Boston, 1976), p. 23.

[117] Ibid, Morton.

Swamps and wetlands off the coast of New England were frozen hard that winter, as the previously-mentioned Kingston, Rhode Island note about British troops fighting Narragansett Indians in 1675 relates.

Spin-off effects from a volcano like Krakatoa in 1680 could have taken the following forms over the North American northeast (and naturally, England and central Europe)—since the wind patterns from the Asian basin climb over this region and the North Atlantic. Additionally, the severity could have been heightened by cloud cover and cold that probably assisted in heightening coastal cyclonic activity. On June 12, 1682 a tornado ripped through Connecticut, leaving a swath of destruction "28 miles long, up to a half mile wide."[118] A month later on July 6, 1682 in the British North American outpost town of Springfield, Massachusetts a hailstorm with stones in some cases nine inches in diameter rained down, damaging houses and destroying crops.[119] A year later in August, the "Hurricane and Flood [of] 1683" occurred in New England, where Connecticut experienced severe blasting down of trees and severe river flooding.[120] In February 1692 the Connecticut River flooded again, not to be outdone in magnitude until the "Jefferson Flood" 109 years later.

Sightings of aurora were becoming more common at this time. A few days after Christmas, 1700, in Salem, Massachusetts (then called Salem Village) the Reverend Joseph Green remarked, "I (was) at home. Warm weather, about 10 at night we saw a white circle like a rainbow in ye north west."[121] So goes the first reported, actual northern light for North America, though the earliest one was believed to have occurred on December 11th, 1719.[122] (Two years almost to the day after the Reverend Green made his diary entry inhabitants of Bergen, Norway observed "a particularly active display" of northern lights on New Year's Day.[123])

* * *

---

[118] Ibid, Ludlum, p. 51.

[119] Ibid, Ludlum, p. 61. Structural note on thunderstorms and hailstorms: hailstorms are summer phenomena characterized by cumulonimbus cloud stacks (very large) and strong vertical currents. Severest among these up and downdrafts in cumulonimbus create hail.

[120] Ibid, Ludlum, p. 41. The Connecticut River bisects a few New England states and rose 26 feet above its normal level, and terrible sea tides as high north as Dover, New Hampshire.

[121] Ibid, Rizzo and Schove, p. 397. Reporting on *Diary of Rev. Joseph Green of Salem Village*, communicated by S.P. Fowler (Historical Collection of the Essex Institute, Vol. VIII, Salem, Mass.) (Essex Institute Press, 1868), p. 219 and cf. p. 215.

[122] Ibid, Rizzo and Schove, p. 397.

[123] Ibid, *The Northern Light,* p. 7.

The weather of the Puritan period of New England finds itself in legend and fiction quite often in nineteenth century America (as weather as a topic does in eighteenth century England). Accurate, semi-fictional accounts of disasters in adverse weather or related tragedies had been in literary vogue since the days of Defoe. A question that could be asked is: was this tendency to describe terrific storms, plagues, etc. in literature just a spontaneous effort shared by a common people (Englishmen and colonial Englishmen)? Or was it a collective, unconscious drive to sublimate the terrible weather and the often destructive side effects the Maunder Minimum or the Little Ice Age years caused?

In a short story from Nathaniel Hawthorne's *Twice Told Tales* (1837) called "Old News," this descendant of Massachusetts puritans makes a semi-fictional visit to the past through colonial newspapers—many of which were still extant in the archives of young Hawthorne's Salem, Massachusetts of the early nineteenth century. The story is about how the written word in these "dingy half sheets" sometimes "prove(d) more durable ... than most of the timber, bricks, and stone of the town where they were issued."[124] It was also about how these papers contained all that was contemporary in the goings-on in farming villages, trite or otherwise—to include the farmer's prognostications on, and descriptions of, weather. In a sense Hawthorne is quite correct about the words proving durable, while providing a glimpse into the New England of the mid-1600s, weather-wise (albeit in his own words), via his 1830s short story. Significantly, he voices a very common (eighteenth, as well as) nineteenth century New England opinion that

> It is certain that winter rushed upon them [Massachusetts Bay colonists] with fiercer storms than now, blocking up the narrow forest-paths, and overwhelming the roads along the sea-coast with mountain snowdrifts; so that weeks elapsed before the newspaper could announce how many travelers had perished, or what wrecks had strewn the shore. The cold was more piercing then, and lingered further into the spring, making the chimney-corner a comfortable seat till long past May-day. By the number of such accounts on record, we might suppose the thunderstone, as they termed it, fell oftener and deadlier, on steeples, dwellings, and unsheltered wretches. In fine, our fathers bore the brunt of more raging and pitiless elements than we.

[124] Hawthorne, Nathaniel, "Old News. 1. The Colonial Newspaper," *Twice Told Tales* (Walter Black, 1837).

The "thunderstones" Hawthorne referred to were hailstones. And although the "many accounts" of them he may have perused in antique archives are now dust, parts of their memory survive. On May 9, 1695, Judge Samuel Sewall witnessed one of these "accounts" of a violent hailstorm in Boston:

> And about 2 PM a very extraordinary storm of hail, so that the ground was made white with it, as with the blossoms when fallen; 'twas as big as pistol and musquet bullets; it broke off the glass of the new house about 480 quarrels (squares) of the front.[125]

New Englanders fought against rough winter snow storms in the "hard winter of 1697–1698" known as "The Severest of 17th century"—a century already referred to as being "fiercer," storm-wise, than Hawthorne's Victorian-era America. In the Boston area, the following occurred:

> The terriblest winter for continuance of frost and snow, and extremity of cold, that ever was known; 31 snowstorms from November 20 to April 9; Charlestown, Mass., ferry frozen for six weeks; 42-inch snow depth reported at Cambridge (Mass); reputation for severity survived for many years.[126]

Many were the times winters experienced in North America earned a sort of enduring fame. Take this next episode toward the Maunder Minimum's end in c. 1720—an incident in those slower times handed down in local legend, and even forming the basis of a children's book about a hundred years later. The "Great Snow" of New England in 1717 was characterized by punctuated extreme weather (sharply mild—sharply severe) activity repeated in a later cycle, the Dalton Minimum, when volcanic activity added to similarly disturbing erratic cold/hot weather (Mt. Tambora in 1851–1816). First, heavy snows shifted to warm, fine weather in February, 1717, "lulling New Englanders into believing the worst was over." Between February 27 through March 7, four snowstorms suddenly struck the region, with three to five feet of snow, wind whipping it into drifts ten to twenty-five feet high. The resultant severe weather had people trapped in houses for long periods, cutting off most utilities and services such as they then were.

---

[125] October, 1692: "*Dover, N.H., 'First Snow,' Journal of Rev. John Pike*" (p. 121). Others from the *New England Weather Book* by David Ludlum; 1976, 148 pp, Houghton Mifflin Company, Boston), p. 61.

[126] October, 1692: "*Dover, N.H., 'First Snow,' Journal of Rev. John Pike*" (p. 121) Another from the *New England Weather Book* by David Ludlum is: September 1696: "Salem, Mass., A black frost. Ye ice on ye side of my house as thick as window glass [John Higginson]."

Wild animals were trapped in drifts, easy prey for predators, and farmers' stock, destroyed. But not always: animals in some cases were covered over so quickly with snow, they were literally shielded from any more ill effects and were dug up after weeks' confinement. As the Reverend Cotton Mather reported in one incident:

> The Poultry as unaccountably survived as these ... Hens were found alive after seven days; Turkeys were found alive after five and twenty days, buried in ye Snow, and at a distance from ye ground, and altogether destitute of any thing to feed them. The number of creatures that kept a Rigid Fast, shutt up in Snow for divers weeks together, and were found alive after all, have yielded surprising stories unto us.

Early April saw an end to these events.[127] Presumably, it was a welcomed relief.

The debate on human modification and destruction of the Earth and atmosphere and environment—as pointed out earlier in the Dutch case—is not new. The Right Rev. Cotton Mather believed in 1721 that "our cold is much moderated since the opening and clearing of our woods, and the winds do not blow roughly as in the days of our fathers."[128] In so stating, he reinforced the carried-over memories of his ancestors (as with Hawthorne) about the "colder weather" puritans, for instance, endured. Other than simply putting punctuation marks to the end of a long cold and wet spell now called the Maunder Minimum, Mather mentions a key phrase: clearing of the woods, or, deforestation to be exact. Looked upon with pride in some corners of the British Empire, literally being believed to make way for warmer air and milder climates to enhance agriculture,[129] it was sometimes viewed with suspicion and reviled as a sign of moral turpitude in others.

However, bad weather was not over just yet. In North America, cyclonic activity was still reported well into the earlier part of the eighteenth century, as several accounts of a violent hurricane affecting Charleston, South Carolina and areas

---

[127] McCain, D.R., (Internet) *Life in Early America* "The Worst Winters." ("A little ice age bedeviled eighteenth-century Americans.")

[128] Fleming, J.R., *Historical Perspectives on Climate Change* (OUP, 1998), p. 24.

[129] Fleming points out in a chapter called "The Great Climate Debate in Early America" that there was much contention between American colonists and Europeans to justify agricultural and natural-beauty greatness in America. The great plains had not yet been opened up, and the bountiful grain lands harnessed. What occupied Americans, then, was the clearing of native forests to suit the purpose. Which they believed would enhance life and especially agriculture. Even flower types were considered "degenerate" under antique analysis, and this did not endear Americans to European opinion.

northward could testify. The following remark indicates the storm's great violence as it struck the Cape Fear section: "ships were driven from their anchors far within land, particularly a sloop in North Carolina was drove three miles over marshes into the woods." This one was between September 16–17, 1713.[130] A direct connection can almost be made to the severity of this storm and one that had preceded or perhaps had been connected to it in New England at this time, on August 30, 1713:

> New London Hurricane—violent storm of rain and wind on the thirtieth was followed by a Hurricane which blew down several buildings & fruit trees such as hath not been known. It blasted or withered ye leaves & like frost, though warm weather.[131]

---

[130] Internet, National Weather Service, Newport, N.C. "Eighteenth century Hurricanes Impacting North Carolina."

[131] Ibid, Ludlum, p. 41.

5

Surveying the Maunder Minimum

In the previous chapters, the weather, "meteorological" (to include aurora borealis, or the virtual lack, thereof) and solar phenomena (few sunspots, faculae, etc.) noted by witnesses, and by paid—and amateur—observers, underscored a number of abbreviated events. These solar events occurred between circa 1620 and 1720, during which Earth's weather (and other geophysical forces) acted in an accordingly mysterious and disruptive manner.

Galileo noted that:

> "mysteries profound, concepts sublime" abound in the cosmos, yet remain blind to human eyes despite how hard it is studied by the best minds.[1]

Such was the case in studying the Maunder Minimum. We can blame it on that time not being so "information oriented," and less on the fact that the observers and even theorists then were no less able than they are now to observe and record.

Recorded traces concerning what occurred between about 1620 and 1720 on the Sun remained disparate and in many different private archives and libraries, at the mercy, such as the times were, of wars, conspiracies, fanatical personal jealousies and other disjunctive occurrences. Weather recording was not undertaken in any coordinated fashion in Europe. In China recording was regular, as we have seen. But for some periods during the Maunder Minimum interval, this recording was halted for reasons that we do not know.

These observations, measurements and even ideas and abstractions, where recorded, were not collated for years. Indeed, they are still in the process of being

---

[1] Galileo, in *Letter to the Grand Duchess Christina*, on the "mysteries of the heavens" notes: "Within (the heaven's pages) are couched mysteries so profound and concepts so sublime (that) vigils, labours, and studies of hundreds upon hundreds of the most acute minds have not pierced them even after continual investigations for thousands of years." We may well note this same trouble now ... even with many more minds than he dreamed, not to mention money, and equipment. (Ibid, Drake, 1957)

collated and collected. Some, as it has been stated elsewhere, are irretrievably lost.[2, 3] This is one of the reasons why solar astronomer J.A. Eddy revived what the Maunders had first called for: a refocusing on the anomalous nature of the Maunder Minimum interval—and only after a lapse of so many years. We find throughout this story of the Maunder Minimum lapses of many years intervening between random "discoveries," many of which were re-joined in a coherent sequence only at a much later date.

* * *

As uncoordinated and random as they occurred, events in the years that span this minimum were collected and analyzed for oddities regarding weather, atmospheric (electrical) events, sunspot disappearance, and other phenomena. As well as appreciating the sunspots for what they were, we also saw the shift away from magical interpretations of northern lights around Halley's and Römer's time. And as early as 1801 (regarding spot phenomena alone) the year 1670[4] concluded what the English-German astronomer and planet discoverer William Herschel[5] later called "the first of five very irregular and unequal periods" on sunspots' "disappearance" (however, a better word is diminution).

The first period covered roughly 1650–1670. Herschel's next better verified "irregular and unequal sunspot period" began in December 1676 and ended in April, 1684. A shorter, third "spotless" period began in 1686 and lasted until 1688, when the careful Paris Observatory director, Giovanni Cassini, saw no spots at all. A fourth "spotless" period lasted from 1695 to 1700. Herschel's fifth and final stretch of a "spotless" Sun went between 1710 to 1713 from which time onward, spots in relative abundance reappeared on the Sun's surface. About this same time (the latter period) the cold that gripped Europe and elsewhere gradually diminished. Aurora borealis increased. Ostensibly, the weather became more "normal" in northern and north central Europe, as it has been in the past. Presumably, it grew more fair generally throughout the Northern Hemisphere as a whole, as the agricultural

---

[2] Ibid, Hoyt and Schatten.

[3] This book is a start of a voyage of discovery: not one to start and end. One never knows when the lost letter, mislaid observation, or missing relict, will redirect the course of all the history amassed. We also recommend very useful/helpful collections, e.g., in "The Gallery of Natural Phenomena" by Christopher Chatfield: http://www.phenomena.org.uk/home.htm (we thank Sallie Baliunas for sharing this URL).

[4] 1650–1670: this one not clearly ascertained.

[5] Herschel, W., *Philosophical Transactions of the Royal Society of London*, Vol. 91, pp. 265–331 (1801).

and economic recovery in the Chi'ing Dynasty bore out. More positive impacts of better climate were also demonstrated by the filling up of the American colonies with immigrants and their increasingly healthier agriculture and fishing industries. Cold weather receded, somewhat, as did its recording in art, letters, diaries, and political treatises and scholarship.

The disparate and diverse recorders and observers did not know it in the years between circa 1620 and 1720, but what they found with these fluctuating sunspots and painted in their pictures, and noted in letters to their lords or acquaintances, were one of solar cycles' calling cards. This interval is well within a decreasing solar activity cycle which is located in an already-present period of non-linear cold usually referred to as the "Little Ice Age." The Maunder Minimum in such circumstances owed its condition perhaps to the compounding effects of yet *another* low solar activity phase of a prior minimum: the putative "Spörer Minimum" of A.D. 1420–A.D. 1530.[6]

We adopt Herschel's notation of five steps to help relate the non-linear climatic changes occurring together with the anomalous solar activity period called the Maunder Minimum. In noticing the consistency—and sometimes, minor inconsistency—of these events, we can tie in the weather observations and spot non-observations made around this paradigmatic five-step event from the last chapters in graphical format (Tables 1 and 1a, below).

The question of the Maunder beginning much before 1630 is still open, since many physical reports of sunspots, for one, are lacking prior to 1645. Some observations from the 1620s, as shown in the Chinese observations made by the Jesuit astronomer Schreck around 1628, reveal a relatively active Sun. It is revealed even if, for instance, the consequent weather and other events—Alps' snow collection, etc.—in this timeframe (1620s) seems to show that a weakening in activity was somehow going on.

Sunspot observations in the Maunder Minimum timeframe are summarized in Tables 1 and 1a. An improvement is gained in the sense that only the large spot-groups are counted. This happens whether the Maunder is counted from 1620–1720, or from 1645–1710, by Hoyt and Schatten in a 1998 paper[7] using a modified R. Wolf's technique for deriving sunspot numbers. What is gained is a homogeneity

---

[6] More research is necessary to fully understand the richness of the history of the Sun's activity with regard to this time.

[7] Hoyt, D.V., Schatten, K.H., "Group Sunspot Numbers: A New Solar Activity Reconstruction," *Solar Physics*, 1998, Vol. 181, p. 495.

Table 1. A list of Maunder Minimum recorded odd-weather phenomena and solar phenomena (H&S indicates Hoyt and Schatten) (*Continued on the following page*)

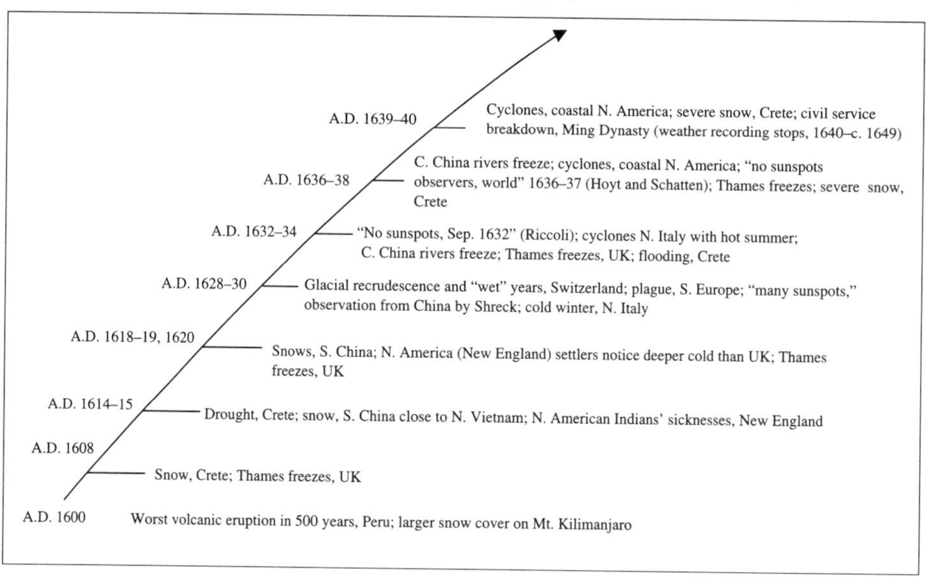

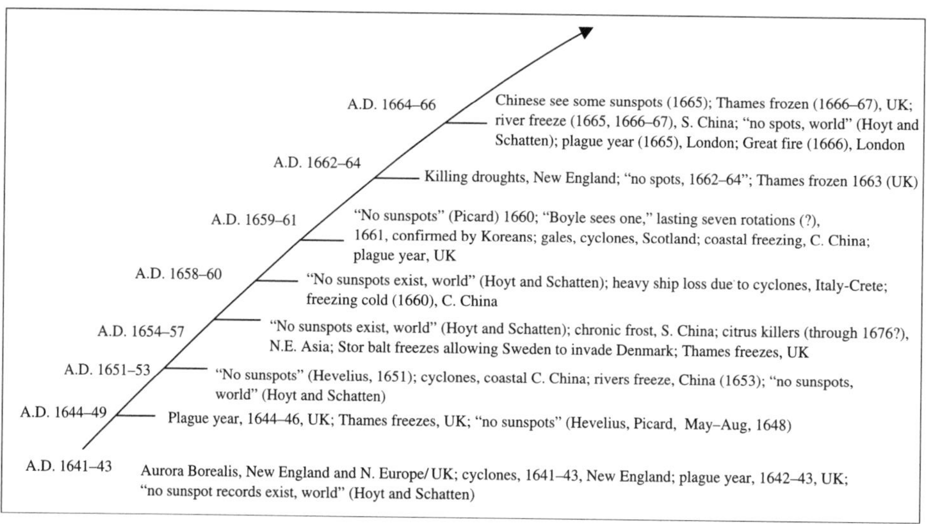

in the basis of counting the spots.[8] The following facts were found after they broke down the quality[9] of these observations made in the entire Maunder Minimum period:

---

[8] Hoyt and Schatten's efforts were aimed towards discerning long-term, inter-decadal- and centuries-long changes of the largest-scale solar magnetic field through a better representation of the spot groups (while R. Wolf's original counting method also weighed in contributions from smaller spots).

[9] Ibid, Hoyt and Schatten, 1998. They broke down the quantity/quality—with recounts—into these categories (after Wolf): Zurich-recorded observers; new non-Zurich recorders, effectively-new

Table 1. *(Continued from the previous page)*

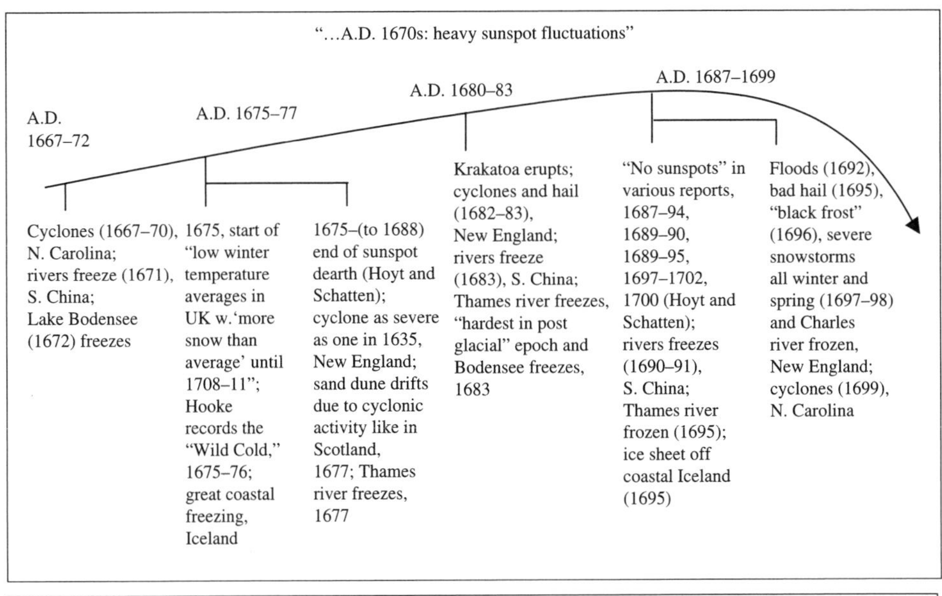

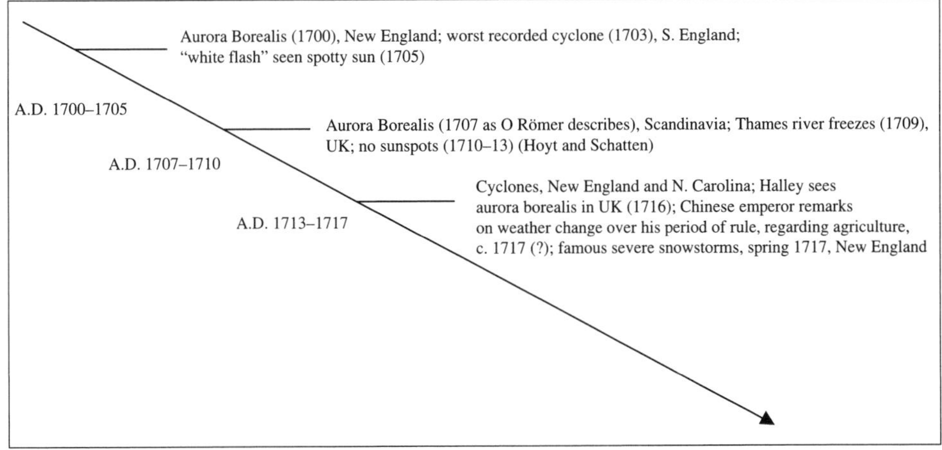

- Spare and incomplete sunspot observations, 1610–1644.
- Complete or nearly complete record for sunspot observations, 1645–1727.

They concluded that the search for observations over these timeframes has not been as "exhaustive as [they] would like."[10]

(continued)

observers; enhanced observers; partially recorded observers; corrected observers; vague observers, summary observers, misplaced observers; lost observers; unknown observers, and poor observers.

[10] Ibid, Hoyt and Schatten, p. 495. What Hoyt and Schatten's new work confirmed was that the Sun was well-observed and that only few spots were seen during c. 1645–1710. It is important to note that

Table 1a. Reliable observers' inability to view and record sunspots and other solar phenomena—or beginning once again to see them—within and around the Maunder Minimum period (some climate anomalies are listed for reference to cumulative severity)

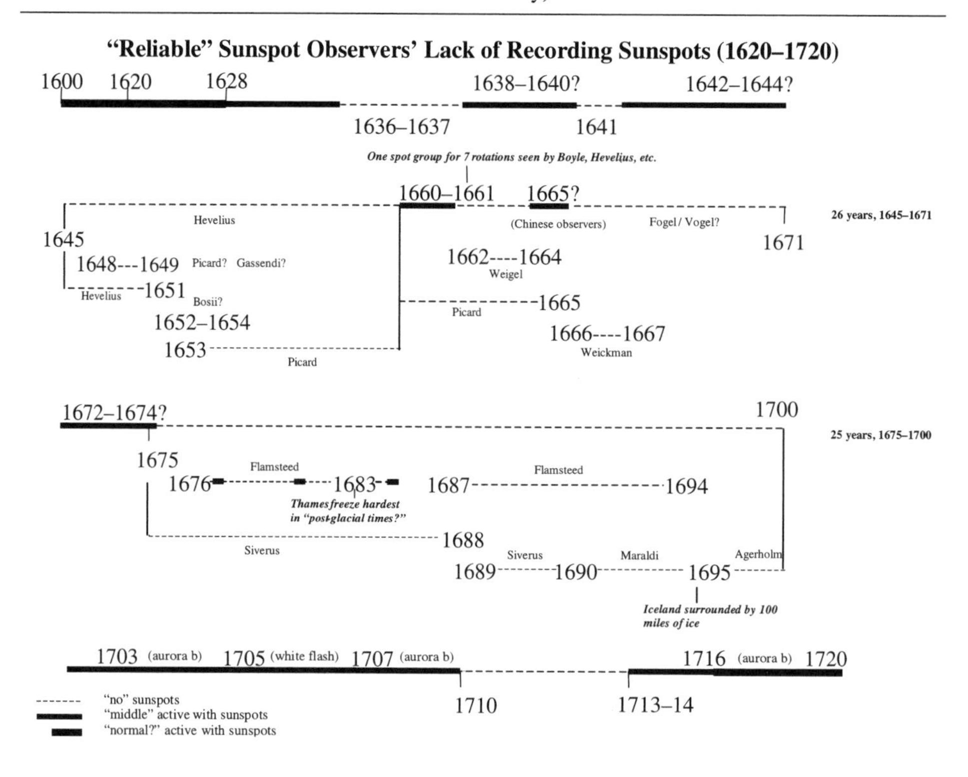

The 1620–1720 interval is summarized in Table 1a, outlining the names of sunspot observers and mostly the absence of spots in this period.

Table 1a confirms that reliable observers saw no spots at times. From 1628 on, we see a gradual diminishing of spot observation but with two actual, one-year observational gaps around 1636–1637 and 1640–1641. The drop in observed sunspots after 1644 is remarkable and pronounced. Several astronomers with reputations for observational care see little or nothing in this regard for 15 years.

(*continued*)

occasional challenges to the claim that the 11-year solar activity cycle (not *spot cycles*, per se) was ineffective during the Maunder minimum still needs to be closely studied. For example, Gleissberg and Damboldt (*JBAA*, Vol. 89, pp. 440–449, 1979) have researched three contemporary Maunder Minimum publications (Otto von Guericke's *Experimenta Nova,* published in 1672 in Amsterdam, Athanasius Kircher's *Iter exstaticum coeleste* [Celestial rapture journey] published in 1671, as well as Michael Ernst Ettmuller's dissertation *De Maculis in Sole visits* [On spots seen on the Sun] defended on August 2, 1693) and available auroral observations to raise objections to the original suggestion that the 11-year solar cycle was completely inoperable during the Maunder Minimum.

In the fifteenth year, a spot group is seen—an enormous one—that lasts several rotations of the Sun (seven months) before going out of sight. But the dearth of sunspots runs for an entire 26-year period from 1645–1671 before a miniscule three-year period of reoccurrence emerges again (1672–1674). But from 1675 to 1700—another 25 years—they are missing for the most part; again with only a few scattered sightings. Several astronomers wonder deeply about this, some remarking on it in diaries and in correspondence, Flamsteed to the effect that this was never the case "in the day(e)s of Scheiner and Galileo."

A human generation is about 25 years long. For all intents and purposes, the Sun for some reason reduced its activity for 51 years: *circa* two human generations. If we focus on the sunspots' absence as a single measuring stick of reduced activity, subtract five years (1660–61; 1665; 1672–1674) then we see 46 years of reduced solar activity. In this timeframe, virtually no aurora at polar latitudes were witnessed (except possibly a few, such as in 1643): in fact, aurora was unknown to two generations, for the most part. And for two human generations, sunspots were also for the most part gone.

Also worthy of note in passing are the "hard freezes" on the Thames River and Bodensee, not to mention the many miles of creeping ice around Iceland in the latter 25-year segment (1675–1700). Again, the *cumulative* nature of the cold in a *gradually worsening state* is pointed out across the Northern Hemisphere. If the first 25-year segment is compared to the latter, then the worsening in climate effects is definitely visible in the second segment. The logical implication of this is that the cold had, indeed, *cumulatively worsened* in this time frame, versus the earlier segment, the double-heaviness of this effect probably worsening not only by over thirty years of greatly reduced solar activity proven by absent sunspots, but by ash blockage first the Huaynaputina (1600) and later the Krakatoa eruption of 1680. Coincidental to the cold and heightened precipitation is the continued absence of aurora—another sign of reduced solar inactivity.

With the time sequence of 1703–1720, signs of vigorous solar-terrestrial activities re-emerge. These periods were "kicked off," it seemed, by reliably-observed "white flashes" indicating magnetic bursts, aurora at various times (as witnessed with awe by the likes of Römer and Halley). But there was a three-year paucity in spot observation from 1710–1713. The "white flash" (flaring) was so pronounced that it could be observed with fairly weak equipment compared with modern equivalents. By the mid-1700s, sunspot and aurora observation was again common: the subjects plentiful. The bad weather, however, takes some time to attenuate. For example, the fishing industry, negatively affected in Iceland in the mid 1600s, did not return to normal until the mid eighteenth century. If one is forced to use a common metaphor for the entire 100-year period, the Sun was like a blow-torch or lamp that went out, after a few feeble attempts at being re-lit. Then a 51-year period of failure to re-light occurs, only to start "clicking"—powerfully, at times—before it finally re-lights in a robust manner once again.

Additionally, in reviewing the events recapped in the previous chapters and in Table 1 and 1a, lower-latitude auroral events are also seen to have occurred during the Maunder Minimum—lending credence to the counter-observation that the Maunder Minimum did *not* mean a total halt in magnetic activity. Rather, there was perhaps an *overall* decrease in sunspot activity, as well as an *overall* decrease in aurora activity.[11]

But the next question is, "in comparison to what"? Suffice it to say that the aurora record before 1620 and after 1720 was much better represented. After the Maunder Minimum, the aurora record in the Northern Hemisphere is very well remarked upon. We quote physicist Sam Silverman regarding astronomers Ole Römer (1707) and Edmund Halley and others (1716) and the "sudden reappearance" of aurora in the early eighteenth century:

> Growing interest in the aurora in the early part of the eighteenth century, which resulted from the spectacular reappearance of the aurora in 1707 and 1716, followed a relative scarcity of great auroras during the Maunder Minimum ... Observations in the early eighteenth century led to questions about the geographical extent, nature, and temporal variability of the auroras. Typically, such observations were included as part of recorded meteorological notations, though occasionally early astronomers, such as Tycho Brahe in the 1590s, included auroras in their observations.[12]

Silverman goes on to note that in British colonial North America (into the period of becoming the United States, between 1742 and 1793)—well after the Maunder Minimum—a pair of observers in Cambridge, Massachusetts recorded 390 auroras, and between 1786 and 1820, another 198 were observed by an observer in the shipping town of Salem, Massachusetts, filled with skilled optical navigators and mariners as it then was.[13] It can be assumed that the frequency of aurora occurrences at least increased strongly after the Maunder Minimum and it can be

---

[11] Silverman, S., "19th Century Auroral Observations Reveal Solar Activity Patterns," *Eos*, Vol. 78, Number 14, April 8, 1997, pp. 145–152.

[12] Silverman, S., "19th century Auroral Observations Reveal Solar Activity Patterns," *Eos*, Vol. 78, Number 14, April 8, 1997, pp. 145–152.

[13] Home of Nathaniel Bowditch, creator of the first good, practical text for sea navigation. It was used up to recent times.

concluded that this was definitely a sign of greatly increased solar activity compared to at least the beginning of the eighteenth century.[14, 15]

Furthermore, the notion could be advanced that the revival of heightened aurora activity—especially from the 1716 example studied by Römer and Halley—was part and parcel of Stable Auroral Red Arc (SARs) activity. These usually occur at mid latitudes and are "red," and not considered "true" aurora, which are polar and multi-colored. The overwhelming number of brief reports of aurora after 1500 in Korea[16] tell of red aurora ("fire like vapors," etc.), lending credence to the possibility that SARs were being observed, maybe in Maunder Minimum times, in Korea—and *not* the high-polar aurora bands that are indicated during periods of greater solar activity.

* * *

What happened in the timeframe 1675 to about 1700 during the colder phase of the "Little Ice Age" (LIA) interval—which, as stated, is broader than previously thought—and perhaps ill-named, is a case in point of long-term variations in climate. This change is likely due to solar variation in part. And perhaps also by attendant geomagnetic variation and changes in the North Atlantic Gulf Stream.[17]

The coast of Brazil, as climate scientist H.H. Lamb reported in one of his works, is characterized by a geological feature that deflects warm water north into the Gulf Stream. A one or two degree shift in the ocean current, north or south of this, can determine how much warm water goes up the Earth's Atlantic coast from there, and how much does not. Wind circulation changes accompany these one to two degree shifts and have been recorded by scientists over the last century.[18] Studying ocean bed sediments have also shown that clear shifts in ocean currents[19]—and how much

---

[14] Unsophisticated or uneducated people in the west were just as prone to fanciful explanation as their eastern and southern cousins, and still are. We assume elites in the west and east at this time had ideas of scientific explanations of such phenomena.

[15] Ibid, Silverman. Data provided by Silverman. Interpretations of the way of describing these events—in the east and west—is our own. We note a possible bias that less skilled observers, observed, prior to 1700, and made fewer written reports, in the North American colonies. But the consensus seems to be "fewer aurora to observe" in the Maunder Minimum.

[16] Ibid, Zhang, *B.A.A. Journal*, 1985, p. 205.

[17] Lamb, H.H., "Climatic Variation and Changes in the Wind and Ocean Circulation: The Little Ice Age in the North Atlantic," *Quaternary Research*, Vol. 11, (1979), p. 5. The authors make no direct connection here between Sun and internal geomagnetic Earth activity.

[18] Ibid, Lamb, 1979, p. 8.

[19] Also to be taken into consideration is tectonic movement. If the Earth's core sends heat up through cracks in the ocean floor, this heats water that is carried up and over the mid-Atlantic coast. From

warm water goes north along this coast—have occurred repeatedly, in the "great cold climate stages of the last 225,000 years."[20]

Given that warm surface water can abruptly stop being pumped northwards at the same time as colder, polar water ceases to sink and move south, it is not hard

Fig. 17 Woodcut of the "Great Fire" in London, September 1666. Superstitious predictions of fire due to the arrangement of the number "666" helped spur efforts by the Lord Mayor to take precautions that year. Climatic caprices, however—combined with combustibles—sealed the fates of many Londoners. (Reprinted from Heywood Gould's Christopher Wren [Franklin Watts, 1970]).

(*continued*)

the 1675–1700 scenario, not much was apparent from this source, seeing as the water temperature sank enough to drive back certain fish. In the study of global ocean energy budget today, this geothermal heating component is noted to be able to contribute a substantial portion of the energy for mixing deep water globally (that in turn is responsible for the maintainence of the thermohaline circulation). See Huang, R.X., "Mixing and energetics of the oceanic thermohaline circulation," *Journal of Physical Oceanography*, Vol. 29, pp. 727–746 (1999).

[20] Ibid, Lamb, 1979, p. 8 (He cites A. McIntyre and colleagues's 1972 study.) Lamb used "time-series' of maps of seasonal circulation patterns based on qualitative data from different locations." For more on this and other reconstruction methods, see *Historical Climatology*, by Ingram, M.J., *et al.* in *Nature*, Vol. 276, November, 23, 1978.

to explain the drop in temperature by 4–5 degrees in North Atlantic waters around Iceland and the Faeroe Islands in the period from about 1675 to 1700, or, the later Maunder Minimum.

Though not necessarily tied to this drop in oceanic temperature, air currents over the middle (Atlantic) latitudes, forcing north or south in such states in a battle of cold and warm can "block" wind progression, naturally as it does from west to east. This blocking, in turn, can "mark very different, though persistent, spells of weather and great differences in heat and moisture transport"[21] from west to east over great distances. In such quasi-stationary states of atmospheric blockings, sharply-occurring differences in weather were noted. Phenomena like extremely hot summers (as happened in 1665 and 1666 in London, England) can occur "back to back." This was not only in, say, the midst of what was called the "Little Ice Age," (LIA) but also on top of a low solar activity period like the Maunder Minimum, which the "Great Plague" (1665) and "Great Fire" (1666) years are hemmed in by. All this could (and does) happen, followed the next year, by, for example, extremely cold summers since a strong atmospheric wind blocking event introduces intermittent fast-moving, north-south flowing winds.

Sixteen hundred and ninety five—probably the coldest winter in England over all these years—was followed by a mild year, which repeats the 1665–1666 pattern, just described. The late Jean Grove's words on the unpredictable and erratic nature of climate come immediately to mind.

A point to be made in this chapter is that many weather and climate events connected with Maunder Minimum-sunspot diminution could be linked to reduced solar brightness and other associated solar wind effects. (Sunspot diminution, it cannot be stressed enough, is just *one* sign of what may occur in a weakened solar activity phase.) By contrast, overall conditions with more sunspots and increased brightness imply the *opposite extreme* for world climate. A matter not to be neglected is the internal geophysical "forcings" on the Earth's climate system like those by changing oceanic circulation and by occasional large eruptions of volcanoes.

* * *

If the Maunder Minimum solar/geophysical data shown and discussed in this and in earlier chapters seem arbitrary and circumstantial, anecdotal, or trivial, this negative claim should not be ignored. It is reasonable[22] to assume that the contents of

---

[21] Ibid, Lamb, 1979, p. 2.

[22] Ibid, Jones and Bradley, *Climate variations of the past 500 years*. They are quite emphatically sceptical about eye-witness accounts and next-best-data renditions (that is, proxy data cooked up with the statistics of your choice), regardless of source, and with good reason.

Tables 1 and 1a, and the information on long-term climate change, are coincidental, given the complexity of the issues involved.

The data re-encapsulated in Table 1 and 1a on top of our weak understanding of climate drivers and responses show the vast gap in knowledge between long-term Earth climate cooling and heating, vis à vis solar aspects, confused as they can be with other inputs (geophysical, such as major volcanic eruptions, etc.).

The search for direct weather and climate events detailing the 1600s and 1700s should continue—even across other recorded solar minima and of course maxima as well. This would only serve to build onto an ever-expanding knowledge base. As illustrated in the previous chapters, there is even relevant climatic information to be gleaned from the world of art, architecture and design—and even literary history in re-building past climate conditions. But seeking solar operating mechanisms that play roles in Earth's climate were—and are—of an even more pressing and vital nature, just as much as the inputs of solely geophysical forces and human impacts to the global climate are urgent.

However, we leave the period of early solar study as it encompassed the actual Maunder Minimum of the seventeenth century. We shall proceed into the future some 150 years to begin seeing how a phenomenon like this extended solar minima slowly managed to be discerned.

6

Maunder's Immediate Predecessors in Delineating Solar Structure and Behavior: Towards Understanding Solar Variability and Sun-Climate Connections

It must be kept in mind that the seventeenth and eighteenth century events shown in Tables 1 and 1a (Chapter 5), were not part of Maunder's general knowledge—though parts of it could have been known to him. There was little effort made—even in his lifetime—to reconstruct or collect any of these data into a coherent framework. Late in his life (the 1920s, when the idea of extended solar minima became clearer), Maunder regretted the vast amount of missing climate data that could round out his formulation of what he understood to be "a prolonged solar calm."

However, Maunder inherited an amount of data as well as concepts and ideas from primarily German and Swiss scientists, much of which he would put to good use. It should first be noted that Gustav Spörer was an interpreter of solar motions and sunspot patterns, and was Maunder's predecessor in delineating the weakening of solar activity during the mid-to-late seventeenth century. Secondly, Rudolf Wolf was another talented observer and interpreter of solar behavior and made tenuous connections between solar activity and eighteenth century climate change. Thirdly, Heinrich Schwabe, after 40 years of observing the Sun, fell upon the fact that it was governed by 11-year cycles, and Maunder was quite aware of these cycles. The German amateur astronomer Schwabe[1] noted the following about the Sun:

> Comparing now the number of groups with the number of spot-free days one finds that sunspots had a period of about 10 years and that they

---

[1] Schwabe became interested in sunspots probably while trying to define them away from an "extra-Mercurial" planet (that is, the putative planet, Vulcan) that he was looking for. Other astronomers, such as Leverrier, were also searching for Vulcan. Vulcan supposedly transited the Sun's face, and its small apparent size would confuse it with sunspots (Meadows, A.J. and Kennedy, J.E., "The Origin of Solar-Terrestrial Studies," from *Vistas in Astronomy,* Vol. 25, 1982, p. 419).

> appeared so frequently for 5 years that during this time there were but a few or no spot-free days ... The future will show whether this period shows some consistency.[2]

The future would come to realize Schwabe's discovery. Schwabe is reported to have watched the Sun daily for 40 years. But this "10-year" lapse was noted by Schwabe after 17 years' worth of viewing the Sun, and the "lapse" came to be known, in Schwabe's honor, as the 11-year Schwabe Sunspot Cycle. With regard to further progress in solar science, Maunder would one day state with conviction:

> Rightly ... was Schwabe honoured for his demonstration of the solar cycle, and for the great advance in our knowledge of the sun ... that rendered [it] possible. But the demonstration due to Wolf and to Spörer of this strange interruption of it, is equally necessary to be borne in mind. The sunspot theory which ignores it stands self-condemned.[3]

C.F. Gauss was probably another German scientist with whose work Maunder was acquainted, though it is certain that Maunder knew the work and data of Edward Sabine (and John H. LeFroy). Sabine was the British soldier-scientist who had access to the far-flung British "watch stations" keeping tabs on the Empire and maintaining her safety on the seven seas (even Gauss was grateful for the data provided by the British watch stations in his own work). Sabine and LeFroy reported on solar magnetic disturbances felt on Earth from stations in Canada and Australia in the 1850s and were aware of it, earlier. They also did work on aurora.[4]

---

[2] "For more than 40 years he observed the Sun daily, and recorded the temperature of the air. After years rich with spots an uneventful period followed in 1833–1834. When, ten years later, there were still hardly any spots to be counted he (courageously!) took up his pen on New Year's Eve, 1843: 'Comparing now the number of groups with the number of spot-free days one finds that sunspots had a period of about 10 years and that they appeared so frequently for 5 years that during this time there were but a few or no spot-free days ... The future will show whether this period shows some consistency, whether the smallest activity of the Sun in producing spots lasts for one or two years, and whether this activity increases more rapidly than it decreases. On April 10 and 11 and on May 10, the Sun was so extraordinarily clear, there being only a slightly overcast sky, that the paler light of its limb stood out very clearly.' " (Schwabe 1844) [from pp. 23–24 of Gunther Rudiger, 1989, *Differential Rotation and Stellar Convection: Sun and Solar-type Stars*, Akademia-Verlag, Berlin].

[3] From *Knowledge,* Vol. 17, pp. 173–176, August 1, 1894: "A prolonged sunspot minimum."

[4] For example, Sabine, E., *On Periodical Laws Discoverable in the Mean Effects of the Larger Magnetic Disturbances – No. II* (from the conclusion, and read before the Royal Society of London, May 6, 1852, pp. 103–124).

Maunder of course had access to the gifted William Herschel's notations on the connections between low crop-yield years and sunspot diminution, and the so-called "five great periods" when sunspots were "absent" in the seventeenth century.

Herschel was an acute observer of the Sun. Or, as the nineteenth century astronomy author—and Maunder's friend—Agnes Clerke put it, "his attention to the sun might have been exclusive, so diligent was his scrutiny of its shining surface."[5] As early as 1798 Herschel was trying to do to the Sun with prisms what Robert Bunsen, Gustav Kirchhoff, Charles Young,[6] Jules Janssen and others, much later, managed to do better. That is, he as well as they broke its light lines down to see what these lines would reveal.[7] Imaginative astronomer and scientist that he was, Herschel used metaphor freely. To him, the Sun's faculae were like "the shrivelled elevations upon a dried apple, extended in length... most of them joined together making waves, or waving lines." And sunspots were "dusky pores."[8]

William Herschel was one of the first to equate perceived phenomena on the Sun with the abundance of heating and cooling it was capable of causing on Earth, and finding some tenuous facts to support this surmise. He believed that "great shallows (spots' penumbrae), ridges (bright elevated, extended features resembling faculae), nodules (bright elevated, but smaller features resembling luculi) and corrugations (less luminous, rough, dark features looking mottled) instead of small indentations (depressed dark features that are extended)"[9] on the Sun would let large amounts of heat grace the Earth. And on the other hand, "pores, small indentations—central regions of dark depressed spots—and the absence of ridges and nodules"[10] meant less heat would touch Earth.

---

[5] Clerke, A.M., *The Herschels and Modern Astronomy* (MacMillan, 1895), p. 83.

[6] See Mark Littmann and Ken Wilcox, *Total Eclipses of the Sun* (University of Hawaii Press, 1991) for more on the history of what light could tell astronomers.

[7] As it was, he managed to filter the light of Sirius (A Canis Majoris) through a prism. (This and the "Ritter's dark rays" section of *William Herschel and His Work*, by James Sime relates this fascinating story [Scribners, 1900].)

[8] Ibid, Clerke, p. 83.

[9] Clarifications of terms directly from Herschel (1801). See also Lubbock, C.A., *The Herschel Chronicle* (Cambridge University Press, 1933). Herschel was an acute solar astronomer, though he believed that the Sun contained a cooler core that may have contained life, like many others in his day. Other good—if rare—readings are *The Herschels and Modern Astronomy* by Agnes M. Clerke (MacMillan, 1895), which gives a detailed look into him in this regard, and *William Herschel*, by Angus Armitage (Doubleday, 1963).

[10] Ibid, Lubbock.

During his recognition of the variability of solar behavior and hypothesized solar structure, he inadvertently picked up the relative absence altogether of spots on the Sun from July, 1795 to January, 1800. He most likely saw the precursor of the Dalton Minimum, perhaps without knowing it. He was perhaps the very first to construct a past record of observed or missing sunspots and found that, in England at least, the absence of spots coincided with high wheat prices. Herschel read his paper on this before the Royal Society. He was completely misinterpreted and heartily ridiculed before that body. Since this comparison only applied to England and the Sun, he was believed to have regarded the entire Earth as having been affected in this way, or perhaps even only specific crops. To many, his points did not ring true.

What Herschel believed was that a periodicity in the Sun's disturbances caused outbreaks of spots. As one biographer—his granddaughter—later said, "to test Herschel's conjecture properly would have required exact information regarding the recurrence of mild or severe seasons, and no meteorological records of this nature were yet available."[11] There were no records (there are hardly any good ones, even now, for the purposes Herschel needed); therefore, his surmising led to laughter in the Royal Society. Years later, in a similar vein, Walter Maunder would utter a lament about the unavailability of weather records to buttress the hypothesis of extended solar minima. Yet a growing database is accumulating, enough to show that both Herschel and Maunder indeed had points to their scientific quests. A solar eclipse of May 12, 1706 was reported by many observers, but to build the argument to support Herschel and Maunder's scientific intuitions, we will settle on Isaac Newton's friend, Facio (or Fatio de) Duillier's observation. This is to show how data—accumulated by others—can lie like dead wood until thrown into the furnace of another's scientific imagination. A sea captain, one Stannyan, reported seeing a "blood red streak" from the Sun's left limb. It was a phenomenon that lasted for at least several seconds. But Duillier described what is now called the "crown," or "solar corona," in the following way:

> The clouds … became red, and then of a pale violet. There was seen, during the whole time of the total immersion, a whiteness which did seem to break out from behind the moon to encompass it *on all sides equally*[12] … This planet [the Moon traversing the Sun's disk] did appear

[11] Ibid, Lubbock, p. 279.

[12] Author's italics. Notice the "all sides equally" observation: a note that comes into play later in this discussion.

> very black ... within this whiteness, which encompassed it about, and whose colour was the same with that of a white crown or *halo*.[13]

Much later, at the end of the century and into the next, Herschel and his friend, Alexander Wilson—probably unaware of Duillier and Stannyan—composed a structural theory based on observation of the Sun. First, there is a luminous solar cloud region seen around the Sun. Adding to this the elevation of faculae, the region could not be less than some thousands of miles deep, Herschel felt. Pithily, he compared the "clouds" in this region to the Earth's aurora. Beneath this region was the solar atmosphere, covered enough to shield the Sun's body, which he thought could contain life (a thought still credible at that time). Herschel and other contemporaries such as Astronomer Royal Airy[14] thought there was a possibility for life on the Sun at this time. The region below may have been cooler than the outer "luminous solar cloud region," too, Herschel thought. To explain spots, this solar atmosphere could be whipped up to a frenzy and generate tornadoes—incredibly powerful ones—and

> A mighty updraft from below rolls back, for a longer or shorter time, the luminous solar clouds. Into the vast pit ... these clouds pour a flood of light on the body and cloudy atmosphere of the sun. The former looks black against the light, but reveal mountains upwards of three hundred miles in height; the latter, with its shelving sides, returns more of the light, and is less black; while the shining matter, rolled back into waves of enormous length and height, is heaped up in fiery storms round the vast gulf. The dark body of the sun is called the macula, or spot; the better lighted atmospheric shield, the penumbra; and the heaped up waves the faculae, which give the sun's surface the roughness of aspect it presents.[15]

We can easily see Herschel connecting the "power storms" hastened up in the "solar clouds" to the "red flame shot up from the burning ocean of the sun's surface to a height of 200,000 miles in a few minutes, rising and falling back into that ocean's bosom in a couple of hours."[16]

The exact numbers in this statement may be wrong, but Herschel was on the right track towards realizing the dynamic nature of the Sun's behavior and various

---

[13] Ibid, Sime, p. 62.

[14] Airy made a structural observation of coronal phenomena he termed "sierras," since he thought these could be mountains (Ibid, Littmann and Wilcox, p. 64).

[15] Ibid, Sime, p. 165.

[16] Ibid, Sime, p. 166.

phenomena. For example, the explosive ejection of solar materials which may then extend to Earth, as well as those time-varying shallows, ridges, nodules, corrugations, pores, and indentations modulating the solar heating of Earth.

Had Herschel seen or found out about the "blood red streak" seen by the sea Captain Stannyan, such observational proof would have strengthened this prototypical coronal theory and, as one author put it, "placed the science of astronomy ... half a century in advance of the position he [Herschel] left it in at his death."[17]

Maunder also knew of British solar astronomer Richard Carrington's[18] observation that dark sunspots "flashed" white, and that these, in turn, recorded a magnetic burst on earth as a then-putative accompaniment like telegraph cable interruption. That is, Carrington in one of his daily sunspot reports actually saw the flash through his telescope while magnetic readings some hours later showed "delayed burst effects" on Earth – without knowing why or how (as discussed before, the British solar astronomer, Stephen Gray, saw a white flash in 1705). Maunder would hardly have known about Gray (though this possibility should not be ruled out).

Richard Carrington,[19] during Maunder's early childhood in the 1850s and 1860s, made a number of observations in his daily routine as an amateur solar astronomer. He elaborated on one of the biggest "clues"[20] at this early date and one that lends credence to the almost forgotten solar astronomer Stephen Gray, who really *had not* been "seeing things" when he saw "white flashes" on the Sun's surface. Carrington also found a case-in-point example of magnetic disturbance *not coinciding* with sunspot occurrence in his study.

On September 1, 1859 Carrington saw what Gray said he had seen two days after Christmas Day 1705. Namely, what long after came to be called a "solar flare," while making his daily sunspot drawing.[21] He received corroborating confirmation

---

[17] Ibid, Sime, p. 165.

[18] From page 425 of *Philosophical Transactions of the Royal Society of London,* 1861, Vol. 151, pp. 423–430.

[19] Richard Christopher Carrington (1826–1875), brewer (like Hevelius had been hundreds of years earlier) and solar astronomer, watched the Sun from November 9, 1853 till March 24, 1861 at Red Hill Observatory (or "Redhill") in Surrey. His contribution to solar physics is enormous, including the discovery of surface differential rotation, the "Spörer's Laws of Spot-zones," the noticing of flare-phenomenon, etc. (see Hoyt & Schatten 1997 book for more). On a personal note, Carrington lived a tragic life, dying about three weeks after his wife Rosa Helen Rodway (considered to have a "dubious past") around December, 1875—possibly due to an overdose of chlorolhydrate (for insomnia). See Chapman, A., *The Victorian Amateur Astronomer* (John Wiley, 1998), p. 41.

[20] Carrington, R.C., *MNRAS,* Vol. 20, pp. 13–15, 1860. See also Cliver, E.W.,"Solar Activity and Geomagnetic Storms: The First 40 Years," *Eos,* Vol. 75, No. 49, December 6, 1994.

[21] Ibid, Cliver, 1994.

of his sighting that day from another amateur, nearby; a Mr R. Hodgson.[22] Comparing his observations with those made at Kew Observatory the same day, Carrington found that a short-lived, but definite, magnetic disturbance occurred at the same time he saw the solar flare. And a magnetic storm of some proportion occurred some 18 hours later. Carrington, in proper cautious scientist fashion, avoided drawing any conclusions from Earth's magnetic variations and the flare's occurrence.

The exceptional nature of the occurrence was not lost on Carrington. Balfour Stewart in a report to the Royal Society quoted him in his own words:[23]

> While engaged in the forenoon of Thursday, September 1, 1859, in taking my customary observation of the forms and positions of the solar spots, an appearance was witnessed which I believe to be exceedingly rare. [He then indicates his equipment was in place, etc.] ... I had secured diagrams of all the groups and detached spots, and was engaged at the time in counting from a chronometer and recording the contacts of the spots ... when within the area of the great north group ... two patches of intensely bright and white light broke out [in A and B, see Fig. 18a] and of the forms of the spaces left white. [He then relates he thinks it accidental: a mere reflection] ... but, by once interrupting the current observation, and causing the image to move by turning the [telescope's] R.A. handle, I saw I was an unprepared witness of a very different affair ... I hastily ran to call some one to witness the exhibition with me, and on returning within 60 seconds, was mortified to find that it was already much changed and enfeebled. Very shortly afterwards the last trace was gone; and although I maintained a strict watch for nearly an hour, no recurrence took place. The last traces were at C and D [see Fig. 18a] ... The instant of the first outburst was not 15 seconds different from 11h 18 m Greenwich mean time, and 11 h 23 m was taken for

---

[22] Being a true inquirer, here is what Carrington said about the importance of R. Hodgson's observation (*MNRAS*, Vol. 20, pp. 15–16, 1860): "It has been very gratifying to me to learn that our friend Mr. Hodgson chanced to be observing the sun at his house at Highgate on the same day, and to hear that he was a witness of what he also considered a very remarkable phenomenon. I have carefully avoided exchanging any information with that gentleman, that any value which the accounts may possess may be increased by their entire independence."

[23] Stewart, B., "On the Great Magnetic Disturbance which extended from August 28 to September 7, 1859, as recorded by Photography at the Kew Observatory," *Philosophical Transactions of the Royal Society of London,* Vol. 151, 1861, pp. 426–427.

> the time of disappearance. In this lapse of five minutes, the two patches of light traversed a space of about 35,000 miles ... Both in figure and position the patches of light seemed entirely independent of the configuration of the great spot, and of its parts, whether nucleus or umbra.

What did not gibe was the sighting of the effect with its delayed "detonative" impact on Earth hours later on September 2. As E.W. Cliver recently put it: "the fact that the Sun could have both prompt and delayed geomagnetic effects [as manifested by Carrington's flare] would be a source of confusion well into the next [the twentieth] century."[24]

But there was not so much confusion at the time of observation. Balfour Stewart continued in his Royal Society report to describe Carrington's discovery of magnetic disturbances which occurred simultaneously with his own observations:[25]

> On calling at Kew Observatory a day or two afterwards, Mr. Carrington learned that at the very moment when he had observed this phenomenon the three magnetic elements at Kew were simultaneously disturbed. If no connexion had been known to subsist between these two classes of phenomena, it would, perhaps, be wrong to consider this in any other light than a casual coincidence; but since General Sabine has proved that a relation subsists between magnetic disturbances and sun spots, it is not impossible to suppose that in this case our luminary was taken *in the act*.[26]

The "luminary" that Stewart and Carrington could only draw and wonder about is shown in the modern European Space Agency-released photograph below taken by the SOHO spacecraft (see Fig. 18b). Though perhaps not of the same magnitude of the one Carrington saw, the phenomena shown in the SOHO photograph is of exactly the same type.

* * *

The systematic and scientific approach to the Maunder Minimum's discovery arguably may have begun with a number of pioneer inquirers, among them, Joseph Fraunhofer and W.H. Wollaston.

These scientists proved that the Sun's lines in prisms were truly *from* the Sun: over 5,000 of them. They found that they were not merely prismatic distortions.[27]

---

[24] Ibid, Cliver, direct quote, December 6, 1994 article.

[25] Ibid, Stewart, p. 428.

[26] Stewart's italics.

[27] Pannekoek, A., *A History of Astronomy* (Dover, 1961), p. 330.

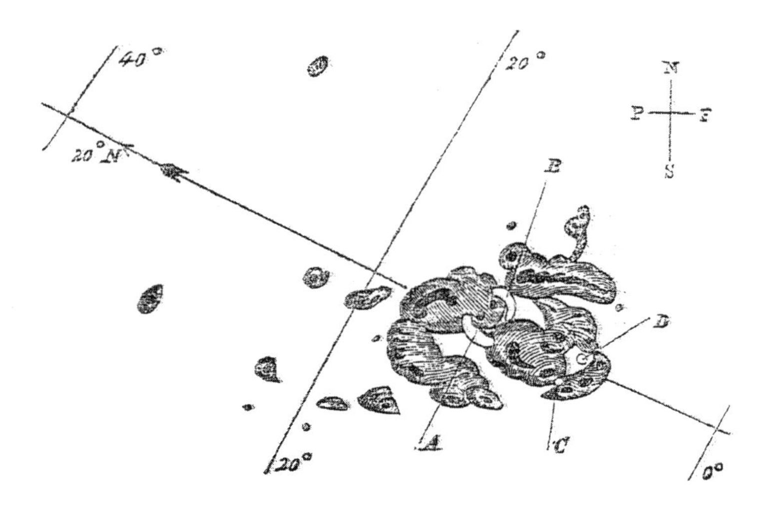

FIG. 1.—September 1, 1859, 11^{h}18^m G.M.T.; sun's diameter = 289 mm (Carrington)

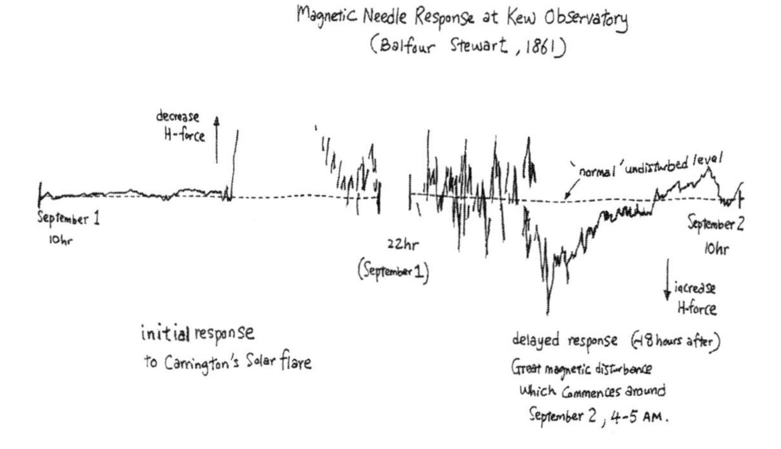

Fig. 18a Carrington's 1860 drawing of a solar flare (light areas marked A, B, C, and D) observed in white light during his daily sunspot drawing on September 1, 1859. The Kew record (Stewart, 1861) of the horizontal magnetic force showing the prompt (September 1) and delayed (September 2) effects of the eruptive flare observed by Carrington. It was further noted that this "singular" solar flare produced "auroral displays of almost unprecedented magnificence" as observed in Europe, America (as far south as Cuba) and Australia. (After Stewart, 1861 and modified from Cliver, 1994)

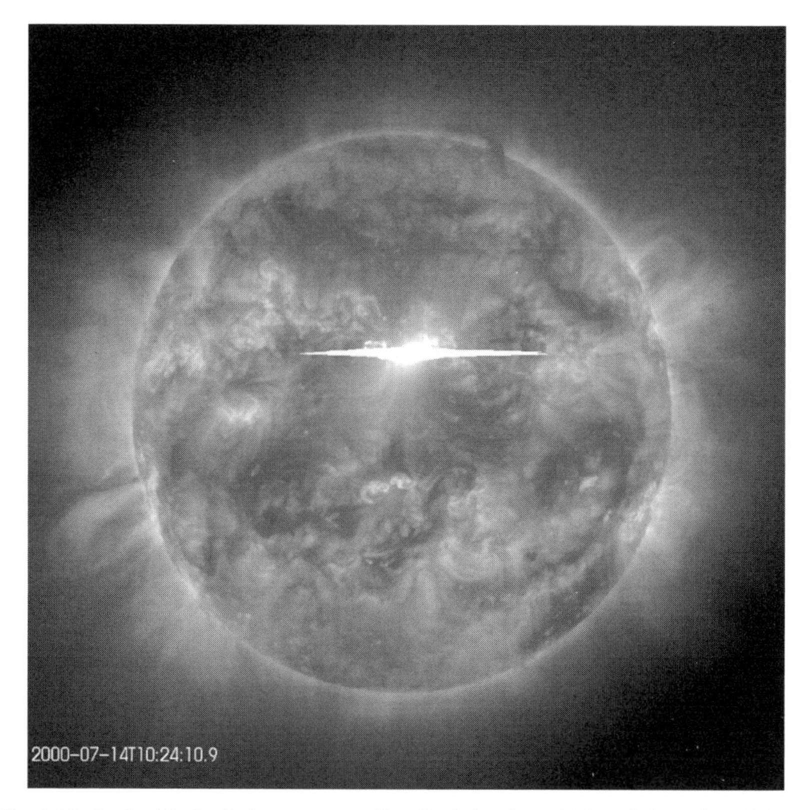

Fig. 18b This is the kind of phenomenon (the flash in the middle of the solar disk) Richard Carrington saw on September 1, 1859, as the SOHO satellite photographed the phenomenon in ultraviolet light 141 years later. (SOHO picture courtesy of the European Space Agency.)

Hence, they took a step away from Herschel's experimenting with prisms—on the right track as Herschel was. Perhaps in the light created by the flame of the newly developed Bunsen burner, Robert Bunsen wrote to a colleague in England:

> At present Kirchhoff and I are engaged in an investigation that doesn't let us sleep. Kirchhoff has made a wonderful, entirely unexpected discovery in finding the cause of the dark lines in the solar spectrum, and he can increase them artificially in the sun's spectrum or produce them in a continuous spectrum and in exactly the same position as the corresponding Fraunhofer lines. Thus a means has been found to determine the composition of the sun and fixed stars.[28]

[28] R.W. Noyes, *The Sun, our Star*, 1982, p. 31.

Other work that would have a solid effect on the course of solar science in these times was interpreting Fraunhofer's lines, in 1860. This led Gustav Kirchhoff and Robert Bunsen to find that each known element—notably in the Sun's case, Calcium (Ca), Hydrogen (H), Sodium (Na) etc.—has distinct spectral lines or positions in the sunlight spectrum.[29]

Also through these years—and more mundanely—Carrington and others also observed sunspot latitude drift towards the Sun's equator. So it became known that, when a solar cycle started, new sunspots developed at mid latitudes. Maunder would later find himself in a position of defending these well-delineated observations and would call these "Spörer's Law of Spot-Zones"[30,31]—probably an early case of an honoured scientist honouring another by using eponyms. Walter Maunder[32] would later start a course toward bringing the "Sun-climate connection" into sharp focus (consciously or not from where Herschel left it) including such concretely-discerned anomalies as witnessed by Schwabe, Kirchhoff, Wolf, Spörer, and the American, Joseph Henry—the latter having found that sunspots radiated less heat than the rest of the solar disc[33] and thus corroborating some of Herschel's "surmises" in this regard.

---

[29] This clarified what Fraunhofer's/Wollaston's "dark (solar) lines" were.

[30] Maunder's name for these in his 1904 paper (*Monthly Notices of the Royal Astronomical Society*, Vol. 64, pp. 747–761).

[31] Probably using Wolf's records, Spörer traced the trend Carrington saw all the way back to 1621 (but note that there is a large gap between 1619 and 1755). This all has direct relevance to modern solar magnetic research.

[32] Analogous to Wolf (and suspected by others). Around 1850, Rudolf Wolf, Alfred Gautier, Edward Sabine and others have noted the relationship between the sun's activity with terrestrial magnetism, as recorded in movements of magnetic needles. In fact, Wolf had used that information to help reconstruct solar activity further back in time until 1784, when direct solar observations were lacking. (Read Hoyt, D.V., Schatten, K.H., *The Role of the Sun in Climate Change* [Oxford University Press, 1997] for more details.)

[33] Joseph Henry, American scientist at Princeton, determined this with a thermopile and a 4-inch telescope imaging the Sun on a wall, indoors, in 1845. Measurements over a period of sunny days confirmed that sunspots were cooler than the surrounding solar disk, or, in his words after twelve measurements were made on the first day, "the spot emitted less heat than the surrounding parts of the luminous disc." (From "Observations on the Relative Radiation of the Solar Spots," *Proceedings of the American Philosophical Society*, Vol. IV, pp. 173–176, June 20, 1845.) Later on, Henry's type of measurements were extended and pursued by, for instance, Samuel Langley and the Jesuit priest, Angelo Secchi. The scientific lineage of Joseph Henry, Samuel Langley and Charles Abbot constituted the Smithsonian Astrophysical Observatory's long tradition of measuring "solar heat" (or "solar constant") that was among the few serious attempts to arrive at a physical understanding in the study of the factors of climate change (see a review by Hoyt, D.V., "The Smithsonian Astrophysical Observatory solar constant program," *Reviews of Geophysics and Space Physics*, Vol. 17, 1979, pp. 427–458).

Consistency in accurate, long-term observation with long-term data accrual was paying empirical dividends for solar science. The records, observations, and measurements now being too many, and too catholic in breadth to lose track of, was having a beneficial effect on this scientific field of endeavor. Walter Maunder would prove to be an exceptional articulator of these data as well as much of his own.

7

Maunder's Early Life and Associations

On November 18, 1882, Queen Victoria was holding a review in Hyde Park. The morning was somewhat foggy, and the sun shone dull and red through the thick air, so that it was easy to look at him. On this occasion there was a great spot on the sun; so big that it caught the attention of the soldiers who were marching across Blackheath, ... they pointed it out to each other.

The Heavens and Their Story (1908), p. 106

He was born Edward Walter Maunder on Chesterfield Street, St Pancras, London,[1] on April 12th, 1851 in the middle of England's Victorian period. He was the youngest of four boys (Thomas the eldest, and as it would turn out, his mentor; George, and one who did not survive, Henry) their father being a Wesleyan minister or as it was called "a minister of the Wesleyan Society"[2]—the Reverend George Peter Maunder.

The England of his time was mid-Victorian, and perhaps ascending to the height of the British Empire. It was a point of English pride at this time to know the "Sun was always rising" somewhere on the face of the Earth in the British Empire. England's global sea-reach had for a long time already contributed to solar science, due to the many "watch stations" that existed in places as far-flung from each other as Canada and Australia. In such places, men like E. Sabine and J. LeFroy collected vital data about the Sun, and the implications this apparently had to Earth's weather (and especially, geomagnetic effects). British colonization of many parts of the Earth—already well advanced in the seventeenth century—was competitive with many other nations. Germany in particular was a major rival to British colonial interests and relationships between British and German scientists

---

[1] Edward [was] a name he never used, and he signed himself as 'E. Walter.' The tradition of dropping the first name continued in his family (personal communication with direct descendant Alan Maunder).

[2] Wesleyanism dates from the 18th century.

Fig. 19 E. Walter Maunder (1851–1928) at the apex of his solar study (Elliot and Fry).

were growing increasingly strained in part due to national political interests impinging upon detached, scientific ones. (A twentieth century analogy of this phenomenon would be the "space race" between the Soviet Union and the United States in the 1950s and 1960s.)

Not much is yet well known about "E." Walter Maunder in his formative years (and in fact, this book is not particularly focused on his and his wife's personal lives). He was, however, in the midst of the debates on Darwinism as a boy. As a member of a relatively poor family he must have come in frequent contact with the British brand of patriotism huckstered on the streets by the penny press from Shoreditch to Whitehall, street urchins maybe selling drawings or tintypes of what their heroes withstood at Balaklava during the Crimean War of 1854–56.

Living in this area at this point in his life, the church fire of 1867 in Croydon must have certainly affected him. Like his father he may, perhaps, have wondered if religion in England wasn't going to yield to the devil. Yet, how these events and others would affect this youngest son of a Wesleyan ministers' Nicene faith is unknown as yet. His lifelong record of actions—as it shall be seen—speaks far louder and gives a definite clue.

Perhaps what the Anglican Church was undergoing at the time bothered his Methodist's conscience much less than the rampant social injustices of his time. John Wesley's creed certainly included confronting and ameliorating the effects of such injustices. Everything from the creation of the Society of the Prevention of

Cruelty to Animals, to Florence Nightingale and a nurse corps in the Crimea having been acts wherein Wesleyans probably took active parts. Wesley's faith was borne of his travels in America many years before, and had its roots in the New England pastor, Jonathan Edwards', deistic world view. This particularly American stamp to Maunder's father's faith could just as well have influenced the English liberal government in its tacit support of the American Civil War's higher ideals (to wit, the end of slavery as a tenable idea).

Like Anglicans, Methodists serve God believing that their commitment is demonstrated through action, good deeds, not by mere words, nor through revelation—through the Holy Spirit. Concomitantly, their faith is not hemmed in by particular distinctions of class, race, or sex, and they go out of their way, ideally, to demonstrate this. To what extent his father, mother or elder brother played in forming his early religious outlook is not known. But his Methodism is congruent with his presumed later good acts, though not necessarily with his ability in speculative and pure sciences. In addition to his work in solar science, Maunder became a scholar of religion and wrote a great deal on it later. He was to become a considerable force in this in conjunction with his solar astronomy. He not only wrote books on religion but also became ensconced in an institute for religious study,[3] where he gained considerable respect.

That he often forced the hand of truth in finding errors in the research record against notables such as William Thomson, Lord Kelvin and Alfred Russel Wallace for instance—and bevies of lesser-knowns—cannot be denied. His public stance—shown *facta, non verba*—on those actions involving women's rights as they affected him and his near ones betrays a Methodist's belief in good *actions*. This is not to imply that Methodism is superior as a faith to all others, simply because one individual seemed to embody its higher ideals so fully.

His early schooling was at the school attached to University College on Gower Street.[4] Like many before and after him, a period of severe illness (in these harrowing times, usually fatal), probably in the mid to late 1860s, could have brought about a religious experience that concentrated his energies. Following this illness for one reason or another he made a map of Croydon by pacing the length of the streets and measuring angles by eye.[5] Perhaps the discipline of having to win out over severe illness had disciplined his mind in such a way as to make him a

---

[3] The Victoria Institute. (Where Newton critiqued Christian theology.)

[4] B.A.A. Hundred and Ninth Annual General Meeting, February, 1929, p. 318.

[5] Obituary of E.W. Maunder, *Journal of the BAA,* Vol. 38, Session 1927–1928, No. 6, p. 230.

diligent observer: a sort of technical precursor of the scientific achievements that would one day come. As to an interest in astronomy, and what inspired this, there is little evidence. However, in a book written by his second wife, Annie S.D. Maunder (and partly by him) years later, Maunder wrote this reflection on sunspots seen when he was fourteen years old, revealing an observational bent for solar astronomy:

> In February, 1866, as I was returning home from school one evening, I saw the sun, low down in the west, shining red through the mist. The sun was dim and red enough for me to look at him without blinking, and I saw plainly on him a round black spot ... It was the first time that I had ever seen anything on the surface of the sun ... The next time that I saw the sun lying low and red in the west, I saw the black spot ... again, but it had changed its place, and was now much further from the centre of the sun. Two or three days later, it was gone.[6]

He entered King's College, London[7] as an "occasional student" on January 26, 1872 somewhat before his twenty-first birthday, studying chemistry, mathematics and "natural philosophy" (the equivalent of what is called physics today). It should be pointed out how primitive this course must have been in comparison to what was available for study and reflection just a few years later. He enrolled again in April (18th) a few days after his birthday and continued the same course of study. In October, 1872 he became a full-time student in the Department of General Literature and Science, being placed in the "Modern Division." His university career continued as an occasional student once again in 1873, where it trailed off.[8]

For a time, maybe in connection with his part-time status as a college student, he found employment in a London bank. This was doubtless a testament to his predominant ability for numbers, if not necessarily the quantitative reasoning he would become adept at (perhaps directly supported by his more thoroughly-educated,

---

[6] Maunder, A.S.D., Maunder, E.W., *The Heavens and Their Story* (Dana Estes, 1908), pp. 103, 105.

[7] He never graduated from college. However, Maunder is listed in the College Calendars for King's College as having matriculated as a student of the Department of General Literature and Science from 1872 to 1873. Beyond that, there are no further details of his academic career (courtesy of Victoria Holtby, King's College archivist, London).

[8] Information courtesy of Victoria Holtby, archives assistant at King's College London (Strand). Maunder's tuition per term was eleven pounds and seven shillings. Referring to physics as "natural philosophy" continued in England's universities for some years, until quite recently at Edinburgh University (noted by Mary Bruck).

math-talented second wife, Annie, later on). It is a strong indication that he had a need to earn a living of his own. As he would ultimately leave the bank and obtain a rare civil service appointment, Maunder, fortunately for him, pursued a science career in a linear path to advancement. At the time in the Royal Greenwich Observatory, where Maunder was to work, it was common practice to hire children talented in arithmetic as "ciphers," or "human computers." After a certain age, however, they were routinely discharged to work in banks, where often enough they were deemed "unfit for business."[9] These children, unlike Maunder, suffered a retrograde career fate.

We do not know the reason for Walter's quitting the bank (or even why he ever insisted on the "E" instead of "Edward"[10]—or how he got the job in the bank). He was addressed by friends and associates as "Walter." But it seems he was following in the footsteps of his family: perhaps there was, during the touch and go times he grew up in, the need for family members to secure "stable employment in the professions," such as these were. University education sometimes being impediments to the very bright rather than an advantage—and class/money barriers in England such as they then were—Maunder's path did not lay there, either.[11] The same was true of his elder brother Thomas Frid, who became an accountant and remained one all his life, supporting Walter not only in being the secretary, but an assistant in the founding of, the British Astronomical Association (B.A.A.) with him and many others.[12] Thomas Frid was its secretary for thirty-eight years and remained so nearly up to his death, outliving his younger, more famous brother considerably. Such a relationship implies not only a brotherly concern, but also a mutual interest in the scientific if not scholarly occupation of his younger brother.

There is, then, a great gap between the crossroads seen here with these two men and a chance for the younger, at about age twenty one, to take part in a civil

---

[9] Meadows, A.J., *Greenwich Observatory: The Royal Observatory at Greenwich and Hersmonceux, 1675–1975* (Vol. 2: Recent History [1836–1975]) (Taylor and Francis, 1975), p. 10.

[10] The tradition of dropping the first name, Edward, continued in the Maunder line for many years. One son, named Edward Arthur, was always called "Arthur," for instance (personal communication with Arthur's grandson, Alan Maunder, direct descendant, October, 1999).

[11] Non-university graduates being placed in assistantships stopped in Christie's time. In the 1890s, Maunder and another civil service appointee of the 1870s, Criswick, were "serving side by side with young university men in their twenties." (See Bruck, below, p. 85). Getting appointed in Airy's time probably guaranteed him a career, as he might not have been eligible in Christie's era (noted by M.T. Bruck).

[12] Obituary of Thomas Frid Maunder, *Journal of the BAA*, Vol. 45, Session for October, 1935, p. 408.

service examination in 1872 held by the London Commissioners. The test was the first ever to be held in England, and it was a model of impartiality that probably agreed well with Maunder's upbringing. Poor as the times were, certain unattached young men of lesser means at least were getting "a chance" at a career in government.

The test was held to fill a vacancy in the Royal Observatory, or "Greenwich." (It was not called "RGO," for Royal Greenwich Observatory, until the 1950s.) It is difficult to say what led Maunder, with no previous family contacts in this science world that are as yet known, and presumably, no expressed urge to be an astronomer except for some youthful observational ability and perhaps from his college studies in "natural philosophy" to this route. There must have been an abiding interest in the field on his part shown by the opportunity given by the City of London, and reflected in his choice of studies at King's.

To dangerously speculate, such odd luck (for soon after Walter's appointment, non-university graduates rarely, if ever, got assistantships again at Greenwich) may have had a strange chain of coincidental events behind it. England was in the midst of an economic depression in the 1870s,[13] still reeling from expensive wars and then suffering depleted markets where the climate may have had a negative effect on her colonies, which meteorologists such as Charles Meldrum, in Mauritius (a "watch station" of the British Empire) were noting. It took Benjamin Disraeli and the financing of the Rothschilds to bring England some financial edge. Queen Victoria courted American billionaires such as George Peabody. We can assume in these times of aggrandized royal splendor an increased focus on the common man (Gladstone); of Irish Home Rule and attacks on the Union Act, and national contrition felt by the survivors of the Crimean War. At the same time, there was colonial strife that would lead to the creation of men of high ideals and strong national temperament like Gandhi in those self-same colonies.

As a consequence of Gladstone's assistance in forcibly legislating social changes in England, there must have been some agitation on behalf of the Astronomer Royal at the time, Sir George Biddell Airy, for more staff. Airy was a recipient of the Devonshire Commission's reports coming in from the colonies and perhaps even the "watch stations" manned by the Sabines and the Meldrums. Airy was known for his "authoritarianism." In such high ranking Englishmen,

[13] Curiously, there was a terrible El Niño current and Southern Oscillation (ENSO) event that may have made matters worse by 1877. There was famine in India, for example. But the event was later proven world-wide. See Richard H. Grove *et al.*, *Nature and the Orient* (Oxford University Press, 1998), Chapter 11.

however, a wish for efficiency and setting a mark for institutional excellence, borne out by the refined nature of their contributions, are sometimes misconstrued as being the acts of an "authoritarian."[14] Walter Maunder—on record as not getting along with the Royal Observatory director—had the following to say about him:

> So, too, his regulation of his subordinates was, especially in his earlier days [he was Astronomer Royal from 1835 to 1881] despotic in the extreme—despotic to an extent that would hardly be tolerated in the present day, and which was the cause of not a little serious suffering to some of the staff, whom, at the time, he looked upon in the true spirit of Pond, as mere mechanical "drudges."[15]

Airy of course, was later to replace John Pond. We get a view of Pond in Maunder's words, as well, since Pond seemed to scarcely have been much better in terms of attitude towards subordinates. However, unlike Pond, Airy was *begged* to assume the post of Astronomer Royal (no longer was the job referred to as "Astronomical Observator" as it was in Flamsteed's time). Airy also designed many instruments, held his professional staff to strict publishing schedules of their work, and was the kind of administrator who, in any case, paved the way for the observatory in becoming a "new thing," altogether, at the time: an "astrophysical" observatory.

Outside of his "factory-like"[16] work methods which involved hiring low-paid human "computers" (a practice not necessarily begun in his stewardship), and his "Victorian moralism" he was personally responsible for extracting the observatory from the vice-like grip of the Admiralty, which was preventing it from becoming a world-class contributor to the knowledge of astronomy in some respects.[17]

The roots of what was to become the driving force for forming solar research bodies at South Kensington and the Royal Observatory, however, sprung partly from the aforementioned globally-spread, colonially-inspired "watch station" notation of weather as influenced by solar activity. This activity was reported by the likes of Meldrum and geomagnetic activity by Colonel (later General) Edward Sabine – the latter whom Airy also apparently did not get along with.[18]

---

[14] See good reference books on such matters by Patrick Moore.

[15] Ibid, Meadows, 1975, p. 10. Pond was Airy's predecessor.

[16] Bruck, M.T., "Lady Computers at Greenwich in the early 1890s," *Quarterly Journal of the Royal Astronomical Society* (1995), Vol. 36, pp. 83, 95.

[17] Ibid, Meadows, 1975, pp. 1–26.

[18] Ibid, Meadows, 1975, pp. 1–26. Whether or not Sabine's being part of the military hierarchy, and by association, being influential in this mutual tension is only speculation.

Terrestrial magnetism was studied at the Royal Observatory even before Airy's time, however, and in the 1830s and 1840s, Airy actually took measures to strengthen the research activity in the field of magnetic or geomagnetic research. At one point, he even hired Michael Faraday to plumb the depths of these problems. But Sabine and others pointed out the nature of the bad instrumentation to be found at the Royal Observatory and continued to work from Kew Gardens.[19] Regardless of the issues involved by the 1860s and 1870s, in part through the auspices of the Royal Commission on Science, solar science and magnetism became a major concern of the stewards of Greenwich. A key member of this commission—one with much influence—was the astronomer and publisher (founder of *Nature* magazine) Joseph Norman Lockyer of whom much will be said later.[20]

The English liberal government under Gladstone pushed through an Education Act in 1870, compulsory education for children, and, most intriguingly, a Ballot Act in 1872.[21] This latter-mentioned act was itself an attempt to reform voting by freeing people of bribery or intimidation, or both; practices common in England at the time, and still prevalent in many parts of the English-speaking world. For fairness was one of the erstwhile-priest-to-be Gladstone's aims in the English body politic. Some of his reforms led to much boat rocking in the Empire.[22] Such consequences of Gladstone's actions did not endear him to a queen who already despised him.[23] It is not far-fetched to see the reforming of the civil service, then, and a move toward open testing for posts, whether or not posts were impartially staffed via test-taking, or not. These reforms were to occur in Maunder's England in spite of what the queen and her court may have thought, including royal commissioners like Devonshire, who had headed the Royal Commission on Science. These were "interesting times," as the Chinese ironically say, in England, depressed as they were, and were characterized by much political agitation and the desire to be free of monarchic rule.

In the June 1st, 1872 report to the Board of Visitors (Royal Observatory), Airy requested for sunspot and solar spectroscopic work to be undertaken at the Royal

---

[19] Ibid, Meadows, 1975, pp. 95–105.

[20] Lockyer, J.N., "Simultaneous Solar and Terrestrial Changes," *Nature*, February 11, 1904, No. 1789, Vol. 69, pp. 353–355.

[21] *History of England,* Ed. J. Burke (Guild, London, 1988), p. 263.

[22] Irish agitation and ascension in Parliament led to much political power being obtained by Irishmen for the first time—an unseen consequence of Gladstone's reforms.

[23] Burke notes that "she [Victoria] could not bring herself to attend his funeral."

Observatory.[24] The move to expand this work would be one that would make the Royal Observatory a "physical observatory, *pro tanto*."[25] This implied a plea for rigor in astronomical or astrophysical sciences, pure or otherwise, as practised at the Royal Observatory. Able administrator that Airy was—for all his shortcomings and unstoppable halting of progress—the Royal Observatory was clearly on the path to modernization.[26]

How many tested for the open post at Greenwich, or how Walter got word of the test, are facts which are still buried in archives or in letters. The posts were openly advertised in the press, so it could have been this way in which Maunder found out.[27] It is known that the initial single-post opening at the observatory became *three* posts—perhaps a testament to how many competed for the job—and three of the best or luckiest aspirants all found jobs at the Royal Observatory. Walter Maunder was the third hired out of the three posts that were filled.

The intended open staff post at the Royal Observatory was filled by the degree-holding Irishman Arthur Matthew Weld ("A.M.W.") Downing.[28] A second tested individual, W.H. Finlay,[29] was made the First Assistant at the Royal Observatory, Cape of Good Hope, in South Africa. Curiously, it was only after a *second* examination was given that Walter Maunder was chosen, and indeed, given a post. Why a second examination was given is unknown. It could have been on behalf of William Christie, Airy's chief assistant at Greenwich, seeking a good solar observer, the qualities needed for such a post apparently possessed by young Maunder, and by none of the others tested[30] (this is purely speculation).

---

[24] The RGO no longer exists as a working observatory (closed in October, 1998). In the Maunders' time, the "RGO" was called either "Greenwich" or the "Royal Observatory." It became known as the RGO in the early 1950s, when it was moved from Greenwich to Sussex (this note, courtesy of M.T. Bruck).

[25] Obituary (E. Walter Maunder) *Nature*, April 7, 1928, No. 3049, Vol. 121.

[26] Airy, no fool of the bureaucrats, tamed not only the Admiralty, but also the Board of Visitors thus earning him the respect of other observatory directors in Europe, like Struve in Pulkovo (Ibid, Meadows, 1975, p. 4).

[27] Personal communication, Mary Bruck. Bruck asserts at this time, civil service openings were openly advertised in the press.

[28] Ibid, Report of the Council to the Hundredth and Ninth Annual General Meeting, p. 314, Arthur Matthew Weld Downing later became Superintendent of the Nautical Almanac Office, and supporter of Alice Everett and Annie Russell as members of the Royal Astronomical Society at a later date. (Ibid, Bruck, p. 90, and Downing's obituary of December, 1917, *JBAA*, Vol. XXVIII, No. 2, p. 69 by Annie Maunder, where she points out his support for her, Everett, and Elizabeth Brown in career matters. Downing had earned a B.A. and an M.A., and was awarded an honorary doctorate.)

[29] Like Maunder, these two became prominent astronomers.

[30] Ibid, Session 1927–28 Obituary.

However, the post Walter Maunder obtained was a *new* one altogether at the Royal Observatory, Greenwich. He was appointed photographic and spectroscopic assistant there on November 6, 1873—the rank being equivalent to "Assistant of the Second Class"[31] and the pay being a then-reasonable £200–300 per annum.[32] (For comparison, the Astronomer Royal Airy had already negotiated an annual pay of £800 in his initial 1835 appointment.) He became the head of the solar photography section there and remained so for his entire Royal Observatory career.[33] Curiously, Maunder enrolled—or had earlier enrolled in—a course at the Physical Laboratory at King's College on November 11, 1873, noting that his parents' address had changed from 80 East India Road, East London, to 47 Lonsdale Square, North London.[34] As a new employee of the observatory he may have been sent there by that institution to learn spectroscopy and photography (photography of the sort employed at Greenwich was then an advanced technical subject somewhat akin to advanced computer graphics programs today).

The need at the Royal Observatory was mainly for the study of sunspots, went the curricula request, and was perhaps a rolling agenda item since Sabine's day as the Royal Society secretary. Sabine made connections between sunspots and Earth's magnetic storms as early as the 1850s[35] and equipment from public sources like a photoheliograph found itself being transferred from the Kew Observatory to the Royal Observatory. A spectroscope was received from a Mr. Browning, a well-known optician, and was attached to the 12¾ inch Merz refractor called "the Great Equatorial." This was used for line of sight work and was replaced by the 28 inch equatorial in 1892 for photographic purposes.[36]

Maunder found himself in a secure and necessary job. He avoided the fate of the "boy computers." (They worked twelve-hour days in Airy's time, at the

---

[31] Ibid, "Report of the Council to the ...," p. 314.

[32] Ibid, Meadows, 1975, p. 8. Chief Assistant's pay was between £500–600, First Class Assistant's was between £320–450. Before 1870s, Airy recommended against annual increments, but during Maunder's time, annual increments of £20, £15 and £10 were introduced to the payscales for Chief, First-Class and Second-Class Assistant posts, respectively.

[33] Crommelin, A.C.D., Maunder Obituary, *The Observatory,* May, 1928, p. 157. What rank or status Maunder had at RGO during his 1913–1918 wartime return is not clear.

[34] Ibid, personal communication from Victoria Holtby, King's College, London, May 28, 1999.

[35] Kelvin, Lord, W.T., "Astro-Physical Notes," *Philosophical Transactions of the Royal Society of London,* 1893, p. 76. This is probably in reference to the May 6, 1852 paper read at the Royal Society by Sabine on "periodical laws discoverable in the mean effects of the larger magnetic disturbances."

[36] Ibid, Crommelin, *The Observatory,* Maunder Obituary, May, 1928, p. 157.

observatory.[37]) The reforming political movements of the day may have created a niche for someone like Maunder and he was quick to fill it, possibly too well. His life's record, in fact, shows an over-compensation in ambitious activities, however equal they were to his ample abilities.

So far as it is known, Maunder was appointed to this rare post on ability alone. He was to be spared the existence his somewhat reserved and retiring brother, known for his "dry humour," accepted. The elder brother could have warmly embraced this luck, living a vicarious existence as the staid supporter of his younger brother's work. For, different as they were, "totally different" in the words of one,[38] they both were "imbued with the same ideals, a lofty purpose and high integrity of principles."

Having a marked talent for public speaking, we can assume that Walter was not the shy and retiring type. He possessed a certain fire; perhaps even recklessness—an intellectual fire, so to speak—that could easily enable him to get burnt, were he not careful. Purely speculative as it may be, it is not unreasonable to say that his "dryly humorous" older brother could have kept a weather-eye on the storms his brother was headed for. Like Stephen Gray in an earlier age, we see in the brother a similar wish to retire from non-astronomy related pursuits. In 1907, Thomas Frid retired from accounting, and for the next twenty one years dedicated himself to his brother's institution and subsequently Britain's—the B.A.A.—for some time after Walter's death, dying almost in his chair.[39] As a parallel to Thomas Frid, though much earlier, Gray had quit dyeing work and literally had himself institutionalized in a pensioner's home merely to continue being an astronomer without distraction.[40]

Walter Maunder's ambitions lay less in the politics of science at institutions such as the Royal Observatory. His interests were founded, rather, precisely on the science, even if his apparent zeal at mid career made waves. Initially he found colleagues, the like of Christie and Frank Dyson, and latterly, in his skilled second

---

[37] Ibid, Meadows, 1975, p. 10 (in Maunder quote).

[38] Ibid, B.A.A., October, 1935 Obituary. One "A.M.N." wrote this obituary.

[39] Ibid, 1935 Obituary. To quote from this personal friend of Thomas Frid: "and not until increasing deafness rendered it difficult for him to carry out his duties did he relinquish the office he had occupied for so many years with such faithfulness."

[40] For a fascinating look into the times Gray and Flamsteed lived, see *Under Newton's Shadow* by Lesley Murdin. Gray had literally fled his life as a cloth dyer, taking various oaths of decency and cleanliness, and admitting poverty, to have himself cloistered in a men's home—where he continued his astronomical investigations, becoming quite well known for them.

wife. He avoided controversy of the destructive kind, simply since few understood what he was doing and so could see no threat to the powers that were. Maunder's reputation and achievements found themselves laid by the way perhaps for this reason. The reputations he courted disaster with, like the indomitable—and fortunately aged—Lord Kelvin, were kept largely within institutional frameworks. The abstruse nature of the work Maunder took part in contributed largely to the lack of recognition he would receive in posterity, and was a major factor in his and his wife's work being forgotten. However, abstruseness was not the only reason for this neglect. The same institutions—admiring his "greatness"—known or otherwise, kept a lid on his reputation through time. More liberal times would see luckier or better-placed men able to capitalize on the work of a path-breaking scientist without too much backward thanks being expressed.

Clearly a man before his time in an England not in the least ready for him, Maunder quite possibly experienced the institutionalized burial of some of his and his wife's more vital works and insights. But this is also speculation. Much of this work was based on good mathematical proofing, crystal-clear reasoning, and pristine logic—and the later scientists who saw this body of work were in many cases anxious to build on the knowledge base the Maunders had brilliantly achieved. As obituaries and encomiums make very clear, Maunder gained "great fame" for his work on "sunspots" as if this was all he did. As stated, there is no reason to assume that Maunder was by any means meek, however much was the "suavity of his manner and a softness in speech"[41] stated in the official, semi-Victorian obituaries or obliging of what the English once referred to as "one's station."

At the same time, he was not particularly "old-boy-network" oriented, as his and his wife's stance on women's membership in the Royal Astronomical Society would show. For example, Annie Maunder was nominated a fellow of the society in 1916, and became one in December, 1916—upon recommendation of F.R.A.S. Walter Maunder, himself.[42] The B.A.A. was founded, in part, to get around discrimination by gender, for one. He was even against, for instance, the word "amateur" in the British Astronomical Association's title, so as to separate them from professional associations. This separation is still a thin line, today. For such acts, his "kindness of heart" truly did "lay deeper than its outward manifestation."[43] But it is beyond doubt that it gained him no sympathy in some important quarters.

---

[41] Ibid, February, 1929, *BAA* encomium, p. 318.

[42] *The Observatory,* Vol. 39, Number 501, June, 1916, p. 256. This reference contains her nomination and proposer.

[43] Ibid, February, 1929 encomium, p. 318.

Christie succeeded Airy in 1881, and it was during this time that solar research was placed at the forefront of scientific activities at the Royal Observatory, the motors of change chugging since at least 1871. Walter Maunder and Christie worked together on observations of displacement in lines of solar spectra for determining velocity in the line of sight. But then, there was the growing importance of stellar photography requiring special attention: literally, the CCD camera work of the day. We can assume that this primitive plate photography, requiring long exposure times and patience and skill in development, was refined by Maunder, directly or indirectly, at the observatory—the developing task being at last relegated to a trained expert. It was a skill Maunder would later transfer to his wife Annie Maunder.

Notably, Annie (A.S.D.) Maunder was employed by the Royal Observatory as one of a team of female "computers" who were placed on par with the dubious position of the male computers—a program that later failed, but which gained her an "in," in a field blocked to women totally, at the time. This was under Christie's time as Astronomer Royal in the early 1890s and Christie may or may not have been sold on accepting the program by the educator Dorothea Beale.[44]

The procrustean attitudes within which women were held in most professions until only recently only skimmed the surface of the profession of astronomer, by comparison. Astronomy is a field that requires not only patience but persistence.[45] This is even shown in the matter of the women plate photographers, for example, at the Harvard College Observatory in the U.S. Though laborious, the photographic assistants' and cataloguers' labors were held in high esteem by those deeply involved in the work, as the Draper Catalogue done by the likes of Williamina Fleming and later, Annie Cannon (for their work in spectral analysis), at the Harvard College Observatory would show.

The fate of the segregated women of the Harvard College Observatory was never the case with the female computer program at the Royal Observatory, which lasted from 1890–1895. A.S.D. ("Annie") Russell was indeed the last woman to leave the program, quitting entirely upon marrying Walter Maunder. At the Royal Observatory, the women astronomers worked openly and interdependently with men, and some often observed at night using observatory instruments. However, the program eventually died out. Notable woman scientists—at one point, even the

---

[44] Ibid, Bruck, p. 85.

[45] Cursory inspections across history show a high involvement of women in proportion to men in this science—and the record is revealing more who have been active, yet silent, all the time. Many are well known, well regarded, and deserving of due credit and fame, as this book will partly show.

science historian and observer, Agnes Clerke—were invited as "honourary members" of the Royal Astronomical Society and offered permanent jobs at the Royal Observatory (Clerke declined). But Annie Russell, who sat for Mathematical Tripos at Girton College, Cambridge, in 1889[46] was also given—as a testament to her skill and value—an "open contract" at the Royal Observatory. The pay was atrocious, the program doomed from the start, and she quit paid work in astronomy altogether upon marriage to Maunder.[47]

Walter's first notable work, done with Christie, entailed the furthering of the work of Pierre Janssen and the influential Joseph Norman Lockyer. Both Janssen and Lockyer found out how to study the solar chromosphere (that part of the Sun's atmosphere lying above the bright photosphere) as well as solar prominences without needing an eclipse. Maunder's work involved measuring those prominences, prominences being eruptions, as they were then known, which sometimes loop and twist off our Sun's surface. Correlations between solar photographs taken at one location were compared to other ones taken at the same time in other locations—something that figures in this book significantly later.[48] And it was in this timeframe that Maunder more or less put into effect "the formation of [a] unique record of the solar spots associated with his name."[49]

[46] G.T. Walker was Senior Wrangler that year, with Frank Dyson, second. Annie was placed 42nd, a Senior Optime rating (second class honours).

[47] Ibid, Bruck, and Mrs. E. Walter Maunder's obituary from December, 1947's *The Observatory* (No. 841), p. 231. One can't say how much she earned from books: one reason why E. Walter's name was attached to "mostly Annie's" book (*The Heavens and Their Story*) could have been the popularity of his name. She of course earned nothing from lecturing, as she did not lecture in public (courtesy of these observations from M.T. Bruck).

[48] At Dehra Dun, for example.

[49] Ibid, E.W. Maunder Obituary.

8

Maunder and the Connection of Sunspot Behavior and Geomagnetism: Resolving "the Fifty Years' Outstanding Difficulty"

Most of E. Walter Maunder's important solar work and its publication was accomplished between 1873 and 1905.[1] Interpreting this data, however, started from about 1895 onward. In some of this perhaps his wife's careful and methodical input is seen as well.

From about 1877 onwards, Maunder began measuring sunspot areas and faculae areas with a position micrometer. It was during this time that he collected data for creating his famous "butterfly" diagrams, referred to as such even by himself. Near this time, he suffered the loss of his first wife, Edith Hannah Bustin,[2] and was responsible for five children;[3] he was not to marry again until 1895, his soon-to-be wife (A.S.D. Russell) at that time being one of the prized assistants then engaged in work at Greenwich.

The Maunder butterfly diagram, which is the fruit of the 1904–1905 analysis of Maunder's work, begs for insight even now. Its value in studying the cyclical nature of the Sun is still being realized. The diagram is created by showing sunspot

---

[1] Ibid, Maunder's *The Observatory* obituary. The writer pointed out that Maunder's principal research stood out as: (1) a re-determination on the position of the Sun's axis, after Carrington; but Maunder used more data to show this, (2) a study of the motion of spot zones in latitude in the course of 11 sunspot cycles, the "Butterfly Diagram" being a graphic proof of this research, and (3) showing the strong tendency for magnetic storms to recur after the period of the Sun's synodic rotation, and that the storms are produced by streams of "electrified" particles discharged along definite stream lines from disturbed areas of the Sun.

[2] Ibid, Maunder's obituary in *The Observatory*. He married Edith Hannah Bustin on September 11, 1875 in Wesleyan Chapel, High St, Clapham, Wandsworth. (Personal communication of direct descendant Alan Maunder, October, 1999, and notation of the same's family tree.)

[3] Maunder had five children; three sons (one dying in infancy) and two daughters (and none with Annie). Their names were (in order of birth): George Harvard, Edith Augustus, Irene Matilda, Walter Anthony (who died in infancy in 1885), Edward Arthur, and Henry Ernest. (Personal communication from direct descendant Alan Maunder.)

centers in latitude per rotation of the Sun. When a spot center fell on one or more days during a certain rotation, at a certain latitude, a line was drawn over that degree. Over time this created, up and down, a long black line, with white gaps showing where no sunspots appeared. Taken together in this observation path, the collected data in "normal" times of black and white takes on something of a moth's shape; not Francis Bacon's bee, which collects from source to source, transmuting something new, altogether: a butterfly.

He first made this plot and exhibited it at Greenwich Observatory on June 6, 1903 (the *Monthly Notices* [of the RAS] reference, however, appeared in June, 1904). This paper[4] if read by modern scientists could almost be looked at as a forerunner of the better work routinely seen in important scientific journals like the *Astrophysical Journal* or the *Physical Review*. That is, with its proliferation of sine curve charts, plotted distribution curves and the baffling Maunder butterfly diagram at the end.

A lot can be recorded over thirty years' worth of patient, careful recording and observation and much wheat separated from chaff: much complacency abolished; many errors unmasked. The paper represented in "both tabular and graphical forms the results of thirty years' working the measurement and reduction of about 9,000 photographs of the Sun, and the latitudes and areas of about 5,000 separate groups of spots." In this paper some of the Sun's most hidden secrets are revealed for the first time.[5]

In the first six diagrams of this seminal paper the plots show how—individually—the two hemispheres, North and South, behave in the Sun (for the hemispheres may change independently). Secondly, these diagrams also illustrate what he emphasized as being "Spöerer's [Maunder's spelling] Law of Spot-zones." It shows that the two hemispheres do not differ much in total spotted area. Assumed as fact in this 1904 paper are Schwabe's 11-year solar cycles and it points out sunspot maxima and minima noted within these cycles, per solar hemisphere, both North and South. The rises in spot latitude are abrupt in both solar hemispheres.

Maunder, perhaps with his wife's assistance, divided the 5,000 sunspot groups into three "classes." This was done to establish a path of understanding the phenomena

---

[4] Maunder, E.W., "Note on the Distribution of Sun-spots in Heliographic Latitude, 1874 to 1902," *MNRAS*, Vol. 64, June 1904, pp. 747–761.

[5] This fact may include E.W. Maunder and Astronomer Royal Frank Dyson's paper, making improved determinations of the inclination angle of the Sun's rotation, relative to the ecliptic—a task which was tackled earlier by original solar researchers like Richard Carrington, Galileo, Scheiner, and others (see, for example, *MNRAS*, Vol. LXXIII, 1913, p. 50).

in some tangible, regular way. The first class was called *undeveloped* (lasting for only a few hours and which are small). The second class was called a *normal spot stream,* or, "*pairs,*" growing into long streams with the first and last spots increasing in size. In this class, the middle spots die out. (This class they deemed to be fairly long-lived.) The third class was a *"giant spot group"*—the significance of which will soon be evident. Those of the third class are sometimes five to six times the size of spots in the second class and can vary greatly in how long they last. Fascinatingly, these giant spots show isolated peaks and spires in these charts. They were due

> to isolated outbursts, not by any means to zones of activity, and are brought about by the purely computational concentration into single-degree zones of spot areas really distributed over many degrees.[6]

The data show that there is usually a period of one sunspot maximum in each solar hemisphere per year, with small exceptions to this rule being observed. Right after a sunspot minimum period, Spörer's Laws indicated that the first members of the new cycle appear at high solar latitudes while the expiring members of the old have not completely died out at the lower solar latitudes.

However, Dr. William J.S. Lockyer, son of the famous J. Norman Lockyer[7] (both as the discoverer of the element Helium in 1868 in the hot gaseous atmosphere [chromosphere] of the Sun[8] and the founder of *Nature* science magazine) challenged the validity of Spörer's curves. The point William Lockyer drove home was that within several spot zones, one arrives at a latitude in which spots may not exist at all. Maunder replied to this by saying that this applied only to a new sunspot cycle's *beginning.* For then, and only then, said Maunder, are there definite and distinct zones of spots in either solar hemisphere.

Other points in Maunder's paper are that in the rise to sunspot maximum, the spots avoid the equatorial region and that the curves (that is, Spörer's) reproduce those of the corresponding curve of the 11-year cycle (the Schwabe sunspot cycle). In these notations, Maunder even noticed that the "eleven-year cycle" was not quite exactly eleven years but varied in length from time to time—an important modification.[9] (Rudolph Wolf as early as 1861 recognized that: "*greater activity on the Sun goes with shorter periods, less with longer periods* [sic.]."[10]) Spots in

---

[6] Ibid, Maunder's June, 1904 paper, p. 754.

[7] They made a father and son team in 1901, writing a paper on Indian monsoons and cyclones.

[8] Proceedings of the Royal Society, October 20, 1868.

[9] Ibid, Maunder's June, 1904 paper, p. 756.

[10] Wolf, R., "Abstract of his results. By Prof. Wolf. (Translation communicated by Mr. Carrington)," *MNRAS*, Vol. 21, 1861, pp. 77–78.

a degree higher than 33 on the solar disk are rare, he noted, seeing from the thirty-year spread of his collected data an agreement in insight noted with fair accuracy as early as Galileo's time.[11]

Maunder published the paper in June, 1903. But the debate and controversy, or perhaps argument, ensued in meeting/counter-meeting in the halls of wherever the B.A.A. met and in the Royal Astronomical Society, as well.

In the correspondence section of *The Observatory* for August 1903,[12] Maunder's open rejection of part of Lockyer's observations would dampen relations with Lockyer, as small as the critique was. The same could be said of William Lockyer's father—an illustrious discoverer.[13] Maunder risked the ire of J. Norman Lockyer. He was no light-weight, having done the politicking needed to acquire funds for programs that at least indirectly assisted Maunder in earning his own living.[14] At the same time, Maunder was originally hired at Greenwich to investigate some of the questions and observations the elder Lockyer had posed.

In the letter, Walter Maunder suggested that Dr. William Lockyer check the data of the Royal Observatory from 1884 to 1903 to see the ample proof of spots occurring in particular zones where Dr. Lockyer said there would be none. Maunder linked his observed phenomena on the lack of sunspots in times of sunspot minima firmly to Spörer's findings. His letter in *The Observatory*, perhaps penned in May, was drafted in response to a May 8 Royal Astronomical Society presentation by William Lockyer. In the meantime, Maunder's formal butterfly-diagram paper, published by June, 1904, went out, where he pointed out the arbitrary nature of Lockyer's observation, locked into detailed graphs, and which was based on ample methodical observations that could be cross checked by any impartial researcher later on.

Colonial rivalry between Germany and England in this period was ugly: the path towards what would be termed "the Great War," was already beginning to take shape. Maunder, in this period of anti-Prussian zeal, dared argue impartial scientific proof amongst his own in England. Truth or not, he was clearly siding with the work of a German scientist, however historical and impartial this association was. As the elder Lockyer, at least, had actual occasion to be rudely challenged at times

---

[11] Ibid, Drake.

[12] *The Observatory*, letter by E.W. Maunder entitled "Spöerer's Law of Zones," August, 1903, pp. 329–330.

[13] An important "discovery" is every scientist's holy grail. Now as it was, then.

[14] Evidence of this is the fact that J. Norman Lockyer sat on Devonshire's commission in the 1870s, and served as the emissary for the importance of solar-terrestrial research from India to Britain. Details of the history of funding solar research are seen in Lockyer's *Nature* article of February 11, 1904.

by German scientists in painful ways,[15] we can only speculate on Lockyer's feelings towards Maunder, bearing his message before the local scientific community.

Scanning evidence in figure seven of this paper, Maunder pointed out that there was no evidence to support Dr. Lockyer's[16] idea of downward moving

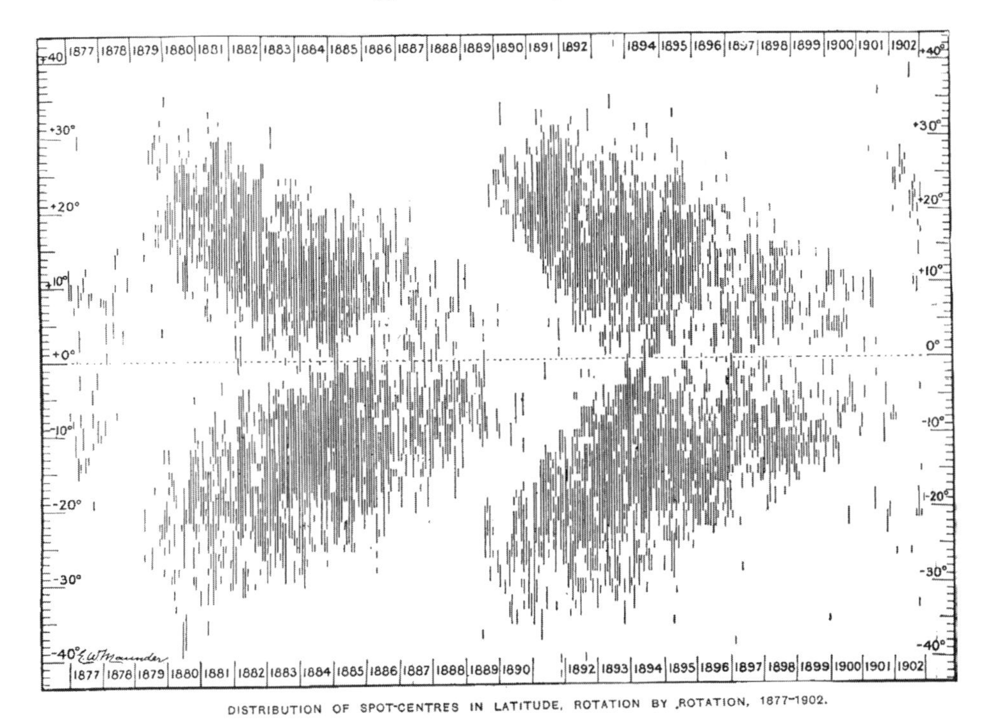

Fig. 20 Maunder's original "butterfly diagram" (1904) for the distribution of sunspots between 1876–1902 (north, above; south, below).[17] The original butterfly diagram is now under the care of the High Altitude Observatory. During the Second World War, Annie Maunder passed on this piece of solar research legacy to Walter Orr Roberts (the first director of High Altitude Observatory) for safe-keeping. (Information/source courtesy of Tom Bogdan[18] of HAO/National Science Foundation.)

---

[15] This is not pure speculation: William's father had contact with German colleagues, and misunderstandings arose, as J. Norman himself pointed out some German peevishness in an apologetic footnote to his 1904 *Nature* article of 11 February. This goes hand in hand with the trends in tensions between the Germans and English at the time.

[16] We may well note here William Lockyer's academic rank, versus E. Walter's (a Ph.D from the prestigious University of Göttingen, contra a civil servant non-degree holder, in spite of being "Superintendent of the Solar Department, RGO"). William Lockyer also served in World War I, first as a lieutenant in the Royal Naval volunteer reserve, and ending with a rank of major in the Royal Air Force in 1918.

[17] Ibid, Maunder's June, 1904 paper.

[18] Private conversation between Tom Bogdan and Willie Soon at the American Astronomical Society's annual meeting in Washington DC, January 2002 (with the pleasant company of Major [Air Force] Paul Bellaire Jr.).

"spot-activity tracks," but that these, indeed, were arbitrary. The eighth figure, the ominous butterfly diagram, showed proof, stated Maunder, of Spörer's Law. It also showed that the southern current of sunspots reached the solar "equator," and crossed it, and that there was only *one* sunspot zone. Maunder in fact actually added to Spörer's findings, much as he had added to the other German's results—Schwabe's sunspot cycles.

Most of the facts just mentioned also happened to buttress observations that Rudolf Wolf in Switzerland made not so very long before. That is, the great low solar activity/no sunspot period in the seventeenth century (at least) corresponded to cold weather in Europe and North America.[19] Around 1850, Rudolf Wolf, Alfred Gautier and Edward Sabine and others noted the relationship between the Sun's activity with terrestrial magnetism as recorded in movements of magnetic needles—something Maunder wished to investigate further. In fact, Wolf had used that information to help reconstruct solar activity further back in time: at least as far back as 1784,[20] when direct solar observations were somewhat lacking. Walter's naïve (for the time) respect for these Swiss and Germans' abilities as scientists was great to say the least. We do not know if he disliked them for being associated with a nation prominent in the Franco-German War and terrible colonial competition and other strife. We can assume he knew the difference between what German scientists stated and represented as it regarded his work and what German nationalism represented, and chose to steer a path between the two.[21]

To Maunder, the laws of spot zones had much more to do with science—matching his own long years of watching and recording them—than with other concerns. It must have been hard for him to defend the rationale of observations as they matched fact, no matter who happened to be the supporters or even initiators—of the said same facts. In Annie Maunder's book, *The Heavens and Their Story,*

---

[19] Ibid, Maunder's June, 1904 paper.

[20] See Hoyt, D.V., Schatten, K.H., *The Role of the Sun in Climate Change*, p. 39

[21] Chillingly—but ultimately a victorious achievement in solidifying and uniting German and English science—a crucial work on geomagnetism, covering the Maunders' hard won research, was achieved by the British Chapman and the German Bartels, during World War II—writing and thinking with difficulty across wartime barriers. Nonetheless, they did it. A stunning and most fortunate achievement—probably for people of all nations. We can note from the majority of the frightened/ashamed demeanors of German scientists' in exile at that time (1940s) that a similar situation could have existed for them in Maunder's. These men were not responsible for the German Romantic Movement, nor for the philosophical roots of German militancy—which inferred by some (such as Karl Popper) was Hegel, Fichte, and others' obscure and pro-annihilation, pro-authoritarian historicist philosophies that obscured those of Schopenhauer and Kant.

written just a few years later, the Maunders (for they each took part in writing it) had this to say about sunspots and the behavior they exhibited. Or the "laws" they obeyed:

> It is very rarely that any spots are found outside the parallels of the sun's latitude that are 35 degrees north or south of the [solar] equator. A very few have been found in a latitude so far from the equator as 40 degrees; one spot has been seen in 50 degrees; but all of these were very small, and lasted for no longer than for two or three days... Nor do the solar spots migrate freely between these limits of 40 degrees north and south of the solar equator. Rather there seem to be narrower zones within which the spots that arise may move, but beyond which they may not transgress. This seems the first law laid upon them. The next law seems to be that in any particular zone on the sun's surface spots do not arise at all times, but only at particular times do spots seemed to be formed. There are long intervals when the zone breeds no spots at all. The third law seems to be that though spots do not stray from one zone to another, there is a bond or connexion [*sic*.] between the different spot-breeding zones, for there is an order observed in the times of breaking out of spots in different zones.[22]

Maunder had more to say regarding terrestrial magnetic activity after much observation and research. For instance, there was an essay he wrote entitled *The Solar Origin of Terrestrial Magnetic Disturbances*.[23] In this essay, he said there was no question as to the existence of ordinary terrestrial magnetic disturbances. Movements to Earth's magnets occur as three kinds, he stated. A daily range, which is a slow westerly movement from circa eight each morning to circa two or three each afternoon, then back again for the remaining 24 hours. Then there is the cyclical inequality of the daily range, which runs through its period in synchrony with sunspot activity. Maunder also noted an annual inequality with "the daily range [being] smallest in winter, and greatest in the summer."

More violent interruptions, on the other hand, are indicated by large shifts in the magnetic needle on Earth. Roughly speaking, this is what Carrington noted on the needle a day after he saw his "white flash" of 1859, which was shown earlier (Chapter 6). For instance, the mean amount of the daily range of magnetic declination at Greenwich, he noted, was about eight minutes of arc; or, about a quarter

---

[22] Ibid, *The Heavens and Their Story,* 1908, pp. 135–136.

[23] *Popular Astronomy,* Vol. XIII, No. 2 (Whole No. 122), February, 1905, pp. 60–63. An article also appeared with this identical title in the *Astrophysical Journal* around the same time.

of our Moon's apparent diameter. In a major magnetic disturbance of the kind Carrington saw, this needle moves over an arc *fifteen times* greater.

In *The Heavens and Their Story*, the Maunders make much of these disturbances and how they are connected with giant sunspots. Very importantly, they noted where these giant sunspots' location on the solar disk occur:

> In the November of 1882, a monster sun-spot easily visible to the naked eye, crossed the sun, and when it was about halfway across, on November 17, a very violent magnetic storm, as these agitations of the magnetic needle are called, occurred, ... ten years later, in February, 1892, a still greater spot ... appeared upon the sun, and when it had passed a little to the west of the sun's centre ... a still more violent magnetic storm occurred than in 1882. This great spot passed off the sun, and returning to the eastern edge, again crossed the sun's disc. When it arrived at the same distance from the centre of the sun, there suddenly broke out again upon the earth a great magnetic storm. Eleven years later, in October, 1903, yet another giant sun-spot appeared ... and there was a magnetic storm, but not a violent one ... but a fortnight later [that is, about 2 weeks] when an important, but smaller, spot had got into the central position on the sun's disc, a magnetic storm burst suddenly ... the most violent that has been experienced in the memory of man.[24]

They go on to make a reference to telegraphic transmissions or messages being disturbed,[25] as well as submarine cables; that is, the cross-Atlantic telephone cable—placed there in part under the auspices of Lord Kelvin. Again, the discussion is very much like the wording found in Sabine's and Carrington's papers for the Royal Society on similar matters of magnetic phenomena (telegraph line disturbances, etc.).

Briefly stated, a "fifty years' outstanding difficulty," as Lord Kelvin described it, grew from E. Sabine's posing of the "difficulty" years before. The difficulty, as

---

[24] Ibid, *The Heavens and Their Story*, p. 187. (E. Walter's name appears on the book, but he states in the preface that it was "almost wholly" Annie's work.)

[25] This was known years before. In fact, Balfour Stewart reported on such interruptions in his "On the Great magnetic Disturbance which extended from August 28 to September 7, 1895, as recorded by Photography at the Kew Observatory" (read November 24, 1861 before the Royal Society). He with others identified the disturbances as having a "throbbing or pulsatory character" which "agreed well with the nature of its action on telegraphic wires." From page 425 of *Philosophical Transactions of the Royal Society of London,* 1861, Vol. 151, pp. 423–430.

Kelvin pointed out, was in proving concretely how the Sun and Earth transacted such cross-magnetic activity—in spite of the work of Sabine. For all the evidence supporting the observations of solar-Earth magnetic disturbances, where was the mathematical proof for the Sun-Earth magnetic link? Kelvin had, in fact, some years earlier done calculations showing why the Sun could not possibly be the responsible source—in spite of observational evidence to the contrary shown by the likes of Sabine, Carrington, and others.

Sabine and other military-politico English imperial government scions such as John Henry Lefroy had an incalculably valuable contribution to make on geomagnetic activity due to the world-wide reach of English "watch stations."[26] Though the British had navigational reasons relating to world trade and military arts for what they did in the main for these watch stations, the data Sabine collected proved useful for later research in solar science. Incoming reports from synchronized observatories worldwide, done in part with German assistance by the likes of C.F. Gauss,[27] for example, also contributed importantly. The record they built was priceless. These men and many others manned the observatories in far-flung locations in the British Empire such as Toronto, Canada; Hobarton, Tasmania; St. Helena in the Carribean; and at Cape Town, South Africa. Here they alighted on such commonalties of magnetic disturbance from Sun to Earth as simultaneous world-wide effects of magnetic impacts from the Sun on recording instruments.

* * *

Kelvin's incorrect assumption, Maunder claimed, was that Kelvin's calculations, pristine as they were, were based on the assumption that the magnetic waves coming forth from the Sun are sent out in all directions—isotropically—through space, at all times. But alas, this was not the case.[28]

---

[26] See S.M. Silverman's "Joseph Henry and John Henry Lefroy: A Common 19th century Vision of Auroral Research," *Eos*, Vol. 70, No. 15, April 11, 1989.

[27] Meadows, A.J., Kennedy, J.E., "The Origin of Solar Terrestrial Studies," from *Vistas in Astronomy*, Vol. 25.

[28] Another Kelvinism, spoken with pride in the years when mathematics first started showing itself invaluable to scientific progress in utilitarian terms: "when you can measure what you are talking about and express it in numbers, you know something about it; when you cannot express it in numbers, your knowledge is of a meagre and unsatisfactory kind." This should be an industrial engineer's credo; however, pure science abides by intuition and observation, as well as by measurement, and by abstraction that will *look* for its measurements (and if finding none, will reject the hypothesis until "next time"). But his statement—commandment-like, as it is—may have shown how Kelvin's scientific judgement could be affected.

Take what Annie and Walter wrote, above, on the February, 1892 disturbance, as well as the October, 1903 one, and (probably) Annie's rather methodical, careful prose as to spot size, per location on the solar disk, per size of the storm. Walter Maunder noted the connection between large sunspots and great magnetic storms and he presented this finding publicly. That was something Sabine had done fifty years before. However, Maunder then rigorously sought a scientific explanation for it.

Most remarkably, he states in *The Heavens and Their Story* that more spots equaled enhanced magnetism, which may lead to a *brighter* Sun:

> A spot like that of February, 1892 [they quantified it as being 92,000 English miles in extreme length and 62,000 of the same, in breadth] is enormous of itself, but is a very small object compared to the sun; and spots of such size do not occur frequently, and last but a very short time. We have no right to expect, therefore, that a time of many sun-spots should mean any appreciable falling off in the light and heat we have from the sun. Indeed, since the surface round the spots is generally bright beyond ordinary, it may very well be that a time of many spots means no falling off, but rather the reverse.[29]

This, then, was the first modern description of more spots equalling a brighter Sun—rather than the opposite (more spots equaling a dimmer Sun). Such an important insight into the Sun's magnetism and its visible light output, albeit only mentioned in passing while searching for the magnetic link, would only bear direct proof from 1980s measurements using satellite instruments.

The path towards reaching his understanding of the Sun-Earth magnetic relationship in this form started out slowly, and involved the process of methodical analysis, coupled with a clear-headed open-mindedness and a stubbornness towards failure in trial and error searching. At this period of his career, Maunder had collected over thirty years' worth of solar activity data. He and his wife were just then starting to analyze those data to answer questions such as those Kelvin and others posed.

Maunder first did a photographic plate analysis of storms recorded at the Royal Observatory, taking him—as he said—ten months to do.[30] Then he made a general comparison between magnetic disturbances and faculae and prominences. It is important to note that at first this research led nowhere.

---

[29] Ibid, *The Heavens and Their Story*, p. 183.

[30] Note this ten months of thrashing around: science is often built this way.

However, seeing a series of four disturbance intervals towards the end of 1886, and another set of similar ones in the 1887 record, something made him think:

> It occurred to me, since this interval was precisely that of the rotation period of the Sun as seen from the Earth, to calculate the longitudes for the centre of the Sun's disc for the beginning of all the magnetic disturbances in my catalogue, 276 in number. It became then apparent that there was a strong tendency for the disturbances to commence when certain meridians were on the centre of the disc.[31]

"In short," he wrote, "the solar rotation period was brought out from one end of the catalogue [with the 276 disturbances] to the other." With the magnetic storm Annie mentioned for February 1892, the great storm began instantaneously with a movement that was simultaneous the world over when the centre of the giant ("monster") sunspot group was exactly 17 degrees past a particular part of the Sun. The group came around again, and, hitting 17 degrees west of the previously noted point, a *second* great storm occurred. Walter Maunder inferred that the magnetic disturbances emanated not only *from* the sun, due to some action, but that an action that, in some way, co-rotates *with* the Sun. From this, consequences such as the emanation being limited only to certain *specific, or limited, areas of the Sun* emerged as stronger observational proof aligned with measurable data. It also showed "that the action does not take place by radiation equally in all directions, like light and heat, but along certain definite lines."[32]

With these new insights based on facts drawn from measurements Maunder was able to disentangle "the difficulty ... so forcibly put by Lord Kelvin ... for his calculation was expressly and explicitly based on the assumption that the magnetic storm was due to magnetic waves emitted by the Sun, equally in all directions in space."

This "forcefulness" on behalf of Kelvin was in Kelvin's presidential address to the Royal Society of December 1, 1892, where first and foremost, Kelvin quotes Maxwell's "electro-magnetic theory of light," and "the undulatory theory of propagation of magnetic force which it includes might hope to perfectly overcome a fifty years' outstanding difficulty in the way of believing the Sun to be the direct cause of magnetic storms in the Earth." Curiously,[33] Kelvin quotes

---

[31] Ibid, *Popular Astronomy*, p. 61.

[32] Ibid, *Popular Astronomy*, p. 62.

[33] Kelvin, Lord W. T., *Nature*, No. 1205, Vol. 47, December 1, 1892.

in his address on the Sun-Earth relation only works from William Ellis and Professor W. Grylls Adams[34]—rightly so, but both researchers were experts in *Earth* science. It is not clear what Maunder had to say about this even if he was in direct contact with Kelvin on the matter.[35]

In this same address, Kelvin detailed the math used to make his assumptions as to the impossibility of connections between magnetic storms and sunspots, which was one of the major problems in the difficulty concerned. He even dates the issuance of this problem to his predecessor Sir Edward Sabine[36] forty years before.

To put this—one of Lord Kelvin's most influential views—into concrete terms, it is necessary to quote his crucial estimate of the energy involved in the Earth's magnetic storm of June 25, 1885:[37]

> To produce such changes as these by any possible dynamical action within the Sun, or in his atmosphere, the agent must have worked at something like 160 million, million, million horsepower [or 12×10^{35} ergs per second], which is about 364 times the total power of the solar radiation. Thus, in this eight hours of a not very severe [geo, or, "Earth"] magnetic storm, as much work must have been done by the Sun in sending magnetic waves out in all directions through space as he actually does in four months of his regular heat and light. This result, it seems to me, is *absolutely conclusive* [our emphasis] against the supposition that terrestrial magnetic storms are due to magnetic action of the Sun; or to any kind of dynamical action taking place within the Sun, or in connection with hurricanes in his atmosphere, or anywhere near the Sun outside. ... [W]e may also be forced to conclude that the supposed connection between magnetic storms and sun-spots is unreal and that the seeming agreement between the periods has been mere coincidence.

---

[34] Adams was a pioneer in obtaining evidence (making measurements) for geomagnetic storms beginning both suddenly and almost simultaneously all over the earth (for example, Adams, W.G., "IV. Comparison of simultaneous magnetic disturbances at several observatories," *Philosophical Transactions of the Royal Society of London*, Vol. 183, 1891, pp. 131–139).

[35] Maunder's title at RGO is stated in the 1905 *Popular Astronomy* article: this is "Superintendent of the Solar Department" of the RGO. Whether or not Maunder was being deliberately ignored is speculation.

[36] Ibid, Kelvin, "Astro-physical Notes," *Astronomy and Astrophysics*, 1893, p. 76.

[37] Kelvin, Lord W.T., *Nature*, Vol. 47, December 1, 1892, p. 109.

The effect of such a powerful argument by such an influential man was clear. It was influential in the sense that it could be argued that it temporarily set-back (except for a few doyens of solar science like Maunder and William Ellis) scientific research into Sun-Earth relationships. The impact of Kelvin's argument can be found penetrating the pages of Sir Robert Ball's popular 1893 book, *The Story of the Sun*, as well as lingering almost 40 years later in George Ellery Hale's influential 1931 paper, *The Spectroheliograph and its Work, Part III: Solar Eruptions and their Apparent Terrestrial Effects.*[38] Thus, Lord Kelvin's comments seemed to be leading toward more years of confusion and set-backs for Sun-Earth studies. But, could Kelvin's broad pen strokes cover all the realities there were to be revealed of a Sun-Earth magnetic coupling?

Most probably not. Speaking in modern terms, Kelvin was not only unaware of the powerful extension of the Sun's magnetic arms through the spread of the solar wind. He was also incorrect on one crucial assumption adopted in his evaluation of the problem. As the solar physicist John Wilcox pointed out,[39] Kelvin had incorrectly assumed that the solar magnetic field strength decreases as an inverse of the distance to the cubic power. The correct functional dependence of such a crucial quantity is less rapid and is more likely an inverse of distance-squared.[40] Another distinguished solar physicist, Eugene Parker, summed up this incidence: "He ignored the suggestion that the geomagnetic variations might be the result of a beam of solar corpuscular radiation,"[41] largely an original contribution of Maunder.

---

[38] Hale, G.E., "The Spectroheliograph and its Work, Part III: Solar Eruptions and their Apparent Terrestrial Effects," *Astrophysical Journal*, Vol. 73, 1931, p. 384.

[39] Wilcox, J. M., "Solar activity and the weather," *Journal of Atmospheric and Terrestrial Physics*, Vol. 37, 1975, pp. 237–256. John Wilcox commented that "These words of an eminent physicist [i.e. Kelvin], stated with the absolute assurance that has not completely deserted the profession today, were correct within the frame of reference in which they were uttered. What Lord Kelvin did not know about, and therefore did not take into account in his calculations, was of course the solar wind which extended the Sun's magnetic field lines out past Earth with the field strength decreasing less rapidly than $1/r^2$ rather than as $1/r^3$ as Lord Kelvin has assumed." (p. 238) In addition, an important earlier criticism of Kelvin is his not making the necessary distinction between world-wide and local changes in the geomagnetic field for his energy estimates, and this was pointed out by Sydney Chapman in *MNRAS*, Vol. 79, pp. 70–83 (1918). We humbly apologise to the reader if our application of modern-day understanding so early on may appear to be an unfair "bashing" of Lord Kelvin. But it seems important to have the record straightened out at this point.

[40] From works by Eugene N. Parker, such as "Dynamics of the interplanetary gas and magnetic fields," *Astrophysical Journal*, Vol. 128, 1958, pp. 664–676 and Jokipii, J.R., Kota, J., "The polar heliospheric magnetic field," *Geophysical Research Letters*, Vol. 16, 1989, pp. 1–4.

[41] From Parker, E.N., "A history of early work on the heliospheric magnetic field," *Journal of Geophysical Research*, Vol. 106, 2001, pp. 15797–15801.

Indeed Maunder had successfully put forward his points on the Sun-Earth magnetic connection in the Royal Astronomical Society early on. He did this in a solar magnetism-to-Earth speech opposing Lord Kelvin's supporters, and accomplished this amidst an atmosphere that promised "severe heckling [but where he] answered his critics masterfully."[42] We do not yet know if Kelvin replied to Maunder's essay in concession or denial. For two years later, Kelvin passed away.

A professor at Manchester University, however, did reply to Maunder. Taking Kelvin's method to defend, Arthur Schuster responded to Maunder "immediately," stating "I cannot, therefore, agree with his somewhat boastful[43] claim that he has rendered clear what Lord Kelvin has called a 'fifty years' outstanding difficulty.' "

Schuster went on:

> He has [that is, Maunder] no doubt, added a new fact [actually, several: chief among these being the persistent, definite source area on the Sun responsible for recurring geomagnetic storms, later to be known as M-regions] ... but mystery is left more mysterious than ever; the facts have become harder to understand and more difficult to explain.[44]

Maunder's presumed boastful summation to the *Popular Astronomy* version of his article, *The Solar Origin of Terrestrial Magnetic Disturbances,* the part to which Schuster probably referred, is as follows:

> First of all, it clears up what Lord Kelvin, twenty years ago, called the "fifty years' outstanding difficulty." The origin of our magnetic storms *does* [Maunder's emphasis] lie in the Sun. Next it introduced a totally new conception of the magnetic action of the Sun. It is not from the whole surface, it is not radiated equally in all directions. But it is from certain restricted areas and along certain narrow lines. Lastly it tends to show that the Sun, though a rotating gaseous sphere, is not, as might be expected, homogeneous, but that certain definite areas of it have already assumed a specific character. And this idea cannot be confined to the Sun alone; it must find its application to stellar physics; and it does not seem possible to forecast how far it may change our conceptions in that field.[45]

---

[42] Ibid, Cliver, *Eos*, December 27, 1994.

[43] Akasofu, S.-I., "A Note on the Chapman-Ferraro Theory," *Physics of the Magnetopause*, Ed. Song, P. *et al.* (American Geophysical Union, 1995), p. 5.

[44] Ibid, Akasofu, p. 5: from Schuster, A., *MNRAS*, Vol. 65, 1905, p. 197.

[45] Ibid, *Popular Astronomy,* February, 1905, pp. 64–65.

Walter did not know it, but he and Annie had accurately described and, of course, observed, high speed solar wind streams emanating from solar coronal holes.[46] But another notation should be made. Namely, the effects of these magnetic regions and even the occasional bursts of these, upon Earth. Determined already was the fact that electric transmission was interrupted. But, how? And to what other extent? Again, the Maunders wrote in their book:

> Can they [sunspots, faculae, prominences, corona changes] affect us here on earth even if we hid ourselves in her depths, shut off by many feet of cold ground from the sight of the sun, from the knowledge of day and night, or summer heat or winter cold?

The answer is, "yes." Even if a bar (steel) magnet is suspended deep in a cellar, far underground, away from light or even noise, far from "the madding crowd" (as the Maunders put it), it will quiver.[47]

> The steel bar—a magnetic needle, it is called—is balanced so that it will point nearly due north and south if undisturbed. But it does not remain undisturbed for long. From about nine in the morning till about two in the afternoon there is a feeble swing of the magnet to the west, during the remaining hours it creeps back. Day by day the magnetic needles swing to and fro, but the extent of the swing is not always the same... Every now and then—we cannot tell when—a monster spot breaks out upon the sun. It passes across the sun's disc into his [that is, the Sun's] unseen hemisphere, and may come again round his eastern rim a second, even a third, fourth, fifth time. We know it to be the same spot... Every now and then, without warning, the magnetic needle swinging gently to and fro in stillness, becomes violently agitated.

They then relate that these agitations correspond to the sunspots passing a certain longitude.

* * *

Returning to Annie Maunder and her lengthy sunspot description earlier, other implications to the matter of solar magnetism are found.

Note what Annie wrote about the October, 1903 storm, where the large sunspot evinced on Earth "a weak storm" that gained in severity only some time

---

[46] Ibid, Akasofu, p. 5.

[47] These quotes are from pp. 183–188 of *The Heavens and Their Story*. The choice of words also betrays a passing familiarity with the works of Thomas Hardy. Or a deeper one with Thomas Gray's *Elegy Written in a Country Churchyard*.

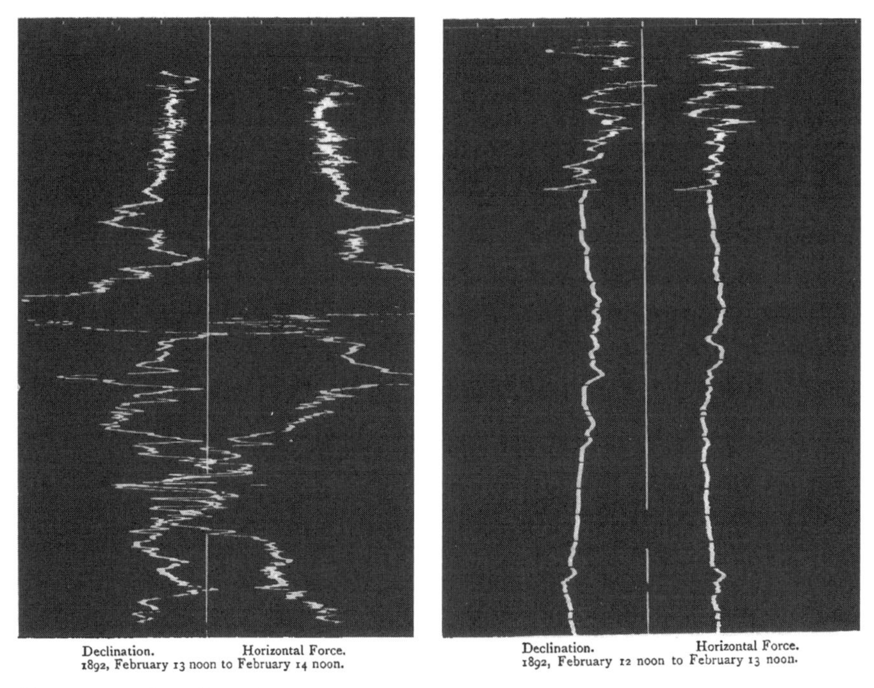

Fig. 21a Trace of the big solar magnetic storm of February 13, 1892, and the readings of the bar magnets on Earth. The register to the right shows the abrupt beginning of the storm and the contrast between the undisturbed (bottom right) and disturbed (to the left) traces. (Reprinted from The Heavens and Their Story, p. 182.)

later. A priest in these proceedings and debates of those "fateful" years starting in 1892, the Jesuit astronomer, Father Sidgreaves,[48] noted that large sunspots do not always produce a large magnetic storm upon Earth. Yes, Maunder answered, *if* the spot is in the wrong place relative to the location of Earth at a specific time, the disturbance probably "often *miss*[*es*] our earth altogether." That is, the disturbance goes right past it, in something of a straight line. And out into space.[49]

In these exciting verbal skirmishes, Maunder had allies such as William Ellis—who more or less pointed out the unfortunately erroneous nature of Kelvin's brilliant assumptions, as well as the pithy ones of Sidgreaves.[50] In the following quoted letter, Ellis (who literally followed in Carrington's footsteps) made this

[48] Ibid, *Popular Astronomy*, February, 1905, p. 62. Father Sidgreaves was a well-known astronomer at the Jesuit Stoneyhurst College in England (noted by Mary Bruck).

[49] A good, intelligent response. But the correct answer lie elsewhere, still.

[50] Ellis, W., "Sun-spots and Magnetic Disturbance," *JBAA*, June, 1901, Vol. LXI, pp. 537–541.

telling description of the Sun's behavior during magnetic storms. We can see in his words similar concern emanating from Carrington's observations many years before:

> Many magnetic storms commence with a very sharp movement of lesser or greater magnitude occurring simultaneously in all elements [hence, a hint for a coherent and global external cause/trigger], declination, horizontal force [picture imaginary electromagnetic lines pushing sideways from the Sun], and vertical force [picture imaginary electromagnetic lines pushing upwards from the Sun], with accompanying earth current. This first movement may somewhat precede the magnetic disturbance or storm, or it may usher in the storm.[51]

By 1904, not only Ellis wrote about connections between aurora and magnetic disturbance but also Annie Maunder, herself. She communicated her results to the Royal Astronomical Society in a provocative study of connections between sunspot activity and secular (that is, century-long or longer) changes in magnetic declination.[52]

In this paper, quoting data, she cited works which suggested that most of Earth's magnetism (95%) arises from its own crust and internal action yet pointed out that 5% of it—from the component that arises from magnetic actions in the upper atmosphere—can be synchronized with certain "critical changes in the sun."

Being able to calculate magnetic solar declination from the present back to the year 1540 at two Earth locations (London and a part of Earth's location that would one day become Baltimore, Maryland, U.S.A.) she noticed that there are century-long secular changes in the magnetic needle's declination. The declination shift from east to west having the maximum east declination in A.D. 1580, switched direction near A.D. 1660 (that is, at zero declination) and reached maximum west declination in A.D. 1810. Such a path suggested to this accomplished mathematician a solar "pendulum shift," rather than a "mere revolution." Naturally, the interesting part about this original study is the notice of the coincidence of critical phases of magnetic declination's "pendulum shift" with the phases of prolonged reduction in sunspots and auroral activity covering both the Maunder and Dalton

---

[51] Ibid, Ellis, p. 539.

[52] Annie Maunder and William Ellis were both published in January, 1904's *MNRAS*, as stated above. Annie is referred to as "Mrs. Walter Maunder," and her paper was "communicated by E. Walter Maunder." We can assume the astronomy in it is mostly hers, and not Walter's. (She was not a member of RAS, yet.)

Fig. 21b Drawing (by engraver-trained W.H. Wesley) of the photograph taken by Annie of the corona, January 22, 1898 (in Talni, India) showing its rod-like rays. The longest ray was six million miles long, the longest extension seen at that time. Agnes Clerke commented: "As regards the corona, Mrs. Maunder with her tiny lens has beaten all the big instruments."[53] (Plate 37 from *The Heavens and Their Story*) (Similar to the fate of the Maunder butterfly diagram, the original of this hand-drawn image is also kept at High Altitude Observatory [information/source courtesy of Tom Bogdan of High Altitude Observatory/National Science Foundation].)

(c. 1795–1820) Minimums. Ellis pushed this latter aspect in his essay in the same issue of the *Monthly Notices*. Ultimately, such a secular solar-geomagnetic relationship adds to the weight of evidence for a coupling of diurnal, seasonal, 11-year time-scales that were previously discussed.

In the *Popular Astronomy* article for 1905 (that is, *The Solar Origin of Terrestrial Magnetic Disturbances*, mentioned earlier) Walter Maunder also noted that in the eclipse of 1898, Annie obtained a photo of the outer corona showing rod-like rays, made up of arches, repelling coronal matter from the Sun. And this phenomenon may provide the ultimate clue for how solar activity variations are communicated to Earth magnetically.

---

[53] Bruck, M.T., "Alice Everett and Annie Russell Maunder torch bearing women astronomers," *Irish Astronomical Journal*, Vol. 21, 1994, p. 284. W.H. Wesley was a business secretary to the RAS, and specialised in interpreting astronomical photographs. He was originally trained as an engraver (noted by Mary Bruck).

9

Studying Aurora ... the Scandinavian and American Connection: Tree Rings, Moisture, and the Missing Sunspot Cycles

The findings shown in the last chapter attempted to clarify and highlight an earlier problem that Maunder noted in an article to the journal *Knowledge* in August 1894, *A Prolonged Sunspot Minimum.* That is, if great geomagnetic storms could be intimately tied to large solar disturbances, then perhaps a reduced terrestrial magnetic variation could find explanation in a much weakened solar magnetism. Agnes Clerke, commenting on Maunder's article a month later in the *Letters* section, noted "there is, besides, strong, although indirect evidence that the "prolonged sunspot minimum" was attended by a profound magnetic calm."[1] She went on to make the following points:

> This evidence [that is, for a profound magnetic calm associated with a prolonged sunspot minimum like in the Maunder] is to be found in the auroral records of the time. For the connection between the occurrence of aurorae and the magnetic condition of the earth is so close, that the absence of the one kind of disturbance may safely be held to betoken the absence of the other.[2]

She makes the point, not entirely accurately, now that discovered diary entries have contravened that in the seventeenth century in England, "not a auroral glimmer was chronicled," whereas they had been abundant in the late 1500s there, and emerged again just in time for Halley to see one in 1716. But we may well note the last sentence in the quote above;[3] that the link between aurora and magnetic disturbance was clearly established.

---

[1] *Knowledge*, September 1, 1894, p. 206.

[2] Ibid, *Knowledge*, p. 206

[3] Ibid, *Knowledge*, p. 206.

In fact, a "connection between aurora and magnetic storms had been accepted since the mid-eighteenth century,"[4] as A.J. Meadows and J.E. Kennedy relate about a Swedish graduate student of Anders Celsius (of temperature scale fame): O.P. Hiorter. Hiorter corroborated his observations on the connection between aurora seen, and magnetic storms experienced, against Celsius'. Thus, a corroboration of observation by independent means of the connection between these phenomena was found:[5]

> Who would have thought that the northern lights would have a connection and a sympathy with the magnet, and that these northern lights, when they draw southwards across our zenith or descend unequally towards the eastern and western horizons could within a few minutes cause considerable oscillations of the magnetic needle through whole degrees? The first time that I saw an aurora to the south and noted simultaneously a great movement of the magnetic needle was on March 1, 1741, in the evening ...when I announced this to the professor Celsius he said he too had noticed such a disturbance of the needle in similar circumstances, but had not wished to mention it, in order to see (whether) I too would light on the same speculation.

Afoot also in Clerke's and Maunder's time were theories arising from the minds of other Scandinavians—albeit, those living after Hiorter. In these years, Anders Ångström (1852) had measured a line in the auroral spectrum. Svante Arrhenius (1900) spoke of auroral theory based on radiation pressure of electromagnetic waves and erupted particles from the Sun.[6] A notable achievement in these proceedings was that of Adam Paulsen,[7] who, working from observations of relationships between northern lights and sunspots, noted that cathode rays—which

[4] Meadows, A.J., Kennedy, J.E., "The Origin of Solar-Terrestrial Studies," from *Vistas in Astronomy*, Vol. 25, 1982, p. 420. (This paper also quotes Chapman's and Bartels's work from 1940.)

[5] To which Meadows and Kennedy—not in adumbration—state: "the relationship between senior research personnel and their assistants has not changed greatly down the years." Hiorter's original paper was "Om-magnet-nålens Åtskillige ändringar" in *Kongl. Svensk Vetenskaps Akad. Handlingar*, Vol. 8, 1747, pp. 27–43.

[6] From "Scientific Auroral Experiments Beginning in the nineteenth century," pp. 85–86 (Ibid, Brekke and Egeland, 1983). Arrhenius expanded on de Mairan's theory. Arrhenius expanded on de Mairan's theory. He published a paper in 1904 on this for the Royal Society, called "On the Electric Equilibrium of the Sun" (*Proceedings of the Royal Society of London*, Vol. 73, pp. 496–499).

[7] The much-awarded and admired Sidney Chapman, who will figure significantly in later chapters on extrapolations of Maunder's and others' work, cited the "indirect assistance" of Birkeland in these proceedings, vis à vis solar activity.

consist of free electrons—make ozone, which in turn makes water vapor condense. He claimed that northern lights, such as they were, "(are) a phenomenon created by the Absorption of Cathode Rays,"[8] and so, could lead to "some auroral cloud formations." Paulsen thought that aurora could penetrate to a kilometer in altitude from Earth's surface.

Members of the Norwegian Aurora Polaris Expedition of 1902–1903 (K.R. Birkeland *et al.*, with official report published in May, 1910[9]) noted "auroral clouds" during that northern expedition. Here, Birkeland quotes "Adam Poulson" as having pointed this out, referring to Paulsen's 1895 paper and also the 1896 paper, *The radiation theory of northern light.*[10] Birkeland, on site in the far north, tested Paulsen's work "in connection with simultaneous magnetic registrations at the same place." He found that the formation of aurora clouds was always accompanied by simultaneous magnetic storms and Earth currents, and mentioned the "corpuscular theory" which had already played a large role in Sun-Earth magnetic theory by 1892—a banner year for solar science.[11]

In work that only reached full appreciation by 1957 and 1958,[12] since the seventeenth century,[13] it was assumed that auroral arcs formed a complete ring around the North Pole. In the Vega expedition to the pole (1878–79) the Swede Adolf Erik Nordensjöld assumed that the center of this auroral ring was near the geomagnetic and geographic north poles, the former being referred to him as "the pole of the northern light."[14] (It has since been named the "auroral oval," and today is seen in composite satellite pictures ringing upper air around both geomagnetic poles.[15])

It should also be pointed out that both the location and shape of the auroral zone depend strongly on Earth's magnetic field. Takasi Oguti synthesized the geomagnetic field records over the past 450 years in work pioneered by Humboldt,

---

[8] Ibid, p. 84.

[9] Birkeland, K.R., *The Norwegian Aurora Polaris Expedition, 1902–03*, Vol. 1 "On The Cause of Magnetic Storms and the Origin of Terrestrial Magnetism." (Longmans, Green & Co.)

[10] in *Nyt-Tidskrift for Fysik og Kemi,* 1896.

[11] Cliver, E.W., "Solar Activity and Geomagnetic Storms: The Corpuscular Hypothesis," *Eos,* Vol. 75, No. 52, December 27, 1994.

[12] And monies supplied for the pursuits in sciences during the International Geophysical Year, or, the "IGY"—1957–1958.

[13] From the work of F.C. Mayer in St. Petersburg (Ibid, Brekke and Egelund, pp. 90–91).

[14] Ibid, "Scientific Auroral Experiments Beginning in the Nineteenth Century."

[15] King-Hele, D.G., "The Earth's Atmosphere: Ideas Old and New" (Milne Lecture, 1984) *Q. Jr R. astr. Soc.* (1985), Vol. 26, pp. 237–261 (see the illustration on p. 240).

Gauss, Airy, Sabine and others into a model of secular changes in the dipole moment and the precession of the dipole axis. Shifts of the geomagnetic field components were also modeled. The result showed that significant changes to the auroral zone had indeed occurred.[16] For example, it was noted that 300 years ago, the auroral zone shifted and stretched towards the European sector, compared to the present concentration closer to the Canadian sector. Oguti's results showed that northern England was in the auroral zone and large parts of England, Denmark, and northern Germany were in the sub-auroral zone 300 years ago. Thus, it cannot be accidental that studies of aurora were rather well developed—even in England and France, in the late-Seventeenth and Eighteenth Centuries. (Such early research can be found in the works of Gassendi, Halley, de Mairan, and Dalton—as well as Scandinavians like Hiorter.)

* * *

But, back to Clerke, Ellis, and Maunder and aurora as it was "interpreted" in England. In the estimate of Scandinavians, England was a "very poor place" to produce experts on aurora, since aurora are hardly so common there as in the north—of course lacking a more scientific description like that of Takasi Oguti. But as testament to Clerke's veracity, judging from written accounts, when in the mid 1700s in Scandinavia northern lights became common once again, old people could not remember any from their youth.[17] As Agnes Clerke's reputation as an archival researcher persists to this day, it is curious to observe the generational lapse in phenomena as witnessed in her descriptions. Or perhaps owing to the long term nature of the lulls in solar activity, relative to human time, whole generations of people can and do miss some of the more common and distinguished solar-terrestrial phenomena like northern lights.

As an observational astronomer Agnes Clerke was self-admittedly a poor one. This did not dampen the efforts of flocks of her male astronomer-admirers to make her a better one,[18] allured either by her fragile charm or her erudition, or both.

---

[16] Oguti, T., "The auroral zone in historic times—The Northern UK was in the auroral zone 300 years ago," *Journal of Geomagnetism and Geoelectricity*, Vol. 45, 1993, pp. 231–242.

[17] Silverman, S.M., "Secular Variations of the Aurora," *Review of Geophysics*, Vol. 30, 1992. Silverman quotes from Lovering, 1867, p. 103. A startled Celsius said, "it is impossible to believe that the skillful observers of the last century, who passed their lives in the Observatories erected for them, particularly at Paris and Greenwich, should not have taken care to transmit to posterity their observations on this admirable phenomenon," and he added with healthy, scientific skepticism: "if it had appeared in their time."

[18] The American astronomer Edward Holden, Henry Reeve, George Ellery Hale, and others, to name a few. A plaintive and highly-cultured woman, educated at home and perhaps repelled by the politics associated with academia and institutions, Clerke refused posts at Greenwich and at Vassar as an astronomer.

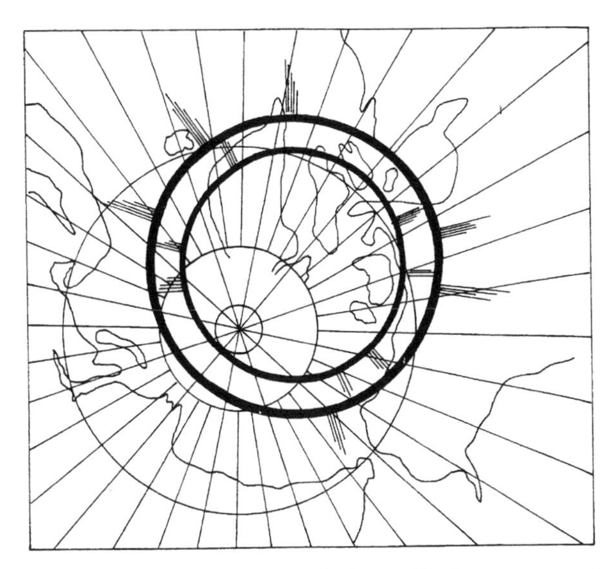

Fig. 22 Nordenskjöld's concept of the auroral ring with its centre at the so-called "pole of the northern light" (figure re-drawn from A.E. Nordenskjöld's Om Norrskenen [1881]). After 1957, it was clear that this point is very close to the north pole, as thought then. (Today, this "ring" is seen as an oval of light in satellite photos.)

Some, like Sir David Gill, succeeded.[19] As a chronicler she was popular and provocative, being almost universally-admired for her many pithy articles on astronomy that served as a "glue" in the science in forums such as the *Edinburgh Review*, based on years of study in, for instance, the libraries of Florence, Italy, where she was a student of Galileo's writings. She was also lauded for her various books—one serving as a benchmark of popular astronomy for many years.

As for the skilled William Ellis, his data-backed insights in another letter to the B.A.A. in January, 1904, one year before Maunder's *Popular Astronomy* article appeared, tie the frequency between auroral appearance to sunspot abundance together. In graphs, Ellis showed that sunspot maximum coincided with magnetic disturbances. The average per year being 94.4 days for twenty spot-maximum years, with fewer disturbances associated with sunspot minimum, or, 55.6 days per year for sixteen spot-minimum years. In Swedish observations made from 1721 to 1877 that he studied, Ellis pointed out in this paper:

> That the observed epochs of maximum and minimum frequency of the aurora during the period mentioned were, with a few exceptions, in

[19] Bruck, M.T., "Agnes Mary Clerke, Chronicler of Astronomy," *Quarterly Journal of the Royal Astronomical Society,* Vol. 35, 1994, pp. 59–79.

> close agreement with corresponding epochs calculated [from the formulae given] indicating a near agreement with the corresponding maximum and minimum epochs of Sun-spots.[20]

Ellis concluded:

> The epochs of maximum and minimum frequency of the aurora thus appear to fall respectively at or near to maximum and minimum epochs of Sun-spots as in our latitude, but follow the Sun-spot epochs a little later in time, and on the whole by a longer interval at maximum epoch than at minimum epoch. There are irregularities, however, in the progression of auroral frequency as well as in that of Sun-spot frequency.[21]

In summary, Ellis admitted that data on polar regions of the Sun and Earth were not available in regard to properly studying "relations existing between solar-spots, terrestrial magnetism, and the aurora."[22] Therefore, in a statement that would perhaps be welcomed by modern-day NASA officials, Ellis even put forward the idea of "plant(ing) observatories on some of the other planets of our (solar) system" to observe such phenomena.

A.E. Douglass, Maunder, and the Connection Between Moisture, Tree Rings and Sunspots During the Maunder Minimum

Interestingly enough near this time, at the Royal (Greenwich) Observatory, studies of Mars's canals went on unabated, and in which Maunder took part. The Italian and American astronomers alike, Giovanni V. Schiaparelli and Edward C. Pickering were fostering interest in Mars' canals.

Maunder confined himself mainly to solar study, yet further work on these oddities continued. Maunder deemed the "Canali," as they were termed, spurious, by conducting simple experiments. Before 1900, Maunder showed that the "apparent regularity of the canals was an optical illusion arising from the presence of minute detail which the retina could not define"[23] However, a great deal of Mars

---

[20] Ellis, W., "The Aurora and Magnetic Disturbance," *JBAA*, January, 1904, Vol. LXIV, No. 3, pp. 234–235. The formula is: maximum epochs plus varied 11-year cycles, times the number of years after the initial epoch.

[21] Ibid, Ellis, 1904, p. 235.

[22] Ibid, Ellis, 1904, p. 236.

[23] Obituary of E. Walter Maunder, from May, 1928's *The Observatory* (No. 648), p. 158.

research and popularisation was taken up in America. To a large extent this was done by retired businessman and well-connected Boston Brahmin/magnate, Percival Lowell.

The only reason to bring up the matter of Schiaparelli's canali or even Edward C. Pickering's "lakes" on the Martian surface is to introduce one of Pickering's assistants at the Harvard College Observatory (HCO). He would become one of Walter Maunder's parallel numbers in the Sun-Earth connection; a New Englander—a Vermonter[24]—by the name of Andrew Ellicott Douglass.

From Lowell Observatory—or more appropriately, Arizona, we appraise A.E. Douglass. While the likes of Lowell, Yerkes, and Carnegie[25] sought the sites to locate telescopes for educational pursuits such as Mars watching, Douglass (with a personal background somewhat similar to Walter Maunder's) had the unique chance to serve as "point-man" for locating both Lowell's Arizona and Pickering's Harvard telescopes.[26]

With HCO director Pickering, Douglass, as junior assistant, was sent to wander in the wilds of South America, seeking after a good "seeing" site for HCO's Southern Hemisphere telescope.[27] To obtain good results through optical telescopes, clear and steady viewing of celestial phenomena are the key. As one source states, Lowell, who needed an assistant of top quality, prompted Pickering to offer his assistant's services. Douglass, perhaps seeing his chance to become an independent researcher, leapt at the opportunity to be Lowell's head scientist.

Douglass was similarly interested in astrophotography regarding the Sun, which was, of course, Maunder's métier—even as an undergraduate in college. Working for Pickering in a side capacity enabled Douglass the relative freedom to pursue many of his own interests, which were often at the forefront of this long-lived, straightforward and rather naïve scientist's mind.[28] Always of a meteorological/botanical bent, Douglass reached both "high" (Lowell, gentleman American Astronomer) and "low" (T.A. Riordan, Irish-American owner of the Arizona Timber and Lumber Company—the latter on record as being Douglass's close friend) in his contacts.

---

[24] To this day, Vermont is known for its libertarianism.

[25] Perhaps following the money-donating example of the billionaire George Peabody.

[26] Ibid, Webb, *Tree Rings and Telescopes*. Both of these were well located.

[27] Arequipa, Peru.

[28] Douglass, born shortly after the U.S. Civil War in sleepy, agricultural Vermont, ended his life in Arizona soon after seeing the first men in space.

It was known up to Douglass's time, not only da Vinci's insight, but in a then-recent U.S. Forest Service report[29] that yearly tree ring growth had been noted by European foresters earlier in the 1800s. These European foresters had made "tens of thousands of counts" of tree rings. It was known that tree-ring types varied between species, and that rings of questionable quality and shape, and other anomalies could be found in old trees. Dutch astronomer Jacobus Kapteyn studied the relationship between rainfall and oak tree growth in the 1880s, but did not publish his results due perhaps to a personally-presumed sense of lack of proof of a connection between the Sun and Earth in these matters. Nebraskan (Nebraska, U.S.A.) doctor A.L. Child[30] observed a relationship between high temperature, heavy rainfall, and tree growth, but without obtaining precise ratios between these.

What Douglass did, in his frequent forays through the breathtaking scenery of Arizona, was to make a correlation between large trees, growing in higher altitudes, and smaller trees at lower altitudes. Here we see Douglass apply what Descartes referred to as possible connections between "clear and distinct ideas."[31] Or perhaps Descartes's fellow-Dutch exile and fellow-victim in the Maunder Minimum, Francis Bacon's, notion of the humble bee, which collects from source to source, and, with a power of its own, transmutes something new from this altogether:

> Because of sparse Arizona rainfall [like on the Colorado Plateau] he [Douglass] attributed the predominant influence in tree growth to precipitation. This relation suggested another link. Because the Sun's heat caused oceanic evaporation and the winds that dropped this moisture on the continents, variations in solar activity would clearly have an effect on tree growth. Annual tree rings should, in turn, reflect these variations.[32]

---

[29] Webb cites George Sarton from an *Isis* essay in December, 1954. B.E. Fernow, "Age of Trees and Time of Blazing Determined by Annual Rings," USDA, Division of Forestry, Circular No. 16 (Washington, 1897).

[30] Webb cites the following papers by Child: "Annual Growth of Tree," *Popular Science Monthly,* (*PSM*, XXII, December, 1882) and "Concentric Rings of Trees" (*PSM*, XXIV, December, 1883). *Nota Bene*: most of this paragraph is extracted from Webb.

[31] Ibid, Gaukroger. *Descartes: An intellectual biography.* This is not to imply that Holland at this time was some sort of intellectual wasteland, so much as it was a freer place at the time. France posed many puzzles for Descartes, personally, as well.

[32] Ibid, Webb, p. 102. Reference to Douglass's scientific paper here is: "Cycles: A Problem in Naming," manuscript dated December 20, 1934, folder 3, Box 67; "Incidents," dated 4/4/39, folder 1, Box 93, Douglass Papers, UAL, A.E. Douglass, *Climatic Cycles and Tree Growth: A Study in the Annual Rings of Trees in Relation to Climate and Solar Activity* (Washington, Carnegie Institution of Washington, 1919), p. 9.

It was providential of Douglass to make these associations as he did between November, 1901 or maybe even somewhat earlier, and on through the 1920s. The scientific intuition was immaculate, and intriguing. But how was he to prove his points?

* * *

In the early years of the twentieth century, Douglass roamed around a really rugged Arizona. Douglass was the eccentric Lowell's "astronomer royal," in a manner of speaking. But Douglass's investigation into the connection between tree-ring growth and decay and climatic variation, rather suggests that, like Descartes, he was at cross purposes in this role: "*bene vixit, bene lui factuit* " ("he who lives hidden, lives well").[33] As Descartes apparently could only do his best work in the calm afforded by an intellectual "desert"—that is, away from intrigue-ridden 1600s Paris, and to Holland—so did Douglass do well, thinking away in

Fig. 23 Andrew Ellicott Douglass (1867–1962) late in life. (Photo by Charles W. Herbert, Courtesy of the Laboratory of Tree Ring Research, University of Arizona [after Eddy, 1978].)

---

[33] Ibid, Gaukroger. *Descartes: An Intellectual Biography.*

the real desert of the Southwestern United States. His goal: the ability to make long-range predictions of Earth's climatic behavior.[34]

Douglass's link to solar astronomy at the time was a direct one. He was informed on the Sun's variable nature via solar radiation measurements of the same, such as those done by Smithsonian Astrophysical Observatory scientists. If his observation of the tall trees in high wet areas/small ones in low dry ones is recalled, Douglass's first hint about the connection between rainfall, size of tree rings, and their connection to others came through reading the Lockyers' 1901 paper[35] on monsoons in India. The paper suggested that pulses of rainfall caused cyclones; one pulse corresponding to sunspot maximum, another pulse, to sunspot minimum.

In 1904 Douglass collected, with the aid of his friend, Riordan, his first tree samples—the first to be used in a long line of crucial examinations linking terrestrial phenomena of this variety to solar events. At nearly the same time (1903) Maunder published an article, *Magnetic Disturbances, 1882 to 1903, as Recorded at the Royal Observatory, Greenwich, and Their Association to Sun-spots*, where Maunder eliminated the element of pure chance as to why magnetic disturbances happened, and demonstrating that they recurred at intervals corresponding to one solar rotation, or, every 27 1/3 days.[36] Later, Maunder recapped these observations in another article tying cyclones from nautical data collected by the aforementioned British, global "watch stations" and others, collected between 1856–1867[37] to solar rotation. In this November, 1909 paper Maunder also noted that sunspots may not be the only factor in tying cyclones to solar rotation, but to what he termed solar "stream-lines" from the Sun. Again, Maunder spoke of specific

---

[34] Ibid, Webb, p. 119. Douglass's work has in part led to studies of another "sub-science": astroclimatology.

[35] This is intimated by Webb in *Tree Rings and Telescopes*, from a paper of Lockyer's to the British Royal Society in 1901, "On Solar Changes of Temperature and Variations in Rainfall in the Region Surrounding the Indian Ocean." However, it may also have reached Douglass by way of papers by C.G. Abbot or by others.

[36] Chapman takes this up in a chapter of his and J. Bartels's epochal (1940) *Geomagnetism*. The writing in this is self-admittedly turgid, though the mathematics and possibilities and insights upon reading it, profound. Considering that the English Chapman, and the very German Bartels, were hacking this out in World War II, it is a miracle anything came of it. Curiously, they credit and quote "Mrs Maunder" in this work, often.

[37] By Charles Meldrum, then the late director of the Royal Alfred Observatory in Mauritius. This is quoted from "Notes on the Cyclones of the Indian Ocean, 1856–1867," by E.W. Maunder (*MNRAS*, November, 1909, pp. 49–62). Maunder did a re-analysis of graphical data.

events—*directed, magnetic streaks*—emanating from particular parts of the Sun, as a link to these events.

But just a few months before Maunder published his November paper, in June, 1909, Douglass published[38] the fruits of his labors in Riordan's lumber yard (and others') from 1904 on. What Douglass found, in Riordan's yellow pine stumps, was rapid tree ring growth in spring/summer weather, and thinner in the fall, as well as *double* rings left there during spring droughts. Careful measurements of these led to a database. He also needed something else: an ability to cross date the rings. In the same article, he claimed strong evidence of cyclic variations in his data: 32.8 years, 21.2 years, and significantly 11.3 years, shown empirically by repetitive patterns in the stump rings. Now, the 11.3-year repetition matched, to his sharp eye, the "same periodic variation found in sunspots."[39] That is, the 11-year Schwabe sunspot cycle.

Larger tree rings would come to be an indicator of excessive moisture (not necessarily rain) and thinner ones, a sign of a lack of moisture. In the course of 15 years (1904–1918) Douglass had amassed a considerable database; he had collected some 230 different tree specimens, mostly in Arizona, but had access to European samples, as well. He had measured 75,000 tree stumps.[40] Long-distance comparisons far into the past, even thousands of years, as rings can determine a tree's age (and some trees can live this long) between these variations showed that some years were drier over longer periods of time than others, and others, wetter—and so, the larger spreads in the tree rings' widths. The accuracy of these measurements was so good he was able to obtain, "in wood," thin ring years proving the existence of Arizona droughts of 1748, 1780, and 1821, hypothetically without recourse to written records.[41]

But the 11-year cycles were boringly repetitious especially in the wetter-climate European samples he was able to examine. Except, intriguingly, for absences of these 11-year cycles in the late seventeenth and early eighteenth centuries. This "mystery" was to remain with Douglass for many years into the future.

[38] Douglass, A.E., "Weather Cycles in the Growth of Big Trees," *Monthly Weather Review*, Vol. XXXVII, June, 1909.

[39] Ibid, Webb, p. 105.

[40] Ibid, Webb, p. 113.

[41] Webb quoting Douglass's 1917 paper, "Climatic Records in the Trunks of Trees," *American Forestry*, Vol. XXIII, December, 1917, p. 733. Curiously—though not related—stalagmites exhibit a similar "ringing" along color: in some cases, like those from caves in Beijing, Gibraltar or even Oman, tan rings in cut stalagmites are signs of wet years, and dark brown ones, signs of dry ones.

Dismissed from Lowell Observatory he became a professor at the University of Arizona[42] until 1922. In this year Douglass made contact with Walter Maunder—now long retired from the Royal Observatory, but as active as ever in the British Astronomical Association (B.A.A.) he founded earlier—in correspondence and article writing.

In parallel, yet unbeknownst to one another, Douglass and Maunder both hunted for the Sun's or Sun-Earth's operating mechanisms. As Maunder with his wife and colleagues graphed sunspots into "butterfly" diagrams, Douglass came up with a methodology called the *periodogram* at the HCO where he worked for a time after a European stint. This was a photographic summary of sunspot crests: with a sunspot record as a test, he did a multiple plot, or, a series of curves, one under another, cut out in white, and pasted them on black. Each curve was identical, but placed ten years to the left, in each successive line. If data from which the curves were constructed showed a period of ten years, the line of crests would slant to the right or left, successively.[43] It was subsequently used to see if longer sunspot cycles than the 11-year Schwabe cycle existed.

In the research funded in part by the Carnegie Institution, Douglass wrote the 1919 paper that Maunder—apparently quite catholic in his research reading—eventually became aware of. This was titled, *Climatic Cycles and Tree-Growth: A Study of the Annual Rings of Trees in Relation to Climate and Solar Activity.* It was this study that showed the results of 15 years' worth of Douglass's aforementioned research, and in which Douglass called attention to the flattened sunspot curves noted between the 1650s and 1720s. Sequoias and yellow pines showed the 11-year cycle very well. Clearly, the cycle was *absent*, as recorded by these two species in this mid-1600s, early-1700s timeframes. Douglass—true to form—left the possibility for what could have been causing it at the conjecture, "some possible interference for a considerable interval about the end of the seventeenth century."[44] Up to this time, he may have had no idea that this interference was caused by the Sun.

---

[42] The outrageous claims of one Dr Thomas Jefferson See so enraged Douglass, he bluntly listed the untruths claimed by him in this controversy. The political heat generated was bad, and See was fired. But Douglass (honest to a fault) later mildly and reasonably suggesting to a third party that Lowell's own work (which George Hale had rejected) was more "literary" than scientific, brought about his own dismissal. He subsequently became a professor at Arizona University. (See *The Explorers of Mars Hill, 1894–1994,* by William L. Putnam [Phoenix, 1994] for more on this fascinating story.)

[43] This entire paragraph paraphrases Webb, pp. 111–112.

[44] Ibid, Webb, p. 122.

So, Maunder came to this problem from a completely different direction and of course, also knew of this odd missing record in sunspots from this time, from years of studying the relative absence of sunspots in Douglass's problematic time period. In February, 1922, Maunder exchanged letters with Douglass; Douglass quoting from Maunder's letter:

> If there is some evidence of well-defined and long-continued abnormality as to the climate of the time it should give us a direct hint to the nature of the connection between sunspot activity and climate; if there is no such evidence, then I think we may safely put any such connection between sunspots and climate as an error in inference.[45]

Skeptically cautious as he was, Maunder provided Douglass with a rare thing. Douglass had secured an impartial, independent corroboration for the absence of the 11.3-year cycles that had been bothering him—and which both had doubtless been hoping for resolution—for some time. He replied to Maunder, in effect, that tree ring and sunspot records were tied together somehow in the absence of 11-year cycles, and that the cycles were related to weather and to cosmic causes.[46] As excited by this unexpected linkage as Douglass was, Maunder could hardly have been less so, since Douglass informed him that this information—derived from 3,200 years of sequoia growth—gave great weight to the history of solar variation over long time periods,[47] quite long time periods, indeed.

Maunder read a letter from Douglass at the April, 1922 meeting of the B.A.A. out loud, commenting on Douglass's paper on the prolonged sunspot minimum of c.1645–1715:[48]

> As soon as I obtained the dates from this journal (1645–1715), I recalled that ... the yellow pines which gave me more trouble than any other in trying to work out the action of the sun-spot cycle. I (noted that): "from 1420 to 1620 the second minimum is generally the deeper. For the next sixty years the curve flattens out in a striking manner." Again ... referring to the sun-spot cycle, "sequoias show strongly the flattening of the

---

[45] Letter from E. Walter Maunder to A.E. Douglass, February 18, 1922, Box 64, Douglass Papers, UAL.

[46] Ibid, Webb, p. 123.

[47] Ibid, Webb, p. 123. This was probably from the letter Douglass wrote to Maunder in reply—on March 23, 1922—to the one Maunder sent on February 18.

[48] *Journal of British Astronomical Association*, Vol. 32, p. 223, April, 1922.

> curve from 1670 or 1680 to 1727." Again, "taking the evidence as a whole, it seems likely that the sun-spot cycle has been operating since 1400 A.D., with some possible interferences for a considerable interval before the end of the 17th century." The period from 1659 to 1727 shows the flattening referred to ... it seems to me, therefore, that you [Douglass, referring to Maunder] have brought to light a very important corroboration of the relationship between solar activity and terrestrial conditions, for I presume that these tree variations are related directly to the weather.

Douglass was looking for solar signals in yellow pine tree rings' growth widths and found hints of regular 11-year cycles there[49] and now had a solar scientist at least in some position to cautiously confirm that he was "on to something." In addition, Douglass also noted an interval from 1660–1730 in which the tree ring growth seemed odd, but for which he had no logical explanation. Apparently, Maunder had given Douglass the necessary added foundation for wondering if the Sun's activity might not be the cause behind the tree ring-width variability. Yet, not knowing what Douglass had already faced (that is, the "yellow pine syndrome") Maunder ventured to suggest that Douglass search his tree ring records for a lack of solar variability in those times.

In a May 29, 1922 lecture given in connection to the Centenary of the Royal Astronomical Society,[50] Maunder stated the following:

> Another feature is suggested by this diagram [here he referred to the butterfly diagram], but will be better brought out in a different connection. It is that the southern hemisphere appears to encroach slightly on the northern at the equator, or at about the time of the close of one cycle and the beginning of the next. The southern influence seems to cross the equator. *Since the origin of the solar spots lies within the Sun, and the northern and southern spots show differences in their behavior, we must conclude that the Sun is not symmetrical in the constitution of its interior. If then we assume, as the basis of any investigation, that the Sun is symmetrical in its internal constitution, we are making an assumption contrary to the evidence supplied by the behavior of its surface* [emphasis by the authors]. If there be this clear distinction

---

[49] Not the 22-year ones, yet: this was developed later and is still under study, today.

[50] Maunder E.W., *Monthly Notices of the Royal Astronomical Society,* Vol. 82, pp. 534–543, 1922.

> between the two hemispheres, is it possible that one of them might go out of action for a time, and if so, what would follow? This is not a mere oratorical question; the event supposed has actually occurred. Dr. Rudolf Wolf, of Zurich, demonstrated, and Spoerer developed that demonstration further, that in the latter part of the seventeenth century and the beginning of the eighteenth the northern hemisphere of the Sun failed for many years to produce a single recorded spot. Was there any effect during the same period recognized here on the earth that could be plausibly associated with this sterility in the Sun's northern hemisphere?[51]

Also in this lecture, Maunder described his "butterfly" diagram work from 1904—interestingly enough, making Galileo-like justifications of what the spots were, and where they came from:

> This diagram has been familiarly called the butterfly diagram, as it seems to suggest three butterflies [by 1920, he or with assistance had made three: it was two, in his original 1904 paper] pinned down to a board with their wings extended. Heads, bodies and legs have disappeared, but the outstretched wings remain. Each pair of wings is distinct from the next; there is a clear V-shaped gap between each of the two specimens. Here again the first deduction is reinforced from an altogether different set of facts. The solar cycle is one. This diagram further suggests that the origin of the solar spots lies within the Sun, not without. They come from below the surface; they are not impressed upon the surface by some exterior influence; neither by planets, nor by meteors. *No exterior influence could invariably begin a fresh disturbance in a high latitude simultaneously on both sides of the equator* [emphasis by the authors].

During the lecture, Maunder directed questions on what implications no spots in the solar northern hemisphere and spots in the solar southern hemisphere could have had.[52] He cited Agnes Clerke's 1894 September comment on his paper, *A Prolonged Sunspot Minimum*[53] of August, 1894 for the overall auroral light-sighting absence at this time and Douglass's work in ring counting-moisture

---

[51] Italics gratis the authors. By May of 1922, Maunder had been aware of Douglass's connection to his work for some time, and intimates the possible weather connection.

[52] *Monthly Notices of the Royal Astronomical Society,* Vol. 82, 1922, p. 537.

[53] Ibid, Clerke, A.M., *Knowledge*, Vol. 17, 1894, p. 206.

connections[54] for a terrestrial connection to solar activity.[55] Concerning her linkage between "prolonged sunspot minima to profound magnetic calm," Maunder went on at some length, making the linkage acute:[56]

> There is strong, although indirect, evidence that 'prolonged sun-spot minimum' [here, Maunder refers to the period of time Eddy later termed the "Maunder Minimum"] was attended by a profound magnetic calm. This evidence is to be found in the auroral records of the time. For the connection between occurrences of aurorae and the magnetic condition of the Earth is so close, that the absence of one kind of disturbance may safely be held to betoken the absence of the other.

Among other things, Maunder lamented the missing (or the uncollected, or even neglected) climate records that he felt could prove his points on the "long period of solar calm,"[57] and tellingly speculated on refined Sun-climate possibilities:

> Did terrestrial magnetism sympathize then with the long solar calm? Theorists have traced all kinds of connections between the solar changes and our cyclones,[58] storms, droughts, and floods. Were those years marked off by any special meteorological peculiarity? For if there be any connection between the sunspot period and weather, so marked and so long-continued a minimum must surely have produced a most notable effect.

---

[54] Between pages 216–218 of "*The Sun, Our Star,*" Robert Noyes provided an insightful recount of Douglass's discovery of this new scientific tool, which is now the basis of the modern field of dendrochronology. M.G.L. Ballie's "A Slice Through Time" (Batsford, London, 1995) tells of many more recent developments and imaginative applications of dendrochronology.

[55] An association Maunder had made as early as 1904 or so (as in the *Monthly Notices of the Royal Astronomical Society,* Vol. 65, pp. 2–34, for November 1904).

[56] The connection here was on the "northern lights' " magnetic nature. Maunder, E.W., *Journal of British Astronomical Association*, Vol. 32, pp. 140–145 (1922) (entitled: "The prolonged sunspot minimum").

[57] "This long-continued solar rest suggests many questions... How we regret the absence of magnetic records, and how ... valuable would be a few of those ponderous volumes of rainfall and temperature that, when published by modern observatories and in these days, we are apt to look upon with scorn. The connection between the magnetic and sunspot curves is unmistakable today. Did terrestrial magnetism sympathize then with the long solar calm?"

[58] Hoyt and Schatten presented the Meldrum result as the first figure of *The Role of the Sun in Climate Change* (p. 4) which shows data from 1847 to 1875, or, the correlation between the number of Indian Ocean cyclones with spot group number leads from meteorologist Charles Meldrum's early works (1872, 1874, 1885): Maunder (*Monthly Notices of the Royal Astronomical Society*, Vol. 70, pp. 49–62, 1909) indeed found encouragement, and was cautious, by seeing hints of solar rotation "signals" in cyclone activity (recurrence of cyclone activity on solar rotation timescale of every 27 days or so).

About solar cyclical interruption, Maunder stated that:

> The very fact [it] is so anomalous renders it of more importance, and it may well prove of greatest value in ... understanding ... the solar ebb and flow...Two centuries of a fairly cyclical oscillation would not have justified us in premising a like condition of things in the half century that preceded them. Clearly, however likely we may consider is that the present cycle will run much the same course as its immediate predecessors, we are not justified in prophesying it. *What has happened before may happen again, or even its reverse; for who shall say it is impossible that, as for so long a time the keynote of solar energy was lowered to such a degree, so for an equal time it may not be proportionately raised?* [emphasis by the authors]. If so, we or our descendants may see our Sun as an unmistakable variable star; for the vague traditions which have reached us from the Middle Ages may be realized.[59]

So, unlike for example, Giambattista Riccioli in the 1600s or Flamsteed or Cassini, Maunder was better equipped to make a comparatively thorough and deliberate connection between cold years and absence of sunspots using a long data record. This in spite of a lack of some facts—and the inclusion of certain errors—marred his associations. But it gave him the chance to review this data with impartiality, since 150–200 years had elapsed since the perceived solar activity minimum ended, albeit eroding the freshness of unpublished reports that may have been extant.[60] Long-term and systematic data accrual for empirical study was paying off for those who knew how to use it.

[59] Maunder finishes by saying, "and for a greater or less time the Sun may be so covered with spots as to lose a very appreciable portion of the light of its general surface. In this case we should probably find the faculae and prominences developed in something like the same proportion, and the Sun might well earn the title of a "bright line star" in a sense which we have never known it to do as yet."

[60] In any case, he pointed out that collecting these records was important to further research. He was also able to make some rather startling predictions as to the course his "descendants" would take in establishing what the Sun did, or might be. Notably, even if Cassini, Herschel, Wolf, Schwabe, and others kept very regular and good solar activity records, no one took such a large leap of faith with the data at hand as Maunder.

10

The Family Maunder: The B.A.A. and Astronomy for All

Walter Maunder and family, in the "teens" (1901–1909) changed their addresses. He signed letters to the B.A.A. alternately as 86 Tyrwhitt Road, St. Johns, in 1903, and 69 Tyrwhitt Road, South East London, by 1909.[1] We can assume Maunder's cares increased, with a capable wife, also an astronomer, one of such obvious talent but with little to show for it. How life fared with his sons and daughters—one child dying in infancy—is little known though at least two, Irene and Edith, accompanied him on certain eclipse expeditions.

This was about ten years after he got the idea for founding the British Astronomical Association, in 1890. His circle of associates from the Royal Observatory aided him in the undertaking.[2] The B.A.A. was sort of a "support group" of its time, vis à vis the RAS. It was conceived as

> an Association of amateur astronomers, astronomers for mutual help, who because of their sex or other circumstances, might be precluded from joining the Royal Astronomical Society.[3]

By 1928, at the time of Walter Maunder's death, the B.A.A. boasted a thousand members, many of whom "are contributing observations of unique character and importance to astronomy," testifying to its success.

Annie Maunder, usually seen in print around this time as "A.S.D.," was probably at odds with the Royal Observatory. She had been retained by the Royal Observatory in supernumerary grade,[4] and may have had no plan to quit, were not the number of female "computers" there whittled down.

---

[1] Maunder (and his family) seems to have resided in the Blackheath and Greenwich areas for most of his adult life (A.J. Kinder, B.A.A. historian, in a personal communication).

[2] Ibid, Maunder's *Nature* obituary, April 7, 1928, p. 546.

[3] Ibid, Maunder's *Nature* obituary.

[4] Personal communication, Mary Bruck.

Fig. 24 Annie ("A.S.D.") Maunder (1868–1947) as a young woman. She was a solar astronomer with a special talent for solar photography. (Courtesy of John McFarland, Armagh Observatory.)

The young Annie was living at 16, The Circus, Greenwich with her eldest sister, Hester, before marriage.[5] On October 31, 1895, Annie Russell quit the Royal Observatory and married Maunder, soon thereafter, on December 28, 1895 at St. Mark's Presbyterian Church.[6] Marriage terminated any further chance of official employment in such a government institution due to strictures of the time. Her last female fellow and friend at the Royal Observatory, Alice Everett, a talented co-worker, first unsuccessfully sought a job as the assistant at Dunsink Observatory. Everett failed at this in spite of being glowingly promoted by the likes of astronomer (also a husband-and-wife team member—William Huggins) and powerful promoter, Sir Robert Ball. Everett eventually obtained a position in

---

[5] Bruck, M.T., Grew, S., "The Family background of Annie S.D. Maunder (Née Russell)," *Irish Astronomical Journal,* Vol. 23 (I), 1996, pp. 55–56.

[6] Ibid, Bruck, p. 90.

Fig. 25 Annie Maunder (on far right, standing) at Girton College, Cambridge, in 1886. She excelled in mathematics, receiving a Senior Optime (second class honours) rank. Fifth from the left, centre row, is Alice Everett, her co-worker at the Royal Observatory. (After Bruck.)

Potsdam in 1895. Incidentally, Ball also wrote a letter of recommendation upon Annie's father's request for her job application at the Royal Observatory.[7]

After marriage, Annie assisted Walter in his work. However, she often carried out her own work and publication schedule. A daughter of an Irish Presbyterian minister—William Andrew Russell[8]—she chafed at times under the prejudice shown women in being denied membership of the RAS through these years, referring to them later as "blackballing"—and so revealing that she suffered not the injustices of the times, gladly, but stoically.[9] Indeed, Agnes Clerke was never[10] granted membership of the RAS, though she did gain honorary status, this being earned on merit. Women were not allowed in as members of the RAS until January, 1916; Annie Maunder being one, and solar scientist

[7] As related by M.B. Ogilvie, "Obligatory amateurs: Annie Maunder (1868–1947) and British women astronomers at the dawn of professional astronomy," *British Journal for the History of Science*, Vol. 33, 2000, pp. 67–84 that Robert Ball's recommendation letter arrived at RGO after the Observatory's secretary H.H. Turner had offered A.S.D. Russell the job.

[8] Like E. Walter, A.S.D. was also a child of a minister. The Dills (her mother) came from a prominent family of ministers. Annie's immediate siblings became quite prominent, in themselves, her brother Samuel becoming an astronomy professor, her sister Hester, a physician (whom she kept contact with in India). Another brother was an educator and a third, a banker (Ibid, Bruck and Grew).

[9] Ibid, Bruck, p. 90.

[10] Ibid, Bruck on Clerke, p. 75.

Mary Proctor[11] being another; the two were some of the first. The first woman scientist allowed into the Royal Observatory on what would be considered by modern standards an "equal status with men" was Flora McBain in 1937.[12]

There were, then, other concerns Walter Maunder turned himself over to in the years preceding his retirement from the Royal Observatory in November, 1913. He continued to publish diversely—like Annie[13]—mostly within astronomy, well

THE HEAVENS

AND THEIR STORY

BY

ANNIE S. D. MAUNDER

HONORARY FELLOW OF THE ROYAL ASTRONOMICAL SOCIETY OF CANADA

AND

E. WALTER MAUNDER

SUPERINTENDENT OF THE SOLAR DEPARTMENT, ROYAL OBSERVATORY, GREENWICH
FELLOW OF THE ROYAL ASTRONOMICAL SOCIETY
FOREIGN ASSOCIATE OF THE SOCIETA DEGLI SPETTROSCOPISTI ITALIANI
HONORARY FELLOW OF THE ROYAL ASTRONOMICAL SOCIETY OF CANADA

AUTHOR OF

'THE ROYAL OBSERVATORY, GREENWICH, ITS HISTORY AND WORK'
'ASTRONOMY WITHOUT A TELESCOPE,'
'THE ASTRONOMY OF THE BIBLE

WITH EIGHT COLOURED PLATES
AND THIRTY-EIGHT ASTRONOMICAL PHOTOGRAPHS
AND FIFTY-ONE OTHER ILLUSTRATIONS

DANA ESTES AND COMPANY
BOSTON

Fig. 26 Cover and title pages of the Maunders' book *The Heavens and Their Story* (notice the initials). Inside sheet—according to her husband—of mainly her book (right) (After Bruck.) (Dana and Estes edition, 1909.)

---

[11] *The Observatory,* Vol. 39, No. 496, January, 1916, p. 48. Proctor established a solar observatory in New Zealand, and her story on this rude awakening is related in the May, 1914 RAS bulletin (No. 474), pp. 189–191 in particular. One of her benefactors was, like Annie's at the RGO—Sir Robert Ball.

[12] Ibid, Bruck on Lady Computers, p. 92.

[13] Annie did book reviews and text analyses, for example, for the RAS organ, on topics like Zoroastrian Star-Champions (May, 1913) and quasi-religious studies with an astronomical bent, much like Walter. To wit, the date and place of writing the Slavonic book of Enoch (August, 1918). She did a book review on the observatory of Jai Singh, in India, and was not shy to point out how far the Indians digressed from actual European-style observational work at this time, versus demonstration, show, and astrology (*JBAA* Correspondence, No. 4, September, 1914). She was also the editor of the *JBAA* twice: in 1894 and in 1917.

after this time. One of these publications, *Are the Planets Inhabited?* was published shortly before his retirement and reviewed in August, 1913,[14] where Maunder points out, using an argument about the lack of temperature-equalizing winds, and low mean temperatures, why Mars had no life as in "our own world."

Throughout Maunder's life, in the capacity of what now may be called associate or contributing editor, he also worked for the now long-defunct *Knowledge*, and the still-extant *Nature*, magazines, as well as the B.A.A. journal and others.[15]

He may have developed his literary skills during his editorship of *The Observatory*, the official organ of the Royal Observatory. Christie, who founded it in 1877, had asked for Maunder's help with its editorship in 1881. For the next six years, Maunder carried out most of the work[16] on top of an observation and expedition schedule.

Interestingly, his editorship of *The Observatory* may have earned him the contacts that helped him drive the founding of the B.A.A.[17] For example, the prominent amateur and first measurer of starlight using photoelectric means, W.H.S. Monck,[18] pointed out the general dissatisfaction with the Liverpool society.[19] This in turn prompted Maunder to reply, "you will allow me to say that a Society (upon the lines that Mr. Monck has suggested) is now in process of formation, and that I shall be pleased to receive the names of those who would like to join."[20] The title "British" was selected, instead of any mention of "amateur." (The addition of "royal" would have been a great honor, had it been bestowed upon the B.A.A.) This institution itself, including its name, was in deference to the many talented individuals who could make valuable contributions to astronomy, even if they possessed perhaps not the right credentials, funding, or luck.

However, the egalitarian aims of the B.A.A.'s founding have already been pointed out. The first general meeting of the B.A.A. was held in the Hall of the Society of Arts, John Street, Adelphi, London, on October 14, 1890. The original

---

[14] Reviewed in *The Observatory* for August, 1914. Book published by Harper & Brothers, London and New York, 1913, pp. 342–343.

[15] Ibid, *Nature* Obituary, April 7, 1928, p. 546.

[16] Ibid, "Report of the council…" February, 1929, p. 317.

[17] Ibid, B.A.A. Obituary, May, 1928, p. 229.

[18] Monck—a lawyer by trade, but trained in science at college—first measured starlight in this way in August, 1892. Monck also discerned a variety of the yellow-type star, which hinted at giant and dwarf stars ("The Monck Plaque," News and Comments, *Irish Astronomical Journal*, Vol. 18, 1989, pp. 122–124).

[19] *English Mechanic*, July 18, 1890.

[20] Ibid, B.A.A. Obituary, May, 1928, p. 229.

membership listed 283, with a budget of £200, 2s, as Annie Maunder related many years later.[21] Here the election of president and other officials was undertaken and head of sections (Meteoric, Solar, Lunar, Spectroscopic and Photographic, Coloured Stars, Variable Stars, Double Stars and Jupiter) voted. Walter's idea was to assign different people to different tasks according to their interests and abilities. Assignments to various study sections still continues at the B.A.A. and allows those interested in one aspect of astronomy more than another to place full concentration on specific subjects and tasks.

The First Ordinary Meeting was held on November 26, 1890 in Holborn. The third meeting was held at Barnard Inn, which contained a Medieval bye-law that restricted men from wearing facial hair, on pain of fine. From a 1942 perspective, Annie—remembering what a bewhiskered age the late 1800s had been—remarked on this with some levity. From 1892, the B.A.A. meetings were moved to Gower Street in London and then to Sion College through World War One. During World War Two, the RAS gave the B.A.A. "shelter."[22]

Walter Maunder became the first editor of the *Journal of the British Astronomical Association*. Thus, where he had a tenuous platform from which to reach the people, he now had one with much more room in which to move about. He was willing to take up disputes or discussions on more esoteric matters in astronomy. These ranged from ancient astronomy, where he published on the recognition of Aries as being the first zodiacal constellation,[23] to biblical matters concerning or paralleling this field and the history of astronomy as well.

In 1900, he published *The Royal Observatory, Greenwich: A Glance at its History and Work*, signed with the rank of Fellow of the Royal Astronomical Society ("F.R.A.S.")[24] he had held since 1875. The book was praised in a review as being "most welcome"[25] since it encapsulated some valuable British history, as in, for example, all the astronomers royal up to that time, starting with Flamsteed, who was not known as "Astronomer Royall [*sic.*]," but "Astronomical

[21] Report of the Ordinary General Meeting, *J.B.A.A.*, October, 1942, p. 286. The founders of the B.A.A. would not have aspired to the addition of "royal" in their institution's title (noted by Mary Bruck).

[22] Ibid, *J.B.A.A.,* October, 1942, p. 287. Annie Maunder remarks here that the RAS had originally seen the B.A.A. as some kind of rival, but had long since recognized the valuable scientific source it had become.

[23] Crommelin, A.C.D., Maunder Obituary, *The Observatory*, May, 1928, p. 159.

[24] Fellow of the Royal Astronomical Society. Remarkably, the book about RGO is available in html format at http://atschool.eduweb.co.uk/bookman/library/rog/index.htm

[25] "New Books and Memoirs," from *B.A.A. Proceedings,* Vol. XI, No. 1. Maunder's book was published by The Religious Tract Society of London in 1900.

Observator" in those Restoration times. In this tome, Maunder mentions that the observatory "was founded and has been maintained for distinctly practical purposes, chiefly for the improvement of the eminently practical science of navigation."[26] While covering each novelty that every new astronomer royal introduced at the Royal Observatory (Halley, a transit and large quadrant; Bradley, a clock-driven equatorial, etc.) the book also discussed the work at Greenwich, up to that time department by department. A reviewer stressed the welcomed non-mathematical nature of the book, as well as its readability—something impartial readers, reviewers, and listeners were to write and say about Walter Maunder's writing and speaking across the years. His speaking engagements were well attended. In a time when speeches rivalled modern-day rock'n'roll music concerts, Maunder's talks were in demand and keenly anticipated.[27] He also kept up a healthy correspondence with his friends, serious researchers, and even casual inquirers in astronomy.

Published from the *Knowledge* editorial offices in 1902 was *Astronomy without a Telescope,* a book acknowledged as a classic in one of Maunder's obituaries.[28] The book also contain much of his papers on constellations and much on the pleasures of visual astronomy without need for excessive technical tools.[29] In one quote from *The Heavens and Their Story,* Maunder relates:

> For the heavens are telling stories of interest, stories of wonder, if we but have the eyes to see and the ears to hear. It is not necessary to be a rich man, and to build a great observatory, in order to become an astronomer. There were great astronomers before ever the telescope was invented; there have been astronomers even in our own days, there are some still living, whose work needs no other instrument than their eyes.[30]

---

[26] Forbes, E.G., Greenwich Observatory: *The Royal Observatory at Greenwich and Herstmonceaux, 1675–1976, Vol. 2: Origins and Early History (1675–1835)* (Taylor and Francis, 1975), p. 23. He quotes Maunder from page 316 of Maunder's book. The notion of RGO "supporting the practical science of navigation" became questionable during Airy's term as Astronomer Royal, even though it was founded in part to support this.

[27] For a parallel look into how public speeches and lectures drew crowds from the time of the U.S. Civil War and somewhat after, read Garry Wills's Pulitzer-prize winning book, *Lincoln at Gettysburg* (Touchstone, 1992).

[28] Rather in the vein of Richard Hinckley Allen's *Star Names: Their Lore and Meaning*—a book written around this time, and which continues as a basis for certain astronomical scholarship (usually of an historical nature) even today.

[29] Ibid, "Report to the Council ...," p. 318.

[30] In the preface to the 1908 edition (Dana Estes, Boston), pp. 7–8.

As previously noted, he was a biblical scholar of merit, admired personally by priest and perhaps rabbi,[31] alike. He was respected enough that he was asked to prepare the astronomical section of the *International Standard Bible Encyclopedia* at one point—which is claimed to have prompted him to write his *Astronomy of the Bible: An Elementary Commentary on the Astronomical References of Holy Scripture* (1908).[32]

He shows his aptitude as a scholar of religion in two back-to-back comments appearing in the B.A.A. journal's *Correspondence* section for October and November, 1909. In these broadsides, Maunder reveals a great deal about himself.

In an October, 1909 counter-thrust to a review arguing about the dates of Genesis in a book by the Rev. F.A. Jones,[33] Maunder took a typical approach. Jones's apparent attempt, using the methods of an expert on the orient, a Mr. A. Hamilton, was to "reduce to a common basis the mystic, if not mythical, chronologies of the great nations of antiquity."[34] Maunder's contentions (he had several in this one letter, alone, with Jones) dealt in part with the realm of separating what could be symbolic periods from actual chronological ones.

He used this logic in defending the Christian side (as well other religions, whose chronological methods Jones wanted to yoke together) in the next letter of November, 1909. This letter invoked the "lesson taught by the White Knight to Alice when she came 'Through the Looking-Glass'; namely, what a thing is, and what its name is, and what it is called, are three different matters, and not one and the same."[35]

Like the October matter, at issue in the November comment was a very complicated topic being fairly lightly commented upon and easily misconstrued, according to Maunder. Typically, Maunder's tone in the review evades any references to such political agendas that may have lain beneath the surface of such books, if this was indeed the issue—and not pure scholarship. But in any case, we see Maunder holding the course on scholarship, and religious scholarship, in particular, and avoiding the politics of the matter.

---

[31] Obituaries, book reviews, and encomiums mention Maunder as being respected by ministers, priests, etc. Given the time of these, Judeo-Christians, only, are implied by these remarks.

[32] Available online at www.cwru.edu/UL/preserve/stack/AstroBible.html.

[33] J.B.A.A. *Correspondence*, No. 414, October, 1909, pp. 390–393. The book in question was *A Comparison of the Biblical Chronology with that of Other Ancient Nations* (1909).

[34] Ibid, B.A.A. Correspondence, October, 1909, p. 390.

[35] Ibid, B.A.A. Correspondence. No. 415, November, 1909, pp. 424–427. "Ancient Chronology." Note also that the author of *Alice in Wonderland*—Charles Lutwidge Dodgson, alias "Lewis Carroll"—was a mathematician.

Maunder's tone in the November letter is lighter,[36] despite the issue at hand—hence the *Alice in Wonderland* quote—perhaps since it was a query into his own research in ancient timekeeping.

At issue were a number of topics, the questions posed by W.H.S. Monck, the answers, provided by Maunder, who betrayed more than a passing familiarity with ancient chronologies, as shown by the October letter's content.

First, in response to whether or not seconds and minutes were known before the Christian era, the answer intimated was, since mathematics to this effect was developed by ancient Greeks, approximations into minutes and seconds were probably done in some way. Another question was how the Jews "later brought their 360 (day) year into accordance with the Roman calendar," to which Maunder replied "the Jews never used a 360 day year," but like the Assyrians and Babylonians, used a "natural, luni-solar year" of "354 (+/−) days, or 384 (+/−) days … and still do." He also criticized Monck for thinking that the Indians of the subcontinent (Hindus) did not reckon in time units smaller than an hour, if not exactly in minutes and seconds as we know them, and then lists a table from the *Siddhanta Siromani*. Maunder was meticulous in his appreciation of these questions, and sensitive enough to note a time called "murta," (beginning) with Prána (a respiration = four European seconds). Maunder's religious understanding—regardless of creed or sect—is exceptional, since he openly defended Hindi, Hebraic, Iṣlamic, and other faiths' theological scholars' or holy men's reckonings, something the rather militant adherents of Christianity at the time had a tendency to lambaste or ignore. Commonality could be found, he suggested: but not in the ways many of these commentators and scholars knew, or perhaps wished.

It is interesting to note that Maunder also critiqued—as did perhaps many a physical scientist, then—Alfred Russel Wallace's contention concerning the Sun's and Earth's "privileged location" (i.e., center) in the universe.[37] In a 1903 comment on Wallace's article in the *Fortnightly Review*,[38] after critiquing the great

---

[36] Perhaps even jocular. Monck was certainly a good acquaintance of his, if not a friend. Taking Monck's lead, Maunder started the B.A.A.

[37] Michael J. Crowe, in *Modern Theories of the Universe: From Herschel to Hubble* (Dover, 1994) gives an indirect discussion of the controversy Alfred Russel Wallace stirred up in the early part of the 1900s with his cosmology.

[38] B.A.A. Journal, 1903 (probably summer) "Letters Communicated to the Association," *The Earth's Place in the Universe* (Maunder's title), pp. 227–234 . Wallace also published this in the *Independent* (see Crowe, p. 199). *Fortnightly Review* was: 73 (1903), pp. 395–411; p. 403 according to Crowe. Article by Wallace titled, "*Man's Place in the Universe.*"

naturalist's fixed notion, Maunder concluded this logical invective against mysticism by appealing to the metaphysical essences Wallace himself must have unwittingly included (or purposely inveigled) in his article:

> And supposing he [Dr. Wallace] had proved all his points, and that the Earth were at the exact centre of the universe, what then? What significance could it have? The Sun is not the centre of the Earth's orbit; the Earth is not in the centre of the [solar] system; yet neither fact, so far as we know, in any way lowers our moral or intellectual standing or impairs our physical comfort.[39]

So it is also seen that Maunder had no fear of bringing to bay matters brought up by the well-knowns in other scientific fields[40] whenever they lumbered over into his own—in addition to noting errors in other astronomers' or theologians' work.

Annie Maunder also kept up a vigorous schedule. As well as her own previously mentioned book, *The Heavens and Their Story,* her work appeared both in RAS publications (as previously shown, under Walter's and the Astronomer Royal W.H. Christie's auspices) and in *Knowledge.* One article was on the polar rays of the Sun's corona, where her studies of rays around the southern polar region revealed characteristics like dark streaks and where the notion that the solar corona is a "body of three dimensions" was given added emphasis.[41]

We see a great deal of official (Royal Observatory) expeditionary work, and "unofficial" (that is, B.A.A.) travels busying the Maunders in addition to routine observation work between 1890–1910. As part of solar work, Maunder and Annie obtained much experience in eclipse work in studying phenomena in the chromosphere and corona. Under the auspices of the British government, with other astronomers, such as Jesuit astronomer Father Stephen Perry, they went to Carriacou, West Indies, on August 29, 1886. A "non-official" (that is, not with Greenwich, but the B.A.A.) trip was to Lapland, in the Norwegian part, in 1896, where weather ruined the view, and in which his newly-wedded wife accompanied him. A note[42] gives a view of the B.A.A.'s activities and social fabric of the

---

[39] Ibid, B.A.A. Journal, 1903. *The Earth's Place in the Universe*, p. 234.

[40] Whether this reveals Maunder's positions on evolution, or personal feelings toward a man who gave up a belief in Christianity, is impossible to say. Given Maunder's nature, such assertions are doubtful.

[41] B.A.A. Journal, 1902, Vol. XII, No. 4. "Astronomical Publications," p. 191.

[42] Ibid, "Report to the council ...," February 1929, p. 317.

Fig. 27 The scientifically unsuccessful—yet apparently socially rewarding—B.A.A. solar eclipse party in Norway, 1896. Standing: second from the right, E. Walter Maunder; centre, with bottle, Annie S.D. (Russell) Maunder. (Printed with permission of the B.A.A.)

time: "the expedition, though failed from a scientific point of view, was a great success from a social point of view."

In one expedition to Talni, India, in January 1898—in which the Maunders "made their own private arrangements, hampered by 'no restrictions whatsoever, having received absolutely no financial help from any public body,'" Annie Maunder obtained a photograph of a "coronal streamer several degrees in length,"[43] using "Sandall triple-coated plates." This translates to a streamer "10 million km in length, or, 14 solar radii"[44] (see Fig. 21b in Chapter 8).

Annie had received a grant from Girton College for the purposes of studying the Milky Way, and with the funds obtained, had made her own 9-inch short-focus camera with a 1.5-inch diameter lens covering a wide field. The camera turned out to be ideal for reaching the outermost regions of the corona. This knowledge betrayed a good mathematical understanding of optics, and was probably not primarily due to hunches or mere luck, but rather, to calculation. Agnes Clerke even

[43] Ibid, Crommelin, A.C.D., Maunder Obituary, *The Observatory*, May, 1928, p. 158.

[44] Bruck, M.T., "Alice Everett and Annie Russell Maunder torch bearing women astronomers," *Irish Astronomical Journal,* Vol. 21, pp. 281–291, 1994.

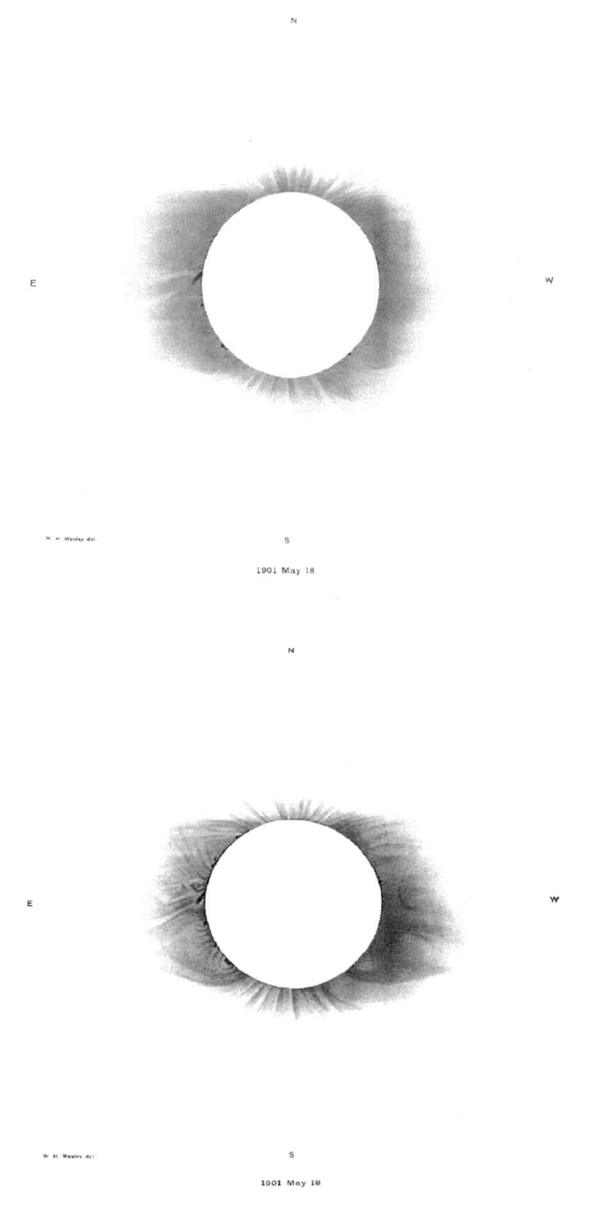

Fig. 28 The official May 18, 1901, Pamplemousses, Mauritius, Royal Observatory eclipse picture of the Sun taken by Walter Maunder (top) versus the unofficial photo taken by Annie S.D. Russell (bottom). The Sun's activity is near minimum, showing a roughly symmetrical, outer solar corona. The two well-defined polar plumes are also clearly displayed. The streamer belts and prominences (darkened extensions within the inner corona region especially near the east limb) near the low-to-middle latitudes imprinted particularly well in the image taken by Annie Maunder (bottom panel). (From Dyson and Christie, 1925–1929.)[45]

---

[45] Ibid, Dyson and Christie, *Memoirs of the RAS*, Vol. 64, 1925–1929.

reported that "as regards the corona, Mrs Maunder with her tiny lens has beaten all the big instruments." This achievement was largely forgotten.[46]

Other expeditions were to Algiers (with the B.A.A.) in 1900, and an official trip (Greenwich) to Mauritius the year later, during the eclipse of May 18, 1901.[47] Here, Maunder was sickened by malaria.[48] However, he obtained the requisite number of successful photos, all the same (see illustration in Fig. 28). The scientific skill of the unofficial Royal Observatory member of this Mauritius expedition[49]—namely, Annie—in photographing the solar corona is undeniably clear (see the drawing of the photo; Fig. 28).

At the invitation of the Canadian government in 1905, Maunder was again stymied in his efforts to take good photographs, the sky playing similar tricks to ruin his work in Canada as it had in Norway a few years before.

---

[46] Ibid, Bruck, p. 287.

[47] Ibid, "Report to the council ...," February, 1929.

[48] Annie was not ill from malaria, herself, though she was curious as to its cause, then. Quinine and Epsom salts, used for treating this disease, was suggested to her as useful in making photographic plates more sensitive (from the *Girton College Review*, "An Eclipse Expedition to the Island of Mauritius" [1907], p. 12). (Personal communication and quote provided by Jean Grove.)

[49] Dyson, F., Christie, W. (Astronomers Royal), "IX. Drawings of the corona from photographs at total eclipses from 1896 to 1922," *Memoirs of the RAS*, Vol. 64, 1925–1929, pp. 363–389.

11

A Particle Theory for the Sun-Earth Connection

When the staff emptied out of Greenwich to man the trenches in World War I, Walter Maunder was recalled to the Royal Observatory due to a staff shortage. Retired from the observatory, he worked at the Victoria Institute studying questions of religion and science.

For an indeterminate period through this war Annie Maunder returned as well—as a volunteer.[1] Frank Dyson was now the Astronomer Royal and had brought onto the staff—still at Cambridge, but "full of much promise"—the young Sydney Chapman.[2] Chapman was set to work editing *The Observatory* magazine much as Maunder had done. (This appeared to be a sort of stepping stone for certain astronomers' careers at the Royal [Greenwich] Observatory.[3]) Chapman was to make many contacts here besides the Maunders. He also met with the likes of Arthur Eddington, Vesto Slipher,[4] and George Ellery Hale while at the Royal Observatory.

A conscientious objector in a war responsible for an enormous brain drain,[5] Chapman was granted exemption from serving, and worked at the Royal Observatory throughout the war. The time there marked a return for Chapman after having quit as an observer and Senior Assistant at the Royal Observatory before the war. His status at the Royal Observatory this time round was perhaps not

---

[1] *The Observatory,* Vol. 60, No. 515, July, 1917, p. 255. Along with a discussion on polar solar "flecks," it is revealed Mrs. Maunder was a volunteer at the RGO through the war. Some date her return as 1916.

[2] Sydney Chapman: 1888–1970, Elected F.R.S. 1919. *Biographical Memoirs* (Headley Brothers, November, 1971), p. 54.

[3] Ibid, Maunder B.A.A. Obituary, May, 1928, p. 232.

[4] Ibid, *The Explorers of Mars Hill.* Vesto was one of the brothers who Lowell recruited in Douglass's place, after Douglass was dismissed for making statements about the director that the director deemed offensive.

[5] Physicist Harry G.J. Moseley (whom had worked with E. Rutherford) comes to mind, whose death at age 27 in World War I (on August 10, 1915 at Gallipoli Peninsula) was a tremendous loss to science, according to many who knew him.

as lucrative or positive as the first time. Chapman's status at the Royal Observatory during the war is not clear, but prior to this, it had been one "attractive both financially and as regards status."[6] Chapman could have been one of those "men in their twenties" of which rank and pay far surpassed Walter Maunder's. Staff as low as it was, this peripatetic and mercurial mathematician—touted as one of the best ever working on questions about the solar-terrestrial relation—could not have missed working or maybe even socializing with E. Walter, together with Annie, in off hours.[7] Even Hale most likely had either met or knew both of the Maunders.

In any case, by 1918 Maunder was no longer formally employed by the Royal Observatory, though the questions he posed certainly remained. So Astronomer Royal Dyson called Chapman's attention in a deeper way to what Maunder had spent his career doing. The work in question was the 27-day recurrence of "geomagnetic" storms—quite a toss back into time (1905) at this late date. But perhaps Dyson—like Airy in past times—did not want fifty years to elapse as had been the case with Sabine, and let the trail of the Sun-Earth magnetic linkage grow ice cold once again. This time it was relatively fresh, and E. Walter's and Annie's rather abundant contributions were laying about the observatory, the R.A.S., and of course the B.A.A., and exuding profundity. In fact, the refiners of Sabine's observations were still alive to relate details and help with matters, if unofficially and of course, if necessary.

This is an important connection, from Maunder's perspective, back to his seminal 1904–1905 papers on the subject. As we saw, Maunder had classed Sun-earth magnetic storms—like William Ellis had—according to size. He found that the biggest storms were associated with large spot groups. The geomagnetic storms occurred when a certain area of the Sun; that is, between 19 degrees east and 47 degrees west solar longitudes, passed into Earth's view.[8] Such findings were also amply brought to bear by Annie in her part of the popularisations found in *The Story of the Heavens*. So also were the lines of force from the Sun imprinted in certain straight or distinct lines, as we recall (in Chapter 8), wherein Maunder showed Kelvin and Schuster being correct perhaps in math, but wrong in the assumption of an isotropic and uniform solar corpuscular emission. However, here, Maunder joined with the previously-mentioned work of the Swede, Svante Arrhenius, and perhaps others like the Norwegian, Birkeland. Herein lies the "corpuscular theory," extant since the 1890s and early 1900s.

---

[6] Ibid, Sydney Chapman, p. 54.

[7] Curiously, Chapman has several mentions of Annie ("Mrs. Maunder") in his seminal 1940 book on geomagnetism with Bartels.

[8] Cliver, E.W., "Solar Activity and Geomagnetic Storms: The Corpuscular Hypothesis," *Eos,* Vol. 75, No. 52, December 27, 1994. E. Walter Maunder's original paper is in *Monthly Notices of Royal Astronomical Society*, Vol. 64, 1904, pp. 205–222.

The "repetitive nature" of the storms, arising at synodic rotation intervals (tied to a 27-day repetition) could (said Maunder):

> only be explained by supposing that the Earth had encountered, time after time, a *definite stream*,[9] a stream which, continually supplied from one and the same area of the Sun's surface, appears to us, at our distance, to be rotating with the same speed as the area from which it arises[10]

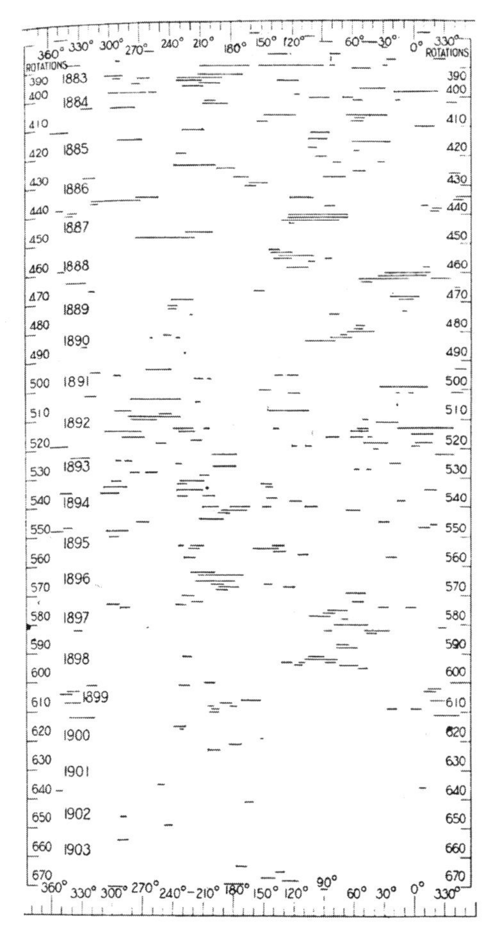

Fig. 29 A figure from Maunder's crucial 1904–1905 paper showing that the 276 geomagnetic disturbances (small, horizontal, broken lines) frequently occurred at consecutive 27-day intervals (this is the rotation period of the Sun as viewed from Earth) when a certain meridian has returned to the center of the disc. Maunder also noted that "the disturbances are not distributed irregularly with regard to the solar meridians, but chiefly affect two or three regions." (After Maunder, 1904–05.[11])

[9] Italics gratis the authors. Notice Maunder's choice of metaphor.

[10] Ibid, Cliver, *Eos*, December 27, 1994, quoting Maunder.

[11] Maunder, E.W., *Monthly Notices of Royal Astronomical Society*, Vol. 65, 1904–1905, pp. 2–34.

and onward, where Maunder saw the non-isotropic "one-direction-only" nature of the storms. And so, the end of the "fifty years' outstanding difficulty"—the legacy of the soldier-scientist, Edward Sabine to Lord Kelvin.

But a crowning glory to these proceedings was Maunder's linking his observations with the particle theories of S. A. Arrhenius. For

> while the coronal streams deemed responsible for [that is, geomagnetic] storms was unknown, the corpuscular hypothesis in the form proposed by Arrhenius that invoked radiation pressure to drive *charged particles*[12] from the sun was consistent with the observations.[13]

* * *

Importantly, we digress to show the background Arrhenius provided Maunder with. Svante August Arrhenius is said to have brought de Mairan's[14] work on aurora "up to date." Swedish Nobel prize winner Arrhenius[15] published as early as 1900 the account that was to figure so greatly in later solar science. The theory, in short, was a description of solar wind *before* solar wind became known as such. It was, basically

> a theory based on radiation pressure of electromagnetic waves and eruptions of particles from the Sun. He [Arrhenius] proposed that during great eruptions on the Sun, large quantities of matter, in the form of droplets or dust, are ejected in a radial direction from the sunspots and are pushed out from the Sun by radiation pressure (A mass transport which today is called the Solar Wind) [*sic*].[16]

Oddly enough, to support the previously-mentioned auroral scientist Paulsen and his claims that the northern lights were "cathode rays" (that is, electron rays

---

[12] Italics gratis the authors.

[13] Ibid, Cliver, *Eos*, December 27, 1994.

[14] Jean Jacques Dotous de Mairan published his main work in 1733. His principle conclusions were that auroral frequency had increased suddenly in 1716 and remained constant, and that it was frequently interrupted in the distant past. He identified 22 such instances from A.D. 500 to 1731, and called their comebacks *reprises* (resumptions). See George L. Siscoe, "Evidence in the auroral record for secular solar variability," *Review of Geophysics and Space Physics,* Vol. 18, No. 3, August, 1980, p. 647.

[15] It is poignant to note that the early Nobel-laureate Arrhenius "proposed in 1908 that radiation from stars could blow microscopic germs from one world to another" (from "Life's far-flung raw materials," *Scientific American*, Vol. 281, No. 1, July, 1999, p. 31).

[16] Ibid, "Scientific Auroral Experiments Beginning in the Nineteenth Century," p. 85.

before electrons were known) it was Arrhenius who could have helped narrow the field on the path to what was later seen as ionized oxygen, here. He did this by noting that, during these "eruptions on the Sun, negative and positive electricity are split into two, and that this division produces cathode rays and X-rays."[17] It should also be noted that Arrhenius proposed that such rays can "ionize the gas through which they propagate" which in turn meant negative electricity is created while leaving the Sun positively charged.[18]

Arrhenius's descriptions are almost frighteningly modern regarding motion and shape. Outward-streaming, negatively-charged particles from the Sun will "hit other celestial bodies, causing them to become negatively charged. They then repel the streaming negative charges from the Sun into hyperbolic paths away from the repelling body."[19] This is, then, a very early reference to the notion of "magnetosphere" around magnetized planets like Earth, Jupiter, and Saturn.

Also notable is that the most rapidly streaming particles can strike the body and increase its charge, and these bodies occasionally discharge as well. Arrhenius and others also believed that these particle bombardment discharges which cause ultraviolet radiation favored certain areas of Earth, at certain times. Arrhenius believed the bombardments occurred on the "dayside" of the equatorial region, but that the flow in the upper air could move such charged particles to other time sectors of the Earth and thereby bring about discharges on the "night side" as well.[20] This ties in with what Sabine, Maunder and others noted regarding "increased (magnetic) activity" at certain times of the day, vis à vis locations on Earth, respective of the geographical observation.[21] It also ties in well with seasonal variations.[22]

Running in parallel, or perhaps competitively, to Arrhenius was the Norwegian mathematician-physicist Kristian Birkeland,[23] who performed a

---

[17] Ibid, "Scientific Auroral Experiments ...," p. 85.

[18] Ibid, "Scientific Auroral Experiments ...," p. 85.

[19] Ibid, "Scientific Auroral Experiments ...," p. 85. A very early reference to the "magnetosphere," in its general description, way before it was so physically and mathematically understood. Note: it is still far from cut and dried.

[20] Ibid, "Scientific Auroral Experiments ...," p. 85.

[21] Ibid, Sabine's May 6, 1852 address to the RAS ("on Periodical Laws ..."), p. 105.

[22] Note Maunder's *Popular Astronomy* article of February, 1905.

[23] See Jago, L., *The Northern Lights* (Knopf, 2001) for a popular account of the heroic life of Kristian Birkeland, especially with regards to his many accomplishments.

physical experiment[24] that underscored the corpuscular theory. It is important to note Birkeland's experimental tendencies and travels, notably to polar regions to test his and others' theories; the nature and insight gained by Birkeland must have been very valuable to Chapman. Birkeland made a model that directed cathode (electron) rays at a magnetized globe to create phosphorescent patterns at its poles that mimicked auroral behavior.[25] The mathematician Carl Störmer[26] later did the

Fig. 30 Birkeland's "Terella" (meaning, "little earth") experiment, where he produced phosphorescent light effects on polar regions of a globe, representing Earth, using cathode rays (Birkeland is at left; his assistant, K. Devik, on the right). This approximated auroral belts around the Earth. Birkeland reportedly proved his auroral theory by 1896—that electrically-charged particles ejected from sunspots on the solar surface are captured by the Earth's magnetic field and directed along the magnetic field lines into the polar regions. As incoming particles reach atmospheric heights they are slowed down by the increasing density of atoms and molecules and in the process the atmospheric constituents become excited and ionized. (Reprinted from *The Northern Lights*)

---

[24] "Terrela."

[25] Ibid, Cliver, *Eos,* December 27, 1994. See also *The Northern Light*, pp. 97–101.

[26] Störmer "mapped the geographical distribution of auroral light." It was found that "the positions of the auroral zones can change depending upon solar activity. During times of great solar activity the zone will move more equatorwards than during quiet solar conditions" (Ibid, *The Northern Light,* p. 89). See also Wilfried Schröder's *Däs Phänomen Des Polarlichts* (Wissenschaftliche Buchgesellschaft, Darmstadt, 1984).

calculations that showed how charged particles are forced to move around the Earth's dipole magnetic field.

* * *

Now that we have an image of the charged particles causing auroral light, we go back to where we left Chapman, considering the Maunders' work.

It was stated that Dyson left Chapman with Maunder's work to further investigate, or to test and eliminate, or confirm, perhaps, the nature of geomagnetic storms and their connection to the Sun's magnetic activity. It was a labor that was to occupy Chapman, off and on, through the 1940s. To some extent, as the record shows, this question was to occupy him for the rest of his life.

The first paper by Chapman in 1918 led to a view of what Birkeland and Arrhenius proposed in a synthesis based on mathematics. The great insights had to survive the test of repeatability: clear, conceptual and mathematical models were the tools used for this. Chapman envisioned geomagnetic storms primarily as systems of electric currents flowing about the higher layers of the Earth's atmosphere. The geomagnetic field variations produced by these currents can be broken down into local, irregular and rapidly-changing parts. However, the most fundamental characteristic of a geomagnetic storm is the world-wide diminution of the intensity of the Earth's horizontal magnetic field as it is detected near the surface.

So what Chapman did was to separate the Maunders'[27] solar-induced geomagnetic storm variations into three distinct components: one representing storm-time variation (D_{st}) that is independent of longitude; one for the occurrence of the storm in local time (S_{d}, "d" standing for diurnal variability) and the last being the baseline geomagnetic component during magnetically-quiet days (S_{q}).[28] This delineation into components seemed to cover all of Maunder's and perhaps Ellis's, the Indian magnetic scientist N. Moos's, and others', observed and recorded phenomena on the matter of geomagnetic storms.

In 1918, Chapman found that storms in local time (S_{d}) are quite different in form, from magnetically quiet days (S_{q}) so that during the storm, the total changes are not simply an enhancement of the quiet magnetic day (S_{q}) alone. Chapman noted that S_{d} had much the same physical form after two days, as on the first day—though,

---

[27] Also extending the Indian magnetic scientist N.A.F. Moos's statistical study of magnetic storms. (Ibid, Chapman obituary, "Biographical Memoirs," p. 64). The following paragraphs paraphrase pp. 65–66 of this same document.

[28] Added complexity that cannot be ignored is the variation of the Earth's self-induced disturbances to S_{q}, which can in turn produce changes in S_{d} and D_{st} much the same in form, although smaller in magnitude as those triggered by the Sun.

only less powerful. Additionally, he showed that storm-time variation (D_{st}) in relation to these (S_q and S_d) at low geomagnetic latitudes can be seen as a *compression of the geomagnetic fields* during the initial phase of a magnetic storm on Earth (the "sudden commencement phase") and as an *expansion of the field*, during subsequent phases of maturation and decay. (To visualize this, recall the sequence of events Richard Carrington, R. Hogdson and Balfour Stewart described, on 1 September, 1859.) From about 1918 to 1927, Chapman's research shed further light on the significance of Birkeland/Arrhenius and others on what caused this compression, which represented the solar-terrestrial connection science interpretations of the phenomena, less the pure mathematics. That is, Chapman refined the conceptual nature of how the phenomena were understood.

The solar connection, between the compression/expansion of the geomagnetic field during the storms, was that the geomagnetic field was believed to be compressed by the impact of a stream of solar charged particles. It must be recalled that in the 1910s the true nature of the solar corpuscular emission was not known.[29] So, like Birkeland[30] and Arrhenius, Chapman ascribed the geomagnetic disturbance factor to solar charged particle emanation from the Sun—but not by electrons, alone; rather, by streams of ionized gas.[31] He was led to accept this complex reality arising from potential dual-processes involved in compression after F.A. Lindemann rejected some of Chapman's suppositions.

At first, Chapman thought that some separation in "sign" was needed to justify the magnetic storms' electrical discharges, but he later concluded that the source of the daytime and storm's variations (S_d, S_q) were controlled by ultraviolet light associated with energetic solar particle streams.[32] Because the particle streams could not propagate cohesively for more than one to two solar radii due to mutual repulsion, Lindemann proposed that a roughly equal number of positively and negatively charged particles (now called "*plasma*") must actually be the

[29] In 1905, Maunder asked the question, "are the particles reaching us from the Sun all of the same size, or, what comes to the same thing, do they all travel with the same rapidity?" (E.W. Maunder, *Astrophysical Journal*, Vol. 21, p. 113.)

[30] Chapman is particularly grateful to Birkeland's indirect assistance in solving these puzzles in papers, notably, his "Historical Introduction to Aurora and Magnetic Storms," *Annales Geophysique*, Vol. 24, 1968, pp. 497–505.

[31] Ibid, Cliver, *Eos,* December 27, 1994. Schuster and Lindemann objected to particles all of one sign. Hence, the shift to ionized gas.

[32] Ibid, Cliver, *Eos*, December 27, 1994.

expelled "cloud of particles" from the Sun.[33] Thus, important advances away from Arrhenius' and Birkeland's initial positions were made at this time.[34]

Chapman proceeded with the theory from his 1918 start, arriving in 1927[35] at further insights into the variations to storms in local time (S_d) and storm-time variation (D_{st}). There are not only external forcing agents to be considered in geomagnetic variations, but also *internal* ones. This was the realization that Earth's self-induced disturbances to the quiet component S_q can—in turn—produce changes in both S_d and D_{st} very similar in form, but smaller in magnitude to those triggered by the Sun. Together, these variations encompassed minor disturbances, which caused a "disturbance geomagnetic field" having characteristics that "do not vary much in form as [much as] its intensity."

Now, he noted that this "disturbance field" was larger and more irregular at higher geomagnetic latitudes than at lower ones. By so noting, Chapman highlighted the disturbance field at the maximal auroral frequency zones near the Earth's poles—like the ones Birkeland described in his Terella experiment. What was inferred as the cause of the disturbance was electric currents flowing in the Earth's ionosphere, concentrating in these maximal auroral frequency zones; namely, the poles. And, why did the currents concentrate here, he asked, pondering Birkeland's experiments and polar observations? They did so because the local charged particles, produced through ionization of the upper air by solar plasma streams, energetic ultraviolet and X-ray lights, are strongly guided by Earth's dipole magnetic field.

Modern Geomagnetic Storm Theory and Delineating the Magnetosphere

A number of incidents converged in 1931 to bring about the first "modern"[36] magnetic storm/solar theory—relative to the Earth—thus making it "geomagnetic," as it included the Sun and Earth fully and properly.[37]

---

[33] Ibid, Cliver, *Eos*, December 27, 1994.

[34] Chapman, for one, could not emphasize enough Birkeland's "indirect" assistance in understanding these proceedings.

[35] 1927 is coincident with Greaves' and Newton's 1928–1929 independent re-confirmation of Maunder's finding. See Cliver, E.W., "Solar Activity and Geomagnetic Storms: From M Regions and Flares to Coronal Holes and CMEs," *Eos,* Vol. 76, No. 8, February 21, 1995, this independent re-test showed evidence for recurring smaller storms every 27 days.

[36] Ibid, Cliver, *Eos*, December 27, 1994.

[37] Charles Chree, who was quoted by Chapman and Bartels in their 1940 book, *Geomagnetism*, praised Maunder for not mystifying his work by using obscure language (see Cliver, *Eos*, December,

To widen understanding especially of the "solar secrets" of those times, let us return to the Sun. George Ellery Hale's 1892 invention, the spectroheliograph, allowed Hale to take the first photograph of a solar flare ever to be published. This device allowed isolating narrow solar emission lines (calcium, hydrogen, etc.) "to build up an image of the Sun by scanning a slit across the disk."[38] Hale called flares "emission phenomena," or "eruptions," and he photographed this publishable one in July, 1892. In these early years, Hale favored the notion that solar magnetic storms could be caused by disturbances in prominences or faculae from anywhere on the Sun.[39]

In Hale's 1931 paper, *The Spectroheliograph and its Work, Part III: Solar Eruptions and their Apparent Terrestrial Effects*,[40] Hale shows some of the thoughts on the subject that were prevalent at the time. On page 405, Hale writes about Chapman that:

> He [Chapman] ascribes auroras and magnetic storms to the action of small temporary spots of very high temperature on the sun, from which blasts of ultra-violet light ionize the gases at high levels in the earth's atmosphere. The ions thus formed descend in spirals around the lines of force toward the northern or southern auroral zones. Chapman maintains that these terrestrial *corpuscles*,[41] with a speed of about 10 km/sec, could not possibly penetrate the atmosphere to the observed lower level of the aurora.[42]

Hale also noted that the "intensely hot regions" had yet to be discovered, but postulated that if they were to be found, that in relation to ultraviolet light emission, they would occur in time and position with the calcium and hydrogen eruptions he described elsewhere in the paper.[43]

(continued)

1994). Chapman, on the other hand, may have unintentionally begun a chain of—if not obscure—then at least, arcane, usage of single, complex words to denote several large concepts: a habit he was praised for throughout his life (he was as compact with math as with words—a disturbing trend for all but those as mercurial as the inventor). Maunder probably never used "geomagnetic" (the start of which uses a classical prefix for Earth) but phrases like "terrestrial magnetism."

[38] Ibid, Cliver, *Eos*, December 27, 1994.

[39] Ibid, Cliver, *Eos*, December 27, 1994.

[40] *Astrophysical Journal,* Vol. 73, 1931

[41] Italics gratis the authors. Note how corpuscular takes on its meaning in relation to charged particles.

[42] Ibid, *Astrophysical Journal,* Vol. 73, 1931, p. 405.

[43] Ibid, Hale, G.S., *Astrophysical Journal,* Vol. 73, 1931, pp. 405–406, reading into his footnotes.

So the "corpuscular theory" is mentioned here, one could say, in a more dynamic sense. It is also modified significantly to contain newer insights into solar physics. And near this time, the first modern theory was proposed for geomagnetic storms by Chapman and his student, V.C.A. Ferraro.[44]

This modern theory can be encapsulated as follows. We have, then, Maunder's directed lines of electromagnetic stream from certain parts of the Sun, striking Earth, and universal acceptance of this except from less important quarters. We know it hits Earth, but, how? Lindemann, though critiquing Chapman, left it to Chapman and—as it would turn out—his graduate student, Ferraro, to wonder about what consequences on the Earth the impact of such a stream of gas would have.[45] Thus, if Maunder had resolved one of the mysteries of the cosmic stream, ending the "fifty years' outstanding difficulty" by finding it emitted as a directional stream from the Sun, Chapman would be left to his own devices to explain *how* this emanation of an electromagnetic force transmuting into an ionized particle stream affected the Earth's polar regions. To do this implies more complexity than a "Terella" experiment. Chapman must have applied the Maunders' observational solar work in coming to the conclusions he soon reached. For Annie and Walter Maunder observed and described the entire process of solar mass ejection as early as 1905 (as noted earlier in Chapter 8) from "certain restricted areas," rather than from the whole surface of the Sun. They had this insight to report:

> There is no difficulty in conceiving the manner in which the solar action giving rise to our [that is, Earth's] magnetic storms may be conveyed to us. The corona has been at some trouble, these many years past, to visualize for us an action which is at least analogous to that in question. For myself, the first hint of the idea came with my first sight of a total solar eclipse in 1886. From that time I have never had the slightest doubt but that at least polar plumes are strictly analogous to the tails of comets; that is, they consist of very minute particles driven away from the Sun by some repulsive force; whether that might be electrical, or due to the pressure of radiation or to so me other cause, is not at this point a matter of importance.[46]

This solar part of the solar-geomagnetic storm connection later came to be known as "M-regions." Today, the eruptive phenomenon and the solar source have been more

---

[44] For example in Chapman, S., Ferraro, V.C.A., "A new theory of magnetic storms," *Nature*, Vol. 126, 1930, pp. 129–130.

[45] Ibid, Akasofu, "A note on the Chapman–Ferraro theory," p. 6.

[46] Ibid, Maunder, E.W., *Astrophysical Journal,* Vol. 21, 1905, pp. 107–108.

precisely linked to "coronal mass ejections" (or "CMEs") and "coronal holes"—namely, a "spotless" area.[47] To iterate Syun-Ichi Akasofu,[48] the distinguished space science physicist from the University of Alaska, Fairbanks: "It is most amazing that Maunder's statement in 1905 is an accurate description of high-speed solar wind streams from corona holes; in fact, he noted the source region of the high-speed streams on the solar disk is spotless!"[49]

However, sighting and description alone are poor sisters to repeatability, even when models are used. Furthermore, as Lord Kelvin insisted about Carrington and others regarding "one-time observations" (Carrington seeing the solar flare and recording eruptions in 1859): "one swallow does not make a summer."[50]

Such healthy skepticism on behalf of Lord Kelvin and F.A. Lindemann—rather than being rude—is good science. Chapman momentarily aside, George Ellery Hale in his important 1931 paper showed Carrington's 1859 observation repeated nearly ten times over seventy years, in anecdotal evidence. That is, solar flares—or more precisely in modern terms, a combination of flares and coronal mass ejections[51]—lead to geomagnetic storms.[52] There was not much to show after Carrington's 1859 event for some 73 years. Apparently, with Hale's help, the flocks of swallows were gathering.

* * *

But back to Chapman and Ferraro and the loaded questions: "we know the storms hit Earth, but how? What do they look like?" To answer Lindemann's challenge, Chapman and Ferraro went back to the basics. They studied Maxwell on electricity and magnetism. They created a model of how the solar corpuscular stream would interact with the Earth's magnetic field while being slowed down in its journey to Earth. They showed that the highly-conducting solar stream would create a force field with unequal repulsion in parts as it wrapped around Earth. It *would* wrap around Earth, since parts farther away would be subjected by weaker

---

[47] Ibid, Cliver, *Eos*, December 27, 1994.

[48] Akasofu, S.-I., a former student of Sydney Chapman, is a prominent auroral scientist who studied the phenomenon for more than 40 years. He has recently published the insightful *Exploring the Secrets of the Aurora* (Kluwer, 2002).

[49] Akasofu, p. 5 of "A note on the Chapman–Ferraro theory" in *Physics of Magnetopause*, eds. Song, P. *et al.* (American Geophysical Union, 1995).

[50] Ibid, Akasofu, p. 5.

[51] For a broad but technical overview on the nature of Coronal Mass Ejection-Flare phenomenon, see e.g., Lin, J., Soon, W., Baliunas, S., "Theories of solar eruptions: A review," *New Astronomy Reviews*, Vol. 47, 2003, pp. 53–84.

[52] Ibid, Cliver, *Eos,* February 21, 1995.

forces. They concluded that a *cavity* would be made around the Earth with the formation of an overall ring of currents. The solar stream protons are guided against the direction of rotation of the Earth, while the electrons go along with the Earth's rotation. In so doing, the solar plasma compresses the geomagnetic field lines, and hence, the strengthening of Earth's "field" (as during the initial phase of a large geomagnetic storm). The conception of a magnetosphere and the corpuscular theory of a Sun-Earth connection was thus formed. The formulation was published as a note in *Nature* in 1930.[53]

What was also being sought after was how the disturbances at various levels of upper air from the magnetosphere down to the ionosphere (that is, within this Chapman–Ferraro "cavity") are interconnected. As the Earth's dipole field lines are threaded, relative to the magnetosphere during daily rotation, a potential is set up to allow flows of currents between the poles and the equatorial plane around the space of the ionosphere and magnetosphere.[54] With disturbances to the equatorial ring current, electrical repulsion of particles with like charges can cause diffusion or drifting of charged particles away from the equatorial ring. They then accelerate along the lines of force of the Earth's magnetic field, ending near the top of the northern and southern polar auroral zones.

Models and mathematics by Chapman and Ferraro showed that the motions of protons and electrons in such a particle stream from the Sun are strongly linked to the Earth's dipole field.[55] The solar stream, much later to be called the "solar wind," and its charged particles was assumed to be a non-magnetized wind. As such, it was difficult to see how the stream could penetrate the Earth's magnetic field. A strong "shielding current" flows on the front surface of the advancing stream, and is called the Chapman–Ferraro Current[56] in their honour. As a result,

---

[53] Chapman, S., Ferraro, V.C.A., "A new theory of magnetic storms," *Nature*, Vol. 126, 1930, pp. 129–130.

[54] Such a system current is known today as the Birkeland Current. A.L. Peratt noted that Chapman critiqued Birkeland's ideas quite sharply (but Chapman apparently scathingly indicted Birkeland and Alfvén since it was not possible to distinguish unambiguously between current systems that are field-aligned and those that are completely ionospheric from a study of surface magnetic field measurements). However, satellite confirmation in 1967 showed that Birkeland's pioneering insights were largely correct. In deeper space, as a matter of fact (between Io and Jupiter, for instance, a five mega-ampere Birkeland Current was measured) such currents might exist on larger scales. (See Peratt, A.L. *et al.*, in *Galactic and Intergalactic Magnetic Fields*, p. 149, eds. R. Beck *et al.*, Kluwer, 1990). For an excellent review of magnetospheric physics, see David P. Stern, "A brief history of magnetospheric physics during the space age," *Reviews of Geophysics,* Vol. 34, 1996, pp. 1–31.

[55] This went against Alfvén's theory on protons and electrons being semi-independent, in this respect.

[56] All of this, in direct quotes or otherwise, is from Akasofu.

the Earth and its magnetic field are completely confined in a cavity with the solar stream pressure merely producing the compression effect sought after for explaining, for example, the onslaught of sudden geomagnetic storms and the later-discovered Sudden Ionospheric Disturbances (SIDs). In other words, the "corpuscular stream" was actively resisted by the action of the Earth's own magnetic field.[57] Thus, Maunder's hidden assistant on this—Arrhenius—foresaw what Thomas Gold would later come to call the "magnetosphere."[58]

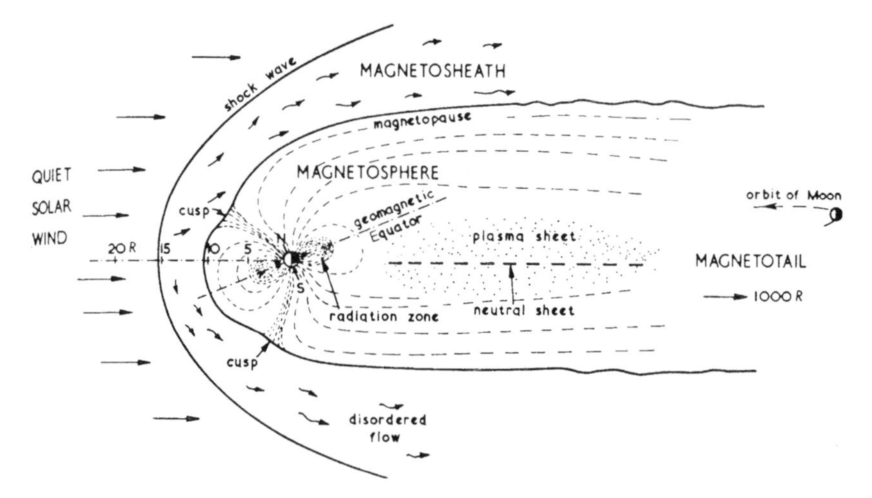

Fig. 31 Simple sketch of the magnetosphere showing elements of its parabolic shape in the magnetosheath. Note that the extension of the magnetosphere is roughly a few Earth radii (up to 10) towards the front bow shock and some 20–30 Earth radii towards the direction of the magnetotail. Larger solar storms pelt the solar wind particles at Earth with all the more intensity. At extremes, when the dynamic pressure of the solar wind drops to really weak levels, the bow shock is known to extend outward as far as 60 Earth radii (R_e).[59] Only in the 1960s[60] was it understood that the access of solar wind particles into the magnetosphere is most effective when the interplanetary magnetic field is southward pointing so as to increase the probability for reconnection with the now, overall, north-pointing terrestrial magnetic field. Other magnetized planets in our solar system may also be similarly pelted. (From King-Hele, 1975, 1985.)

---

[57] Ibid, Cliver, *Eos*, December 27, 1994.

[58] Cliver, E., Siscoe, G., "History of the Discovery of the Solar Wind," from Section News, *Eos*, Vol. 75, No. 12, March 22, 1994. Gold, who coined the term in 1959, was told that the structure was not a sphere, so the term would not be accepted.

[59] Several *in situ* spacecraft detected a shift of the front of the magnetosphere's boundaries to within four earth radii (the geosynchronous orbit is located at about 6.6 Earth radii, or, R_e) upon observing the impact of intense interplanetary shock waves during a large solar eruption. At extremes, when the dynamic pressure of the solar wind drops to really weak levels, the bow shock is known to extend outward as far as 60 Earth radii (R_e) (as observed between May 11–12, 1999).

[60] In the original work of James Dungey, for instance.

The issue of energy transfer across the magnetosphere boundaries was not seriously considered in Chapman and Ferraro's time because it was then believed that only a tiny fraction of the solar wind energy was needed to cause geomagnetic storms. With the benefit of hindsight, Syun-Ichi Akasofu (also a former student of Chapman) related that energy transfer from the solar wind to the magnetosphere "cannot be achieved by non-magnetized plasma." This is because the energy transfer rate is now known to be $10^{19} \sim 10^{20}$ erg/sec, and only magnetized plasma is able to sustain such an efficient transfer process.

Related Sun-Earth magnetic coupling work by Julius Bartels (who coined the term "M region"[61]), W.M.H. Greaves, H.W. Newton and others from 1928–1932 on showed that there were two kinds of magnetic storms: sporadic and recurrent. Large sporadic events begin suddenly, are associated with sunspots, and happen near solar maximum. Recurrent events come from M regions, and usually happen *after* solar maximum. (And are more clearly discerned during years of minimal solar activity.) Additionally, cases where Father Sidgreaves's question as to why solar spots appeared without attendant magnetic disturbances also appeared to be answered. Namely, there was an absence for some reason of a corpuscular or plasma stream emanation at times. This is not what Walter Maunder had postulated, either; namely, that the one-dimensional, linear stream streaked *past* Earth.[62]

Storms, broken down into phases as they were, were hard to theorize. Sydney Chapman could not devise a theory for the "main phase" of geomagnetic storms. Namely, how the solar wind plasma imparted its energy and momentum through this shielded magnetosphere. In the 1960s, J.W. Dungey proposed the "reconnection" theory of magnetic storms. This research describes the transport of a magnetized solar wind plasma across the magnetosheath via the effects of the reconnection between mutually-opposing field lines. Dungey's theory gave improved clarity to solar wind-magnetosphere interactions.

By the 1960s, impulsive solar flaring was observed well with the spectrohelioscope. Still, the major aspect to the mystery regarding continuous corpuscular streams remained. The aspect was this: Where were these "M regions" that couldn't be identified with sunspot groups on the Sun?

---

[61] Bartels, J., "Terrestrial-magnetic activity and its relation to solar phenomena," *Terrestrial Magnetism and Atmospheric Electricity*, Vol. 37, March, 1932, pp. 1–52. See also Cliver, *Eos*, February 21, 1995, and Crooker, N.U., Cliver, E.W., "Postmodern view of M-regions," *Journal of Geophysical Research.*

[62] Ibid, Cliver, *Eos*, February 21, 1995.

12

Our Knowledge of the Sun and Its Variability Today

The search for the origin of the solar corpuscular streams or Bartels's M regions, continued. Even if Annie and Walter Maunder observed these as early as 1905, they were not linked by them to anything like the solar wind. It is time for us to zoom forward to reveal the Sun's "secrets" that were hidden to Hale, Maunder and others then.

To orient us with a modern understanding of the Sun, one can succinctly quote the eminent solar physicist Eugene N. Parker:

> A pedestrian star like the Sun is, in actuality, a physics laboratory we could never build ... The [thermonuclear] core is brighter than ten supernovas at maximum light.[1]

The Sun was formed some 4,700 million years ago and is now located at a distance of about 8 to 8.5 kilo-parsec, or 26,000–28,000 light years from the centre of our home galaxy—the Milky Way. From careful astronomical studies,[2] there is a physical reason to suggest that the Sun was probably born 1.9 kiloparsec (or about 6,200 light years) closer to the galactic center than the present location of our Solar System. Within the Milky Way, the Sun, the solar system planets and our neighboring stars' systems move in an elliptical orbit between about 7.9 and 9.7 kiloparsec from the galactic center, with a period of revolution of about 240 million years. Within the realm of the Solar System, the Sun is about 149 million km, or 8 light minutes away from the Earth (1 astronomical unit [a.u.]).

Materially, the Sun consists almost entirely of hydrogen (71% by mass) and helium (27% by mass). There is a small amount of carbon, nitrogen, oxygen, iron, and other heavy elements. The solar diameter ($2R_{sun}$) is about 1.4 million km, or about 110 times that of the Earth. The interior of the Sun can be divided into three

---

[1] Parker, E.N., *Physics Today*, June 2000, pp. 26–31.

[2] Wielen, R., Fuchs, B., Dettbarn, C., *Astronomy and Astrophysics*, Vol. 314, 1996, pp. 438–447.

main regions: (1) the core from 0 to 0.25 R_{sun}; (2) the radiative zone from 0.25 to 0.7 R_{sun}; and (3) the convective zone as the outermost 0.3 R_{sun}. The core is roughly the region where nuclear fusion[3] of protons into helium takes place to power the entire Sun. The radiative zone is roughly a smooth zone in which the main mode of energy transfer is through the net outward diffusion of photons through continuous absorptions and re-emissions, mostly in X-rays (say photons with wavelengths

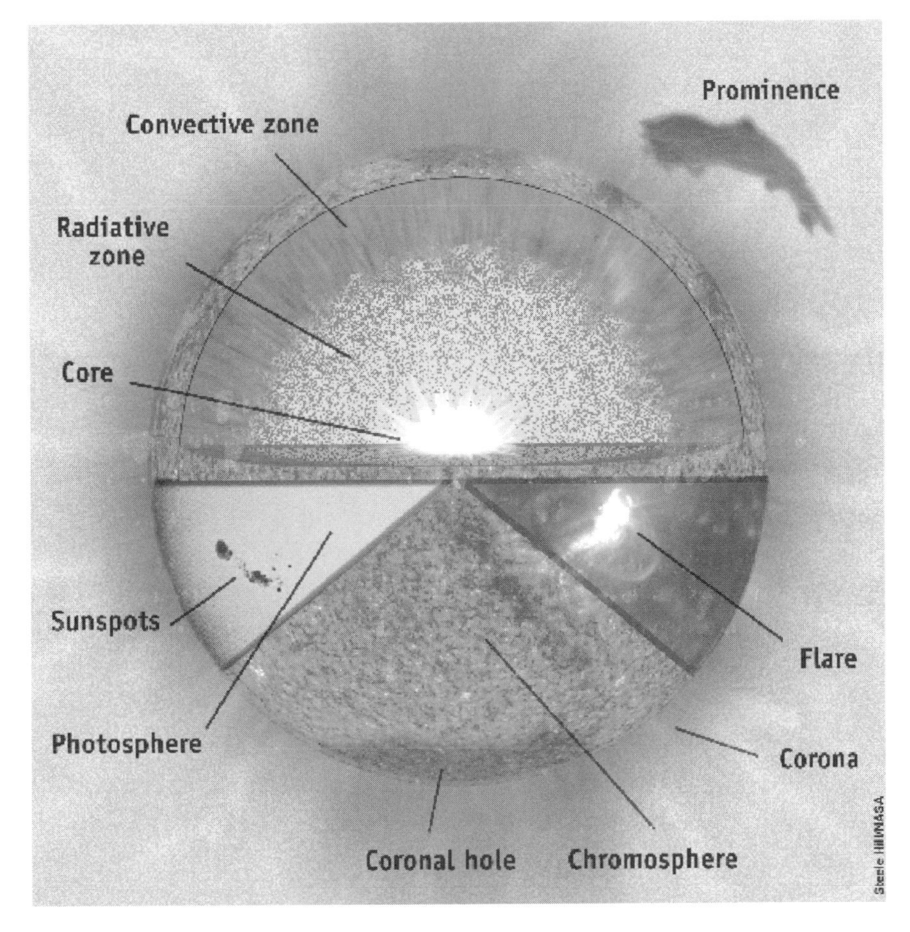

Fig. 32 A cartoon of the solar sphere, illustrating several basic features. (Image courtesy of NASA/ESA/SOHO.)[4]

---

[3] For a long time, the source of the Sun's energy was not well understood. The continual gravitational contraction to account for all of the Sun's energy (more notably argued by Lord Kelvin) is not viable since it would only provide for about 45 million years of the Sun's output at the current luminosity.

[4] Available at http://sohowww.nascom.nasa.gov/gallery/bestofsoho/

10–50 nm) under the high density condition within. The convective zone is the region where the energy transfer is characterized by the flow of hot gases near the bottom of the convective zone at about 0.7 R_{sun} to cooler regions near the surface at 1 R_{sun}.

The temperature of the Sun changes between the surface and the central core. It is about 5,800 K at the surface, rises to 3 million K midway of its radius, and is a hot 15 million near the center. Only the outer surface of the Sun is visible to us. This is the solar atmosphere and it is further divided into three layers: the photosphere, chromosphere and corona.

The Sun emits a vast amount of energy, but only a tiny portion reaches the Earth. This energy comes from nuclear fusion. Since the Sun is now in its stable nuclear-burning phase, it promises to remain at roughly constant radiant energy and surface temperature for a long time. But as will be shown later, this relative stability could be quite deceptive for interpreting past climatic changes. The stability comes from the fact that inward gravitational force is balanced by the outward forces due to gas and radiation pressure.

The vast amount of energy liberated at the centre of the Sun means three things. First, there is sufficient heat pressure within the Sun to resist the inward collapsing force of gravity. Second, the nuclear heat comes in the form of radiation slowly making its way to the solar surface. Only about one part in 2×10^{11} (or 200,000,000,000) of this thermal radiation in the nuclear core leaks out, but obviously, this is more than sufficient to power the light of our star and its surroundings. By comparison, about 3% (or 3 parts in 100) of the Sun's core energy is radiated away in the form of neutrinos.[5] Third, some of the Sun's energy goes into churning and turning gas near the outer layer and its surface, namely, the convective zone. These complicated gas motions, of essentially charged particles owing to the high temperature, can in turn cause the formation of various magnetic field structures. It is rather remarkable that only about 1% of the thermal nuclear energy is turned into the kinetic energy for the convective heat engine in the outermost layer. And only about 10^{-4} to 10^{-3} of that kinetic energy is in turn converted into magnetic form.

---

[5] Note: it may take about 1–10 million years for the core's "nuclear heat" to reach the surface via energetic photons because the core density is about 148,000 kg m^{-3} or about 100 times denser than the surface. This process of diffusion of photons is very slow when compared to that whereby a neutrino takes only three seconds to travel from the core to the surface! Neutrinos do not interact with the dense gas inside the Sun while the photons are forced to be highly interactive with the solar gas. To further marvel at the nature of neutrinos (the famous "ghost particle" conceived by Wolfgang Pauli with theoretical arguments), it is estimated that a neutrino could pass through a wall of lead 1,000 light years thick and not be stopped!

Solar Magnetic Field and Variable Solar Outputs Concerning Historical Solar Minima

Solar activities refer to a host of complex and ever-changing phenomena that originate in the ionized, electrically-conducting and magnetized plasma of the convective zone and solar atmosphere. Most of these phenomena are associated with the magnetic field, which is generally not uniform in its distribution across the solar surface. Superimposed on the general field of about a few Gauss (on the largest spacial scale) are the time-varying high field strength structures of 10–1000 Gauss, most notably those concentrated around sunspot active regions.

Figure 33 shows the sunspot activity cycles, which are the most obvious manifestation of solar magnetism. Usually sunspots come and go in a more-or-less periodic fashion every 11 years or so. But there are also intervals when sunspots drastically disappear altogether from the face of the Sun. This interval is of course now the already familiar anomalous period around 1645–1715 called the Maunder Minimum. Also notable is the relatively depressed solar activity maxima around 1795–1823 known as the Dalton Minimum.

The magnetic field is also likely to be responsible for heating the solar corona to high temperatures of one to two million degrees. Yet the origin of these magnetic fields and their changes are not fully understood.

In the 1980s and 1990s, there came a surprise. For a long time, it was thought that the total amount of energy emitted by the Sun should remain constant over thousands of years. Now, observations of the Sun's total light outputs by numerous NASA satellites since 1978 show that the Sun, in fact, brightens and fades in relatively close synchrony with the 11-year sunspot cycle.

Figure 34 shows this remarkable result as measured accurately by six radiometers. Different radiometers onboard different spacecraft may give different absolute levels of total irradiance, but the brightening of solar luminosity of about one to two parts in a thousand, or 0.1 to 0.2%, from solar minimum to solar maximum is real. Apparently, the explanation for the change in the stream of solar radiant energy rests on the tug for relative dominance between the dark magnetic spots and the bright faculae on the face of the Sun. Dark sunspots inhibit the flow of energy while bright faculae add positively to the solar irradiance. On a rotation timescale of 27 days, very large sunspots (or sunspot groups) could dim the Sun's total irradiance by as much as 0.2–0.4%.[6] Over 11-year cycles, the bright

---

[6] See for example the review by Fröhlich, C., Foukal, P.V., Hickey, J.R., Hudson, H.D., Willson, R.C., "Solar irradiance variability from modern measurements," in *The Sun in Time*, eds. C.P. Sonett *et al.* (University of Arizona, 1991), pp. 11–29.

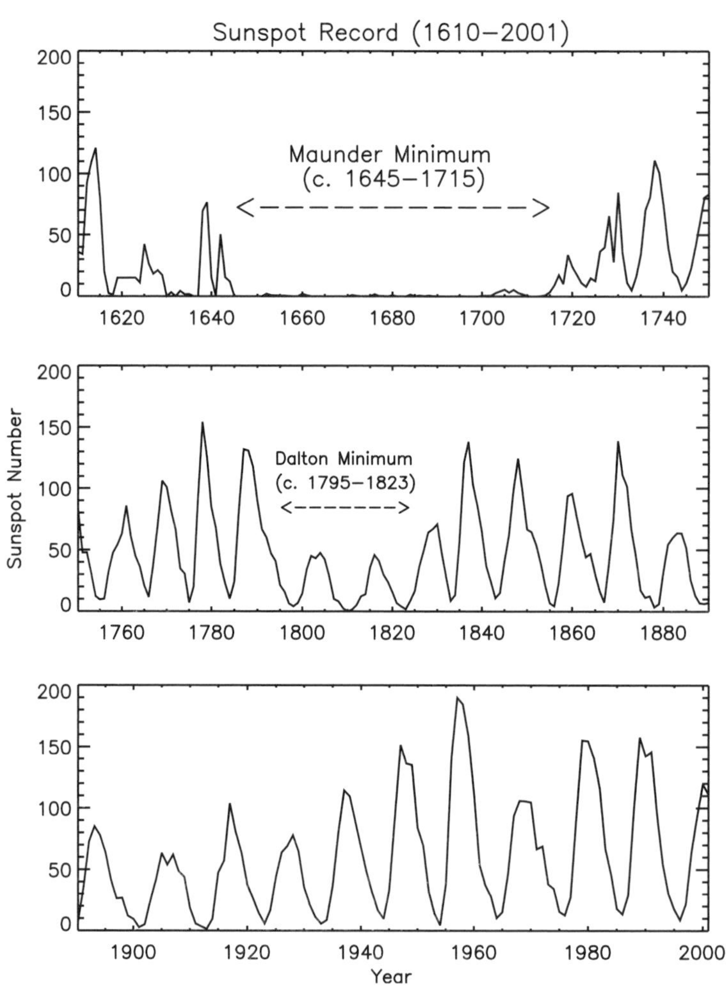

Fig. 33 Sunspot number record from 1610 to 2001. Notable are the regular 11-year cycles and the persistently low sunspot number count during the Maunder Minimum and perhaps during the Dalton Minimum. Even the 22-year cycles of solar magnetism having one low-peak alternating with one high-peak are visible in such a record (starting from about 1855). (Data Courtesy of David Hathaway/NASA/MSFC.)

magnetic components win out, and that is why the Sun is brighter when solar activity is strong.

However, a true understanding of the irradiance changes in relation to solar magnetism remains a matter of continuing scientific research. The question at hand is whether or not the Sun's total irradiance varies on timescales *longer* than 11 years. How much dimmer was the Sun during the Maunder Minimum, compared to the present, for example? There is evidence from solar-star study to suggest that the Sun's brightness during the Maunder Minimum could be anywhere from 0.2% to

0.7% *lower than* present (with a mean of about 0.4%),[7] but the large range of indeterminacy of the estimate is still an ongoing research question seeking resolution.

Solar Winds and Coronal Holes

Moving forward past the Maunders in the study of the Sun brings numerous new and surprising revelations about the nature of the solar corona.

In the 1950s, the now-elderly Sydney Chapman pointed out a remarkable implication of the enormous thermal conductivity sustained by the million degree solar corona. Chapman showed with this fact that the coronal gas must extend far

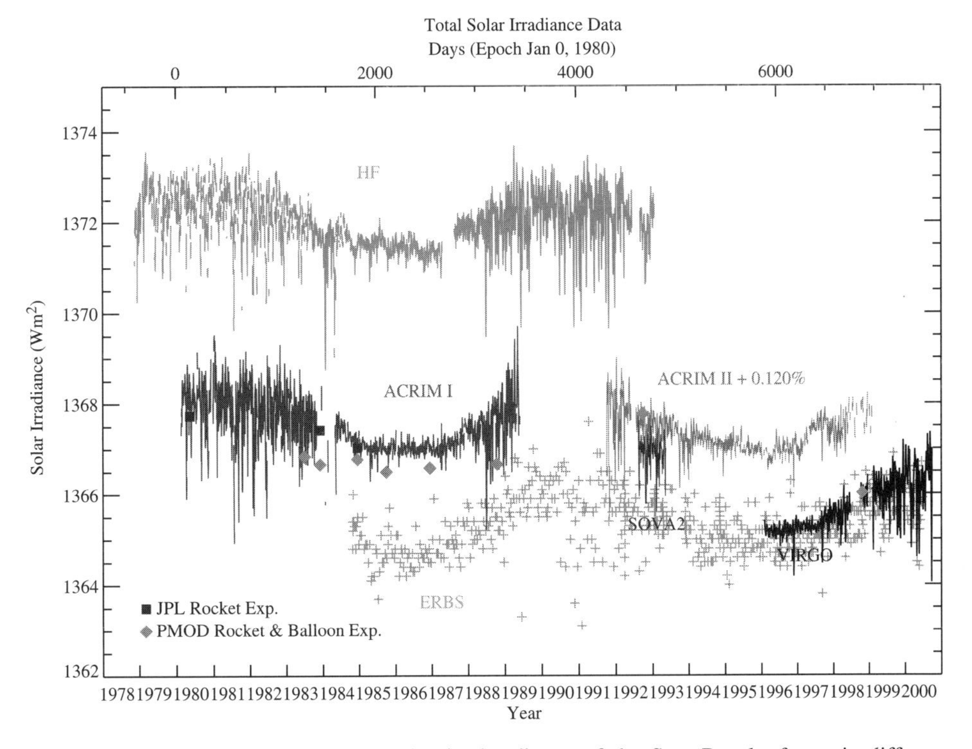

Fig. 34 Records of changes in the total solar irradiance of the Sun. Results from six different radiometers are shown and they proved that the so-called "solar constant" (or total solar irradiance) adopted by most meteorologists and climate scientists is not actually constant. (Courtesy of C. Frohlich of WRC/PMOD and SOHO Virgo Team.)[8]

---

[7] Zhang, Q. *et al., Astrophysical Journal Letters*, Vol. 427, 1994, pp. L111–114, and Soon, W.H., Baliunas, S.L., Zhang, Q., "A technique for estimating long-term variations of solar total irradiance: Preliminary estimates based on observations of the Sun and solar-type stars," in *The Solar Engine and Its Influence on Terrestrial Atmosphere and Climate*, ed. E. Nesme-Ribes (Springer-Verlag, Berlin-Heidelberg, 1994), pp. 133–143.

[8] Available at http://www.pmodwrc.ch/solar_const/solar_const.html

out into space, and it is certainly extended beyond the orbit of the Earth at 1 a.u. In 1951, Ludwig Biermann performed a careful study of the behavior of ionized gas tails of comets, especially on the formation and precise direction of those ion tails as the comets swing near the Sun. It was found that ions in the comets' plasma tails are always *systematically accelerated in an anti-sun direction*, and the *plasma tails are always present in all heliographic latitudes*. Such behavior also appears to persist at all times, regardless of the Sun's activity. To explain the observations, Biermann proposed that there is a continuous flow of solar plasma in all directions.[9]

Ludwig Biermann's proposal meant that this was the weak background flux that "gave rise to incessant minor geomagnetic fluctuations and to faint polar aurorae."[10] There is a problem however: this proposal would contradict the powerful theoretical persuasion of a *static* corona, made by Chapman.

Eugene Parker, studying cosmic-ray modulation in 1955, was struck by both Biermann's and Chapman's arguments. The Sun's hot corona, that fiery solar fringe envisaged by Herschel, as perhaps Maunder could have told us, was *not* static as Chapman had assumed. Parker resolved the dilemma of what this corpuscular stream/plasma from the Sun was doing by stating that it was a hydrodynamic expansion of the solar corona. (Parker is essentially telling us that the gas pressure in the hot solar corona of 1 to 2 million K is sufficiently strong to counteract the binding effect of gravity.) And that was what this particle stream was all about. It was a "solar wind":[11] a gust of solar charged particles blowing into yet another sphere—the heliosphere[12]—where constituents of the solar system reside.

By 1958, Parker had put forward a theory that described the flow of this hot plasma from the Sun.[13] Parker calculated that for the hot million-degree corona

---

[9] See for example, Biermann, L., "On the history of the solar wind concept," in *Historical Events and People in Geosciences*, Ed. W. Schröder (Peter Lang, 1985), pp. 39–47.

[10] Ibid, *Eos*, Vol. 75, No. 12, March 22, 1994.

[11] Artificial satellites, like Explorer 1 in 1958, helped discover Van Allen Zones, or Belts: electrically-charged particles trapped by Earth's magnetic field. The inner belts are mainly electrons and protons. The outer belt is composed of electrons, and it was this discovery that cleared the path for Mariner 2 in 1962 verifying that the Sun was ceaselessly shooting out electric particles in all directions, and in gusts, relative to how active or inactive the Sun is at a particular time in space—which led to calling this outpouring, the "solar wind." For further historical perspectives, see, for example, Karl Hufbauer's *Exploring the Sun* (John Hopkins University Press, 1991). See also David P. Stern (Ibid, *Reviews of Geophysics*, 1996).

[12] Although its exact form is still under close investigation, the heliosphere covers roughly 200 a.u. in its fullest extension (that is, diameter) where the outermost reaches of the solar wind pressure balances the inward pressure from the local interstellar space/wind.

[13] See also a recent after-thought by Parker, E.N., "Space physics before the space age," *Astrophysical Journal*, Vol. 525 (Centennial Issue), 1999, pp. 792–793.

temperature, the solar wind would be flowing with supersonic speed (that is, a flow speed greater than the speed sound waves travel) beyond a few solar radii. Modern measurements[14] show that at Earth's orbit the solar wind velocity, plasma density, and plasma temperature are about 470 km/s, 8 protons cm^{-3}, and 1.0–1.2 × 10^5 K, respectively. Sound waves travel with a speed of 60 km/s in this 1 a.u. medium, so the solar wind is indeed flowing supersonically. The ionic composition of solar wind plasma is about 95% protons and 5% helium nuclei. The solar wind plasma has a high temperature, and is therefore very conductive. It carries with it magnetic field which is ultimately the source for the interplanetary magnetic field. At 1 a.u., the interplanetary field strength ranges from 1 to 10^{-4} Gauss. There are also obvious time-varying features of the solar wind plasma, summarized in Figure 35.[15]

Now that the solar wind was recognised, what was the source on the Sun that brings us the recurrent geomagnetic variation as highlighted by J. Bartels' work? What, then, is the connection between those source regions and the solar wind?

In perhaps what the Maunders, or A.E. Douglass, or others privy to these secrets of the Sun earlier on in the journey could not imagine, it took the space age to find out or double-confirm what M regions are. Coronal holes—literally, holes in this tenuous outermost part of the Sun—were found to be the source of high-speed wind streams and the 27-day recurrent geomagnetic storms (Figure 36a). This was revealed in the 1970s by Skylab. Space flights were the first to observe "dark rents in the solar atmosphere in images of the EUV and X-ray corona."[16]

Coronal holes are, indeed, areas of open magnetic field lines where charged particles stream away from the Sun at high speed. This open magnetic field structure contrasts the more familiar closed magnetic structures at lower heights of the solar atmosphere like those of sunspots at low to middle latitudes, or even some of the non-flaring solar prominences seen protruding from the Sun's limb at higher latitudes. Coronal holes come in a variety of sizes, but are more prominently observed near the polar

---

[14] Gosling, J.T., "The Solar Wind," in *Encyclopedia of the Solar System*, eds. Weissman, P.R. *et al.*, (Academic Press: 1999), pp. 95–122. Materially, we do not have to worry about the Sun losing all its mass through solar wind because this amount is very small, roughly a few part 10^{-14} mass of the Sun is spewed out by the solar wind per year.

[15] Data downloaded from the National Space Science Data Center (NSSDC)'s webpage, http://nssdc.gsfc.nasa.gov/omniweb/form/dx1.htm/. For more descriptions, see for example, Paularena, K. I., King, J.H., "NASA's IMP 8 Spacecraft," in *Interball in the ISTP Program* (*Studies of the Solar Wind-Magnetosphere-Ionosphere Interaction*), eds. Sibeck, D.G., Kudela, K. (Kluwer, 1999), pp. 145–154.

[16] Ibid, Cliver, *Eos*, February 21, 1995.

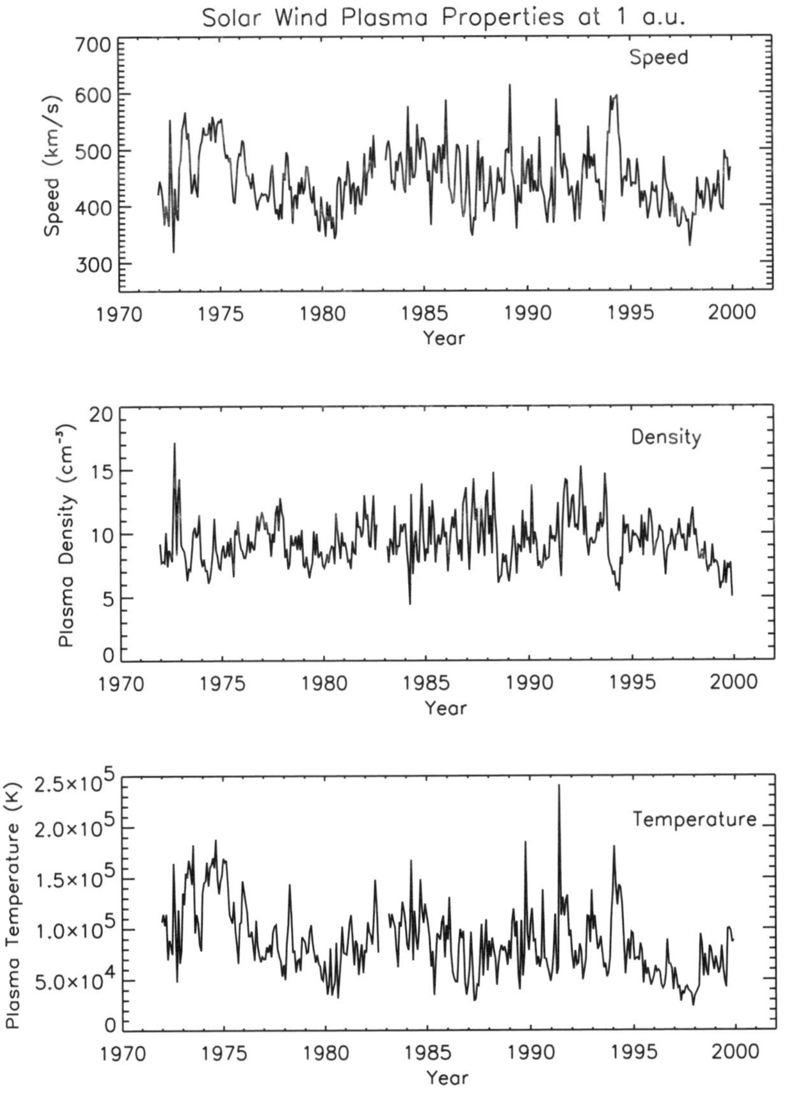

Fig. 35 27-day averages of solar wind plasma properties at 1 a.u. over the last 30 years recorded by a series of *in situ* orbiting NASA satellites. Top top panel shows the wind speed, the middle panel shows the plasma number density, and the bottom panel shows the plasma temperature. (Data courtesy of Joseph King and Natalia Papitashuili, NSSDC OMNIweb.)[15]

regions. The largest individual holes are no more than a few percent of the total solar surface area. The total area of the Sun covered by coronal holes enlarges to about 25% during the declining phase toward solar activity minimum, and shrinks to only a few percent during activity maximum. Apparently, the opened and closed magnetic structures on the solar surface tend to crowd each other out. Figure 36b shows the example of this inverse relationship for total coronal hole area, versus the sunspot number.

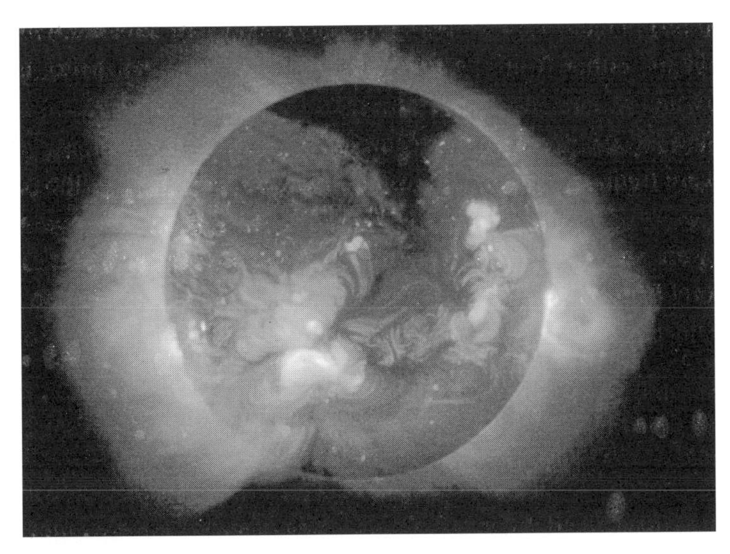

Fig. 36a The Sun in X-ray exposing detailed features of the dynamic corona. Large areas devoid of X-rays, especially obvious near the polar regions, are known as the coronal holes. The coronal holes are simply open magnetic field line areas that extend outward from the Sun's corona into interplanetary space (in contrast to, say, closed magnetic field features like those bright coronal loops at lower latitudes or bipolar sunspot regions down in the photosphere). The coronal holes are also known to be the source regions of a continuous stream of high-speed solar wind. Thus, the secret of M regions is revealed at last. (Image Courtesy of Yohkoh satellite soft X-ray telescope science team. Re-adaptation by David Hathaway, University of Alabama, Huntsville.)

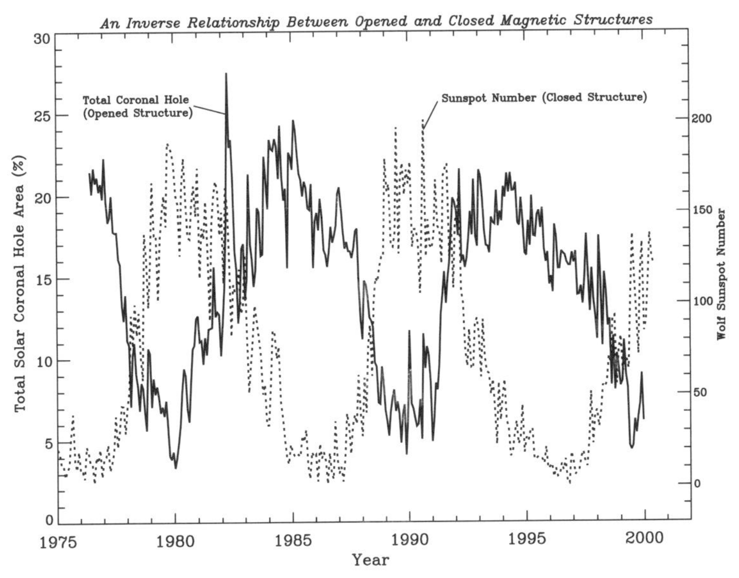

Fig. 36b An inverse relation between opened magnetic field structure (as proxied by total solar coronal hole area) and closed magnetic field entity (as proxied by sunspot number count). (Coronal hole area data courtesy of Y.-M. Wang of the Naval Research Laboratory.)

But the trickiness of the Sun's magnetism does not stop here. There are even more dynamical events: impulsive solar flares and coronal mass ejections.

Solar Flares

Solar flares are indeed complex energetic events normally seen as sudden brightening near the solar surface, and are once again linked to solar magnetic activities. Eruptive phenomena like flares have been known for a long time, as exemplified

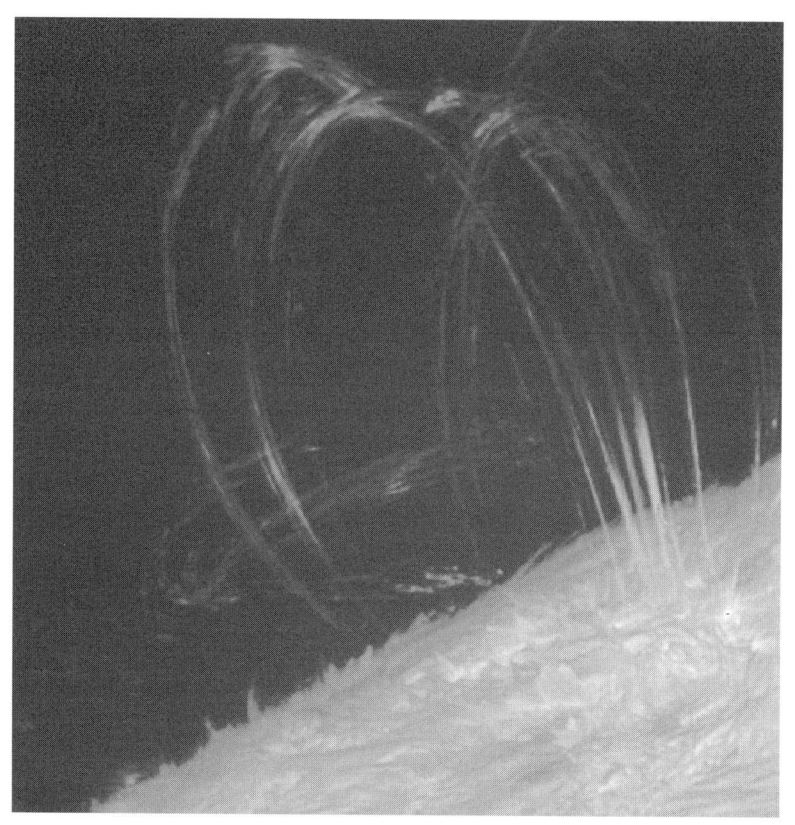

Fig. 37 Trails and loops of hot corona gas after the expulsion of a solar flare near a magnetically active region on June 26, 1992. This post-flare loop system was observed in Hydrogen filter (Hα-filter) by the Lockheed group using the Swedish telescope at La Palma. (Image courtesy of David Hathaway.)[17]

[17] Available at http://science.msfc.nasa.gov/ssl/pad/solar.loops.htm/. Details of this particular flare event of June 25–26, 1992 are described in Moore, R.E., Schmieder, B., Hathaway, D.H., Tarbell, T.D., "3-D magnetic field configuration late in a large two-ribbon flare," *Solar Physics*, Vol. 176, 1997, pp. 153–179.

by Carrington and Hodgson's observation in 1859—or even back to Stephen Gray, in 1705. The study of flaring events have certainly been enhanced by Hale's invention of the spectrohelioscope in the 1920s and 1930s.

Today, we know that flares occur within the magnetically active regions near sunspots, and tend to occur near the boundary between opposite magnetic polarities (that is, directions) of an active region. Flares are also observed to originate out of unstable arches of large prominences in the chromosphere. Solar flares are violent explosive outbursts of energy that send energetic particles, including protons and electrons in kilovolts or more, out into interplanetary space. The radiation accompanying flares is emitted over the full electromagnetic spectrum from gamma rays, X-rays, and radio waves.

Flare events last from a few minutes up to hours, but the effects can be felt for days in interplanetary space. The ejected particles may reach the Earth within a day or more. It was certainly a dream of Walter and Annie Maunder, George Hale, Sydney Chapman, and others, to be able to resolve how flaring events could trigger auroras, disrupt radio wave transmission, and even cause energy surges in high voltage transmission lines. Although flares are believed to be associated with sudden releases of magnetically stored energy, the conversion of magnetic energy into particle energy remains an open question.

Coronal Mass Ejections

Like the discovery of coronal holes, it was also in the 1970s that the Orbiting Solar Observatory (OSO) 7 spacecraft literally saw the explosive *mass ejections* of matter near the boundaries of coronal holes, bringing us a "coronal counterpart to chromospheric flares."[18] Coronal Mass Ejections, or CMEs, as they are now called, were associated with "eruptive prominences" on the Sun more than with flares, and that they even preceded flares in "lift off" in some cases.[19]

Today, these ejected masses are known to contain typically 10^{15} to 10^{16} grams (or about 10 billion tons) of solar stuff while moving between the Sun and the Earth at speeds anywhere between 50 and 2,000 km/s.[20] CMEs are also distinct in that their mass ejecta covers longitudinal and latitudinal extents far larger than the small

---

[18] Ibid, Cliver, *Eos,* February 21, 1995.

[19] For a broad but technical overview on the nature of Coronal Mass Ejection-Flare phenomenon, see e.g. Lin, J., Soon, W., Baliunas, S., "Theories of solar eruptions: A review," *New Astronomy Reviews*, Vol. 47, 2003, pp. 53–84.

[20] Ibid, Gosling and see also Zirker, J.B., "The Sun," in *Encyclopedia of the Solar System*, eds. Weissman, P.R. *et al.* (Academic Press, 1999), pp. 65–93.

regions associated with solar flares. We know that the occurrence of CMEs follows the solar cycle. There is an average of three to four CME events per day at solar activity maximum and one CME event every five to ten days at solar activity minimum. It is estimated that Earth intercepts about 72 CMEs per year during solar activity maxima, and about eight CMEs per year during activity minima.[21] The energy of CMEs is now seen as a major cause of interplanetary phenomena like

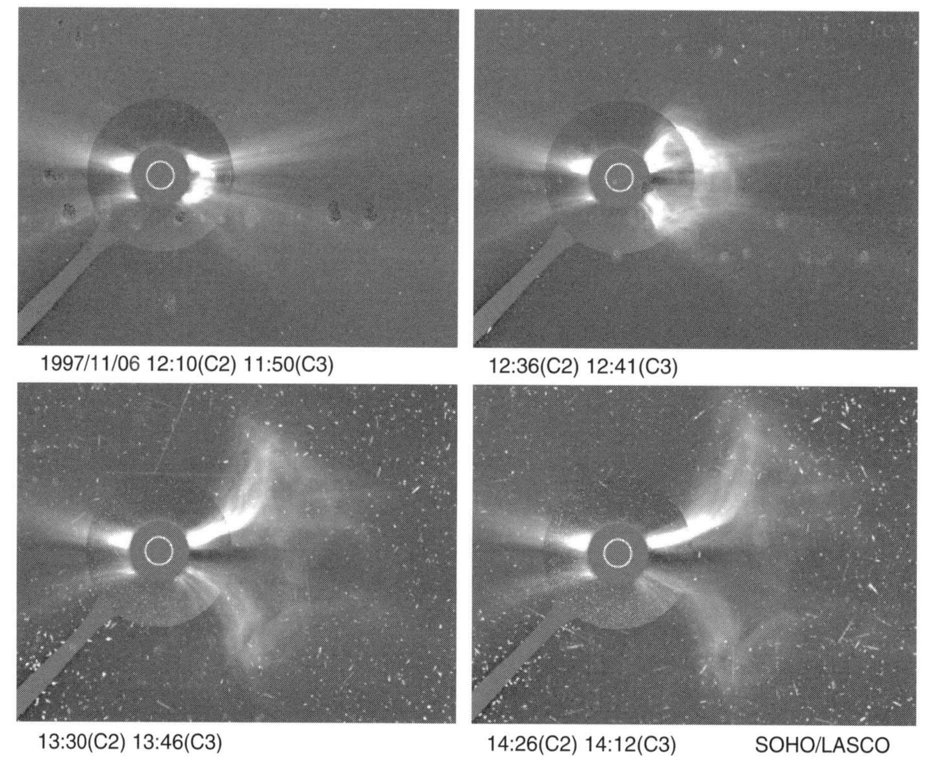

Fig. 38 A wide-angle view of the Sun's extended corona out to 32 solar diameters (or 45 million km, or roughly half the diameter of the orbit of Mercury) showing the development of a Coronal Mass Ejection on November 6, 1997. Ejecta with velocity as high as 1,500 km/s, or 3.3 million miles per hour, are observed. The numerous bright points and streaks in the two bottom panels are caused by impacts of highly energetic protons on the detector. The innermost white circle overlays the occulting disk (to show the fainter corona by blocking off intense light from the solar surface) and shows the actual size of the Sun. (Image courtesy NASA/ESA/SOHO/LASCO.)[22]

[21] Ibid, Gosling, p. 112.

[22] Available at http://sohowww.nascom.nasa.gov/gallery/LASCO/

shock wave disturbances, rather than solar flares. But more research is certainly warranted.

The rational explanation for great geomagnetic storms can now be found in transient blasts of solar energetic charged particles, typified by solar flare and coronal mass ejection events, while the smaller 27-day recurrent geomagnetic disturbances are linked to fixed streams of fast solar wind originating from coronal holes or Bartels's M regions.

In short, today's solar science, with the help of technology, is a vigorous and fruitful pursuit; one with plenty of promise for a deeper understanding of our Sun and related terrestrial effects. As E.W. Cliver aptly put it:

> sunspots, which provided the initial evidence for a solar-terrestrial connection, were eventually discarded in favor of flares and complemented by M regions. In turn, flares and M regions were displaced by CMEs and coronal holes[23]

and the writer admits this is an oversimplification.

Oversimplified, or not, it shows the evolution of thought, study, and proof from the days of de Mairan, Carrington, and the Maunders; Arrhenius, Birkeland, Alfvén, through Bartels, Chapman, Ferraro, Hale, Parker, and others (to name a few) to now.

---

[23] Ibid, Cliver, *Eos*, February 21, 1995.

13

Earth's Atmosphere and Its Story: A Perspective of Past Changes on the Present

We now turn from the Sun and magnetosphere to the lower atmosphere of the Earth. If we look into historical examples of early scientific or fairly-recent religious-philosophical approaches in understanding the natural world, there is often a strict, compositional notion of Earth's atmosphere. In one view of composition, there is the physical structure (Aristotle) and, for example, spiritual growth (Teilhard de Chardin) in the spiritual, metaphysical view. However, in both, there are implications of atmospheric specificity, limitation, and non-change.

For example, Aristotle listed four spheres beyond that of the of the Earth:[1] those of water, air, fire and in the outermost, the celestial sphere. To de Chardin, for example, there were spiritual spheres above the sublunary (that is, beneath the moon's) plane, such as the "Nöosphere," a physical part of the atmosphere wherein "spirit will gain rapidly in its sway against matter."[2]

We now take a leap from these ideas in order to adhere to Lord Kelvin's dictum: namely, that which you can measure, you may have a good or better idea about. Today, the atmosphere has been measured and studied far more than Aristotle could ever have imagined. For many geophysical scientists, Earth's atmosphere is similarly composed of layers. From the Earth's surface out, there first appears the troposphere (or 'a sphere of change' measured at 0–circa 15 km up). From circa 15 km to 50 km runs the stratosphere: a very mysterious place just forty years ago, which contains the changeable ozone layer. Above this to roughly 90 km, we find the mesosphere. From about 90 km to 5,000 km lies the ionosphere, which consists of various distinct plasma layers, in contrast to the layers beneath, the layers being lettered from D to F. And, expanding outward to face the Sun, the previously-much discussed magnetosphere is found.

---

[1] Ibid, King-Hele, D.G. (The Milne Lecture, 1984).

[2] See Gould, S.J., "Our Natural Place," from *Hen's Teeth and Horse's Toes* (W.W. Norton, 1983).

Winnowing in and out of these "spheres" are various "pauses" including the magnetopause illustrated in Figure 31 of Chapter 11. The density of the total atmosphere decreases with height because the air of our atmosphere is mostly a compressible fluid, and because of background gravitational stratification. Solar energy drives the motions of the atmosphere, and the Earth intercepts about half a billionth (0.5×10^{-9}) of the total energy emitted by the Sun.[3] The solar "constant,"[4] as it is called, is—notably—the solar radiant energy that is intercepted by the Earth's outer envelope (that is, outside the magnetosphere). The way the Earth orbits in space means changes occur seasonally by latitude as Earth receives radiant energy from the Sun.

We might reflect on the research of Birkeland, Arrhenius, Chapman, and many unnamed meteorologists and climate scientists when we consider the following points of understanding. Thirty percent of all total solar radiation hitting Earth is reflected back into space.[5] Twenty percent is reflected by clouds; about six percent reflected by the atmosphere, and around four percent reflected by the Earth's surface. At the ground, such albedo effects are manipulated by natural phenomena like vegetation, snow, and ice.[6] Under present-day conditions, clouds are the major reflector of incoming solar radiation since they have global coverage, while ground albedo (reflectivity) effects from snow, ice, and vegetation (rain forests and the like) are confined to limited regions of the Earth. Thus, the rest of the solar radiation (that is visible light or shortwave radiation) is made available for Earth's surface and atmospheric system. Longwave radiation (that is, infrared light) is then the converted mode of energy transfers within Earth's surface-atmosphere-ocean system.

How much of the Earth is heated by the Sun? About 51 percent of solar energy is converted into heat on the Earth's surface while almost 20 percent of solar energy gets absorbed by the atmosphere itself (energy that keeps for example

---

[3] "The radiation budget of the atmosphere," *The Cambridge Encyclopedia of Earth Sciences,* Ed. D.G. Smith (Cambridge University Press, 1981), p. 279.

[4] A clear misnomer because it is now known that the Sun's irradiance on Earth does vary significantly on interannual and decadal timescales.

[5] To appreciate these modern estimates (30%), contrast with the value of 89% reflected sunlight deduced by F.W. Very in 1912. Very was a pioneer in deriving such an important quantity by measuring the relative brightness of the Sun-lit and Earth-lit portion of the lunar disk as well as the neighboring sky (see Hunt, G.E. *et al.*, "A history of pre-satellite investigations of the Earth's radiation budget," *Reviews of Geophysics*, Vol. 24, 1986, pp. 351–356).

[6] Provocatively, other planets (such as Mars) also reflect radiation from their surfaces, outward, in similar manners (that is, albedo effects on Mars, such as ice cap reflectivity).

water vapor, clouds and ice afloat in the atmosphere). This means much shortwave radiation is made available for energy transfers in Earth's intimately related surface-atmosphere-ocean system. Longwave radiation emitted by the Earth is absorbed, re-emitted, and reabsorbed by compounds we are all familiar with: water vapor, CO_2, and material like cloud droplets. By contrast, nitrogen and oxygen are transparent to infrared radiation. In the troposphere—our breathing space—water vapor is the main absorber of longwave radiation. In the mesosphere, CO_2 is the big absorber/emitter, with some net loss of energy through longwave radiation into space.

Cloudless days or cloudless moments on Earth are mostly transparent to longwave radiation. If clear days are a "window," then this "window" can be blocked by clouds or gaseous and particle pollution. The blockage happens, regardless of whether the pollution is natural, like from volcanoes, or is human-made. Longwave radiation is emitted in all directions, then is reabsorbed, and a complex balancing act between short- and long-wave radiation occurs. This provides Earth a mechanism that, were it not in place, would rapidly see a cooling of the Earth's upper atmosphere and a warming of the Earth's surface or *vice versa*. The radiation exchange, in part, is responsible for transferring heat and potential heat around the Earth's entire surface.

You may wonder why radiation fluxes dominate our discussion on Earth's climate system energy budget while experience shows that simple flow or convection of air and water seem to be the dominant factors explaining weather and climate events. The short answer is that electromagnetic radiation is really the only way that will cause the Earth system to gain or lose significant amounts of net energy globally in short timeframes.

Our "living" sphere—the troposphere—contains water and water vapor entering the air through evaporation and other processes, evaporation drawing heat off of surfaces, and condensation releasing the latent heat. As noted before, wetness was a clue to Sun-Earth connections in tree rings as found by Douglass. Water vapor largely follows the "general" air circulation as dictated by differential solar heating and the rotating Earth. The water vapor is transferred pole-wards at latitudes higher than 20 degrees, and equator-wards at lower latitudes. Most of the recognizable warmth is released from condensation processes where northeasterly and southwesterly trade winds converge. Half of this amount of warmth transported within the troposphere originates between 0–10 degrees, north latitude.

Due to astronomical (Sun-Earth orbital configuration), geographical and other factors, atmospheric circulation is stronger in Earth's Southern Hemisphere than in Earth's Northern Hemisphere. So, these geographical features (for example, ocean-to-land surface ratios) affect air flow (or air current), south to north. These features affect water flow in a similar way, south to north, including the previously-discussed example of the deflection of the Gulf Stream by the equatorial ocean surface's counter-current off the coast of Brazil. Dynamically, the air—like water—carries imprints of temperature variations caused by any means in these streams, south to north, to seek energy and momentum balance.

However, due to the Earth's rotation, and due to the different thermal inertias of the world's oceans and its air, complexities arise in all these transfers. Warm air thus does not simply rise in the tropics, heading pole-wards to sink there, and then head equator-wards, again. For latitudes pole-wards of 30 degrees, the dominant carrier of energy comes in the form of transient eddies. The bottom line is that roughly two-thirds to one-half of the total net energy transported pole-wards is carried by the atmosphere (either in the form of the latent heat of condensation or the sensible heat by transient eddies) while the oceans account for the remaining transported energy.[7]

Were we not alternately constrained and given free reign by our elliptical planetary motion about the Sun, a different situation would prevail indeed.[8] At this point we can look upwards again. At heights of over 200 km (the thermosphere, on up) the atmosphere is mostly Sun-controlled. The atmosphere there consists of atomic oxygen (rather than molecules of oxygen). Higher than 500 km, it is mainly helium. The temperature changes at these high altitudes are large, often in tens or hundreds of degrees. The highest daily temperatures are experienced at about 1500 Universal Mean Time (UMT); the lowest, at around 0300 UMT up there.[9] But a lack of extreme ultraviolet (EUV) radiation from the Sun, especially during the lows of the 11-year Schwabe sunspot cycle, can be responsible for very cold nighttime temperatures.

In 1686, Edmund Halley first demonstrated how air pressure decreased with height. Temperature controls the rate at which pressure and density change. Temperatures in the stratosphere rise when ultraviolet radiation is absorbed by the Earth's ozone. In these regions between the stratosphere and thermosphere, nitrogen fills the atmosphere, which is an element important for creating the earlier-mentioned Carbon 14 (^{14}C) isotope. At times however, density varies greatly in

---

[7] See for example, Peixoto, J.P., Oort, A.H., in *Physics of Climate* (American Institute of Physics, 1992), pp. 342–347.

[8] Most of the data in these last few paragraphs were taken from *The Cambridge Encyclopedia of Earth Sciences* (Cambridge University Press, 1981), pp. 279–290.

[9] Ibid, King-Hele, D.G. (Milne Lecture, 1984), p. 241.

Earth's atmosphere—and so does the temperature—linked as it is, sympathetically, to density changes. This consideration of chemical composition changes and heat exchanges is especially relevant for the study of the Sun's impacts on our atmosphere.

Solar magnetic storms can disrupt the Earth's electrically charged upper air and its magnetic properties, as shown earlier. The density can also double in certain areas,[10] such as in the ionospheric sub-layers. Solar disturbances to mundane things like telephone or cellular phone calls and radio transmissions—and to what Annie Maunder described in her book about sea cable communications[11] or what Balfour Stewart mentioned on the disturbance telegraph lines underwent—are thus all seen in a clearer light.

Density variations also occur in the atmosphere in semi-annual patterns. It is no surprise that most hard-to-explain variations in density, temperature, etc., occur in the *upper* atmosphere.

As shown in Figure 31 of Chapter 11, the magnetosphere is the first layer to come into direct contact with the Sun's awesome power, being belted with a cosmic particle spray. The solar wind plasma, some from the eruption of flares and sudden magnetic bursts, conspires to warp, bend, and ultimately penetrate the Earth's magnetic field that is sheathing us. The solar wind-charged particles winnowing their way in through our polar skies and magnetosphere's tail are then able to transfer their energy and momentum. It is this which gives us curtains of aurora in gorgeous greens, yellows, and reds from energy given off by, for instance, the excited nitrogen and oxygen atoms of our air.

"The solar wind plasma near the earth," as Minze Stuiver and Paul D. Quay wrote, "can be considered an extension of the solar corona, and changes in the solar wind properties reflect coronal changes."[12] The corona, as we remember, is

---

[10] Ibid, King-Hele, p. 241.

[11] "[A]t six o'clock in the afternoon of October 31 [1903], a [geo-]magnetic storm burst suddenly, the most violent that has been experienced in the memory of man; so violent that it is disturbed the submarine cables all over the world, and stopped the sending of any telegraphic messages." (p. 188 of *The Heavens and Their Story*). Also, Stewart, B., *MNRS,* 1861.

[12] Stuiver, M., Quay, P.D., "Changes in Atmospheric Carbon-14 Attributed to a Variable Sun," *Science*, Vol. 207, January 4, 1980, p. 11. This quote is essentially a restatement of the remarkable fact first uncovered by S. Chapman (*Proc. R. Soc. London*, Vol. 253 A, 1959, p. 462) that the high thermal conductivity and the hot, million degree temperature of the solar corona must mean that the corona extends far out into interplanetary space (in his model, the electron density and temperature near the orbit of Earth are about 100–1000 cm^{-3} and 200,000 K, respectively) [see for example, Parker, E.N. *Astrophysical Journal*, Vol. 525, 1999, p. 792].

"the vital sign" of an active and dynamic Sun. The magnetic changes in the solar plasma surrounding the nearby planets and interplanetary space actually deflect and modulate the galactic cosmic rays[13] heading for Earth. So the cosmic ray flux arriving at the uppermost part of Earth's atmosphere will actually vary with changes in solar activity. Evidence of this is shown by the connection between the incoming cosmic-ray flux with the actual timing and phase of the solar magnetic polarity (field direction) that vary in 22-year cycles or double the 11-year Schwabe sunspot cycle (which is also called the Hale Cycle). When sunspot activity is low, such as it was in the Maunder Minimum, magnetic shielding is minimal, such that larger cosmic ray flux enters the top-most part of the atmosphere. When sunspot activity is higher—as it is today—these fluxes are lower. Earth's very own magnetic field, with a power of its own, can also modulate the incoming cosmic ray flux and with it, ^{14}C production.

The variations in cosmic ray flux affect the production of Earth's atmospheric neutrons. As ^{14}C is produced from the transmutation of atmospheric nitrogen by adding one neutron, the $^{14}CO_2$ levels in the atmosphere can thus reflect, in turn, changes in the Sun's behavior through its close link to cosmic rays. The record of this atmospheric ^{14}C activity—as noted in tree rings—can give a "printout," as it were, of ancient solar fluctuation recorded on Earth. Yet, for all this, variation of ^{14}C is not due to solar action alone. And ^{14}C production relative to solar activity is not constant with time. Millennial change of geomagnetic field strength is an important factor, as shown by the smooth curve in Figure 5 (of Chapter 2).

To consider the complicated nature of ^{14}C as a radioisotopic time-measuring stick with a half-life decay-time of 5,730 years, Stuiver noted that the amount of ^{14}C in the atmosphere depends not only on the neutrons produced in the upper atmosphere, but also the exchanges of carbon within Earth's reservoirs. These reservoirs are repositories like oceans, the entire biosphere, etc.

Using one model[14] describing long- and short-term carbon and ^{14}C variation, it was found that ^{14}C can reside in the biosphere for about 60 years[15] and stay in the

---

[13] Galactic cosmic rays is a term for the relativistically-moving charged particles coming in from all directions through interstellar space. These cosmic rays are generated and sustained by supernova explosions of dying massive stars. The rate of supernova occurrence for a typical galaxy like ours is about once or twice per century. (Of course, the rate of supernova explosions within 10 parsec from the Sun is much less frequent, roughly once per 10–100 million years.) The constituents of the cosmic rays are about 90% protons, 9% alpha particles (helium nuclei), and 1% heavier particles.

[14] H. Oeschger and colleagues' four-reservoir, box diffusion model.

[15] Ibid, Stuiver and Quay, *Science*, January 1, 1980, p. 15.

deep ocean for as long as one thousand years,[16] thus introducing further uncertainty in the accounting of the ^{14}C budget. Furthermore, relative comparison of modern (circa 1900 to present) levels of ^{14}C to those in the past is now very much complicated by the strong dilution due to releases of relatively carbon-14-free CO_2 from buried hydrocarbons. This human-made ^{14}C dilution factor is known as the "Suess Effect" (see Figure 5 of Chapter 2).

As Figure 31 in Chapter 11 partly illustrates, solar wind and Earth magnetospheric interactions result in electric currents in the ionosphere and magnetosphere (due to motion of charged particles) producing measurable actions on the ground. What comes to a confluence here are the yearly, decadal and century-long changes of solar magnetism and the production of ^{14}C (see the "wiggles" in Figure 5). The relationship between these aspects shows that, during the Maunder Minimum,[17] there was a weakened solar wind-enhancing level of cosmic ray flux, so ^{14}C level was *higher* during this weak solar activity phase. The Maunder Minimum may also have witnessed an unusual incidence of highly weakened interplanetary fields normally supplied by the solar wind, and, having little interaction with the Earth's

Fig. 39 In an ancient manuscript, the Sun is shown ruling Earth's spheres. (After G. Reisch, *Margarita Philosophica Nova*, 1512.) (From D.G. King-Hele, 1975.)

[16] Stuiver, M., Braziunas, T.F., "Sun, ocean, climate and atmospheric $^{14}CO_2$: An evaluation of causal and spectral relationships," *The Holocene*, Vol. 3, 1993, pp. 289–305.

[17] As well as the Spörer Minimum of roughly 1420–1530, and the putative Wolf Minimum of about 1280–1350.

magnetosphere and upper atmosphere, a very much-reduced level of auroral activity occurred, hence the overall "profound magnetic calm" noted by Agnes M. Clerke and pondered by Walter Maunder.

To what degree all these changes played a role in the weather and climate at this time is still under very active investigation. That it may have helped make it colder during the Maunder Minimum is a distinct possibility.

So is the fact that a "Maunder Minimum" could happen again.

* * *

How could these galactic and solar effects actually be tied to the Earth's weather? What are the possible mechanisms for the connection? The specific roles of the Sun come in many forms: both in its radiation and charged particle inputs, as well as its effect on those incoming cosmic rays. However, defining such "forcing" inputs to the Earth's multi-layered spheres is just half of the story. Indeed, the hard questions are tied to the ways in which those Sun-related forcings get played out in the Earth's ocean and lower atmosphere.

First, let us start with the Coriolis effect,[18] in regard to moving air. The deflection effect of this force for moving air parcels on Earth is to the *right* for the

---

[18] This is due to the way the Earth turns, plus a "force arising from the Earth's rotation, deflecting Earth's air to the right in the Northern Hemisphere, and to the left in the Southern" (see (a))—that is, the Coriolis effect. Foucault showed this in an experiment in 1851 (see (b)) after George Hadley's famous arguments based on the conservation of angular momentum (see (c)) and William Ferrel, and others by 1857 (see (d)), had elaborated the general circulation of the atmosphere, but did not account for zonal asymmetry. This is sorted out only much later by Edward Lorenz, Villem Bjerknes, and still later, by Richard Lindzen and colleagues.

Further historical notes:

(a) Self-taught American teacher William Ferrel, developed this idea mathematically in 1858, off Foucault's experiment of 1851 (according to James Burke, "Lend me your ear," *Scientific American,* Vol. 280, No. 3, March, 1999, pp. 93–94).

(b) Léon Foucault suspended a cannonball on 220 feet of piano wire in the French Panthéon in 1851, pulled it to one side with a cord, then burned the cord to release the ball without effecting its motion. As the pendulum thus formed by ball and wire moved in inertial space, a stylus attached to the ball's bottom traced a line that shifted as the Earth turned beneath it. Outside of proving Copernicus, in part, this discovery continued forming the basis for determining how global weather was driven. (Read "And now the weather," by James Burke, in *Scientific American*, Vol. 280, January, 1999, p. 91.)

(c) Edmund Halley attempted a low latitude wind system study in 1686. But George Hadley used words to conceptually describe this motion of general circulation in 1735. Elegantly put, it was thus: "... it follows that the air, as it moves from the tropics towards the equator, having a less velocity than the parts of the Earth it arrives at, will have a relative motion contrary to that of the diurnal [that is, daytime] motion of the Earth in those parts, which being combined with the

Northern Hemisphere, and to the *left* in the Southern Hemisphere. A more playful illustration of this rotational effect would be to consider the deflection of a baseball thrown with a speed of 25 m/s, or about 55 miles per hour, to curve at about 1.5 centimeters to the *right* in the United States—a Northern Hemisphere country.[19] Or, more serious is the World War I tale from battle at the Falkland Islands off Argentina about the British ship gun shells consistently landing approximately 100 yards—some one hundred meters—to the *left* of German ships, despite having already anticipated for the effects of the Earth's rotation.[20] Apparently, the British gunners had done their Coriolis adjustment for 50 degrees *north*, rather than 50 degrees *south*, so that they had mis-aligned their targets by twice the actual distance towards the right, for the target is expected to drift to the *left* in the Southern Hemisphere. Useful to keep in mind for understanding the Coriolis effect in the context of *climate*, is the fact that the parcels of air and water move much more slowly and traverse far greater distances than baseballs or ships' gun shells would. The factor of rotation is, then, much more important in the consideration of climate.

So there is a tendency for air parcels to turn in opposite directions in either hemisphere as the Earth rotates. Besides the general circulation forced by rotation, the weather and climate system are also driven by waves and eddies generated by various forms of inertia, thermal, gravitational, and convective instabilities of the air and sea. As a rule, air is moved or shifted from high pressure zones to low

---

(*continued*)

motion toward the equator, a NE wind will be produced on this side of the equator and a SE, on the other." From Lewis, John, "Clarifying the Dynamics of the General Circulation: Phillips's 1956 Experiment" (*Bulletin of the American Meteorological Society,* Vol. 79, 1 January, 1998, pp. 39–40). For an in-depth discussion of the role of the Coriolis force in atmospheric circulation that arose from the balance of the rotational kinetic energy versus the more traditionally assumed balance of angular momentum, see Anders Persson, *Bulletin of the American Meteorological Society,* Vol. 79, July, 1998, pp. 1373–1385.

(d) For a more mathematical appreciation of the early models of global circulation by first Hadley, and then Thomson and Ferrel, with more modern mathematical models of Bjerknes, Lindzen, E. Schneider, *et al.*, see *Dynamics in Atmospheric Physics* (Cambridge University Press, 1990) by Richard S. Lindzen. A very good view of the picture up to the present (less mathematical; more historical and conceptual) would be John M. Lewis's "Clarifying the Dynamics of the General Circulation: Phillips's 1956 Experiment" (*Bulletin of the American Meteorological Society,* Vol. 79, 1998).

[19] From Persson, A., "How do we understand the Coriolis force?" *Bulletin of the American Meteorological Society*, Vol. 79, July, 1998, pp. 1373–1385.

[20] From Parker, B., "The Coriolis effect: Motion on a rotating planet," *Mariners Weather Log*, Vol. 42, No. 2, August, 1998, pp. 17–23.

pressure ones; that is, due to pressure gradients. The balance between pressure gradients causes global wind, or the so-called "general circulation." A balanced wind that blows along curving isobars is called the gradient wind; if this blows around *low pressure zones*, it is called *cyclonic* (counterclockwise in the Northern Hemisphere, clockwise in the Southern Hemisphere). If the gradient wind blows around a *high pressure zone*, it is called *anticyclonic* (opposite of the Northern-Southern Hemisphere case, just noted).

It is most relevant to recall that, occasionally, the persistent west-east zonal flow due to Earth's rotation can be disrupted by *stronger and more persistent wind flows running north-south.* These north-south flows interrupt the west-east weather flow in the more-land-dominated Northern Hemisphere more easily at times—often causing widely fluctuating weather effects: for instance the extreme year 1666, as recorded in the Maunder Minimum, London, England. That is, blocked weather patterns can help cause extreme cold in locations allowing extreme dryness or even extreme warmth that is uncomfortably unrelenting, forcing droughts and egging on fires (in the 1666 London case).

All such large-scale changes in weather patterns or long-term systematic deviation from the climate's mean pattern could result from either changes in external boundary conditions, such as the Sun's irradiance or volcanic eruptions, or could be due to an internal, non-linear manifestation of atmosphere-ocean circulation dynamics—or both.

Isotopic Recorders of the Sun and Climate: Using the Maunder Minimum To Study Implications for Deep and Deeper Time

Carbon 14

The silent searching of A.E. Douglass is once again recalled, seeing his remarkable connection between the sizes of trees, regarding their topographical locations and the constraints forced upon them by their own biological limitations. What we recall is the wetness that reached those trees (like those growing at higher altitudes under stressed conditions) controlling the size of rings (fat, wet, warm or thin, dry, cold).

Yet, Douglass of course had no idea at that time that isotopes existed: let alone that they "fractionated" during photosynthesis. To quote solar astronomer John Eddy:

> Trees [regarding how much radiocarbon was in the atmosphere at a particular time, such as the Maunder Minimum] keep that record for us, for

> atmospheric radiocarbon (as carbon dioxide) enters their leaves in photosynthesis and is preserved, as cellulose, in new growth wood. Growth rings in many temperate-latitude trees, as Douglass demonstrated, can be identified as an annual diary that extends, in the long-lived species such as the bristlecone pine, for many thousands of years.[21]

Hans E. Suess and Timothy Linick had also noted this effect. Namely, "that ^{14}C in the cellulose present in wood in a given annual tree ring corresponds remarkably well to that in the CO_2 of the atmosphere at the time of the growth of the ring, if one corrects for isotope fractionation during photosynthesis."[22]

We have, here, the isotopic link that Douglass long sought after. It was some time after Douglass's death that Hans Suess, for one, noted and stressed that the century-wide ^{14}C variations in CO_2 relate to solar activity variation. CO_2 is, as Suess noted, a "counting gas," as is acetylene.[23] Counting gases can be used to count radioactive decay, acetylene being the better counter in this case. The isotope of carbon, ^{14}C, was named "radiocarbon" by Willard F. Libby and it was Libby who actually developed the radiocarbon method of dating.[24] As such, Suess was able to correctly date continental glaciation in North America to 20,000 years ago Before Present (BP) using this counting technique.

The story moves from the mid-1940s to the early 1950s, when Libby, part of that "can-do" crowd involved in the Manhattan Project[25] actually found that ^{14}C was a product of cosmic-ray bombardment. One of the goals at the time was to see if ^{14}C was ubiquitous in living things. This took the measurement work over to—of all places—the University of Arizona, where a colleague of Douglass's, Edmund Shulman, kept bristlecone pine samples. There were also sequoia

---

[21] Eddy, J.A., "Historical and Arboreal Evidence for a Changing Sun" in *The New Solar Physics* (AAAS Selected Symposia, 1978), p. 16.

[22] Suess, H.E., Linick, T.W., "The ^{14}C record in bristlecone pine wood of the past 8000 years based on the dendrochronology of the late C.W. Ferguson," *Phil. Trans. R-Soc-Lond.*, Vol. A330, 1990, p. 404.

[23] A colorless but poisonous and highly flammable gaseous hydrocarbon ($HC \equiv CH$) used in organic compound synthesis. See Suess, H.E. and Linick T.W. (1990), p. 403.

[24] For which he was awarded the Nobel Prize in Chemistry in 1960. For the whole history, see Taylor, R.E., Long, A., Kra, R.S. (eds.) *Radiocarbon After Four Decades* (Springer-Verlag, 1992).

[25] Harold Urey was one of the chosen few to lead the effort to separate ^{235}U from the element, uranium. Willard Libby was one of the "two people he hired, in what became a huge, industrial effort" (ibid, Taylor *et al.*, p. 4).

samples reliably dated to 2,000 or so years that were available. Shulman offered his database to the researchers. (The now-elderly A.E. Douglass had left this and had turned his attentions to other astronomical and climate research.[26]) In any case, time variations in the atmospheric inventory of ^{14}C in tree contents was first demonstrated by Hessel De Vries in 1958—and is called the De Vries Effect.[27]

The demonstration was to see if the radiocarbon-dating would, by itself, confirm the already-known dates. But the calculated values deviated too much from the measured ^{14}C ones. Hard questions—some originally posed, as we saw, by Douglass, in part—had to be asked again. Namely:

- Were the deduced tree ring ages (in the samples) correct?
- Do wood samples from different types of trees grown at different geographic locations and having the same tree ring age show the same ^{14}C content?
- Can ^{14}C in the wood samples change through contamination, irradiation, etc?

These were—with reservations and exceptions—"reliably answered," Suess stated. "Fractionation" of isotopes in the photosynthesis process was seen: corrupting values of tree rings being, for example, sap and lignin in the sapwood in older rings.

But, just as true, it was found that cellulose in wood grown at the same time is practically independent of geographic location and altitude, with a few mille[28] (or, a few part per thousand) difference in ^{14}C collected from either the Southern or Northern Hemisphere.

For all this and its use in archaeology ^{14}C dating cannot be expected to be more accurate than the century mark. Also, ^{14}C remains in the atmosphere (as previously noted) longer than expected and will *not* settle immediately downward after its initial manufacture in the atmosphere for several decades. Thus, measured ^{14}C values reflect steady-state concentrations, which depend on cosmic ray production and the rate which ^{14}C finds equilibrium in the terrestrial land biosphere and oceans.

---

[26] Ibid, Webb, pp. 186–187.

[27] De Vries, H., *Koninkl. Ned. Akad. Wetenschap, Proc. Ser.*, Vol. B61, No. 1, 1958, pp. 94–102.

[28] The specific variable that occupies a central role in measurements of radiocarbon is the delta ^{14}C defined by $\Delta^{14}C = (^{14}C_{ref} - ^{14}C_{inv}) / ^{14}C_{inv}$, where $^{14}C_{ref}$ is the chronological reference activity at time, *t*, determined from tree ring counts and $^{14}C_{inv}$ is the true measured activity. This difference is normally then multiplied by a thousand to express the result in per-mille or by a hundred to give it in percent (Sonett and Finney, "The spectrum of radiocarbon," *Phil. Trans. R. Soc. Lond.*, Vol. 330, 1990, pp. 413–426).

So, other than for deviations from the trend of ^{14}C values of dates calculated from tree rings (called "wriggles"), century-to-century variations of ^{14}C in atmospheric carbon dioxide are related to the variations in solar activity. In addition, the variation of ^{14}C is also tied in the overall longer-term trend of changes in the Earth's magnetic field intensity, as previously discussed.[29] When the effect of Earth's magnetic shielding is removed, the resulting spectrum shows cyclical variations on century and millennial scales. It is held that the Sun is a major cause of this variation.[30]

Beryllium 10

Another isotope that can be measured in the firmament is Beryllium-10 (^{10}Be). ^{10}Be, discovered in the 1950s, is formed mainly in the atmosphere by "spallation" of oxygen and nitrogen induced by secondary neutrons. (Spallation is a kind of nuclear reaction wherein several lighter fragments are emitted from an atomic nucleus after being struck by highly energetic cosmic ray particles [mainly protons and bare-helium nuclei].[31,32]) After production in the atmosphere, ^{10}Be attaches itself to either solid or liquid aerosol particles and water vapor there, following the air masses in which it was formed.[33] By comparison, ^{14}C oxidizes to CO_2 as $^{14}CO_2$, and does a complex transforming dance between reservoirs in the atmosphere, biosphere, and oceans—sometimes for long periods before settling. ^{10}Be, on the other hand, does not suffer from this trouble of delays between the time of production and time of being recorded in its terrestrial reservoirs.

Once it is made ^{10}Be reaches the Earth's ground quickly owing to precipitation. The residence time of ^{10}Be in the atmosphere is only one to two years. Literally, like the nuclear explosion wastes that have been stuffed into our atmosphere since

---

[29] That is, tied to the magnetic field's dipole moment. Damon, P.E. "The Natural Carbon Cycle," in *Radiocarbon After Four Decades*, 1992, p. 18.

[30] Damon (1992) notes Charles Sonett.

[31] McHargue, L.R., Damon, P.E. "The Global Beryllium 10 Cycle," *Reviews of Geophysics*, Vol. 29, May, 1991, p. 141.

[32] To readers for whom atomic science is an esoteric mystery, a non-technical, biographical approach to the modern understanding of this science can be read in Richard Rhodes's Pulitzer-prize winning book, *The Making of the Atomic Bomb* (Touchstone, 1988).

[33] Beer, J., Raisbeck, G.M., Yiou, F., "Time Variations of ^{10}Be and Solar Activity," from *The Sun in Time* (eds. C.P. Sonett *et al.*, University of Arizona Press, 1991), p. 345.

the 1940s, ^{10}Be "falls out" and makes, in some cases, direct bee-lines for soil, and, significantly, gets locked in ice. In oceans, it is often in a solution form, and settles into seabed sediments that have made it useful as a tracer in studying island-arc volcanism.[34]

However, for arguments pertinent to this book, ancient ice cores can thus also carry a Sun-activity imprint.

Although sensitive to other perturbations, the key with ^{10}Be is its close connection to atmospheric neutrons. Why does the Sun leave a record of its past activity in these isotopes? This is a complex piece of detective work. By comparing neutron flux at the Earth's surface—like on the mountain tops of the world—against the sunspot number, as a proxy of solar wind plasma conditions in the interplanetary space, year-by-year, one can show a close negative correlation between the two variables. Since ^{10}Be made naturally in the atmosphere ("cosmogenically" versus "anthropogenically"—since people can now split atoms, too) is proportional to the local neutron flux there, it shows an inverse correlation with ongoing solar activity with as little delay as the 1–2 years ^{10}Be residence time in the atmosphere. Once again, contrast this with the ^{14}C story.[35] The main source of short-term ^{10}Be variation is thus interpreted as a result of solar wind interacting with cosmic ray flux.

Seen in this measurable way, ^{10}Be is found in higher concentrations in ice cores when the Sun underwent lower activity—similar to ^{14}C. As such, in ancient ice, we can see when the Sun was more active or not, simply by measuring how much ^{10}Be is in it. In the Maunder Minimum, there was—predictably—quite a bit.

Polar ice is the best for high resolution studies, as it directly samples the atmospheric fallout of ^{10}Be-water vapor-laden aerosol particles and keeps it frozen.[36] It is a lot more stable than ice at lower latitudes, prone as it is to shifting and re-melting processes, which shuffle and mix around the neat deck of annual, time-layered ice cards. With its comparatively regular 11-year Schwabe sunspot Cycle, the Sun, as Maunder well noted, goes through sunspot minima and maxima. The traditional task of quantifying the strength of solar activity by counting sunspot number faces a dead end when there are few sunspots like during the Maunder Minimum interval. In this respect, modern isotopic tracers like ^{10}Be can help trace the historical solar activity with more certainty (see Figure 40).

---

[34] Ibid, McHargue and Damon, p. 141.

[35] Ibid, Beer *et al.* p. 343.

[36] Ibid, Beer *et al.* p. 350.

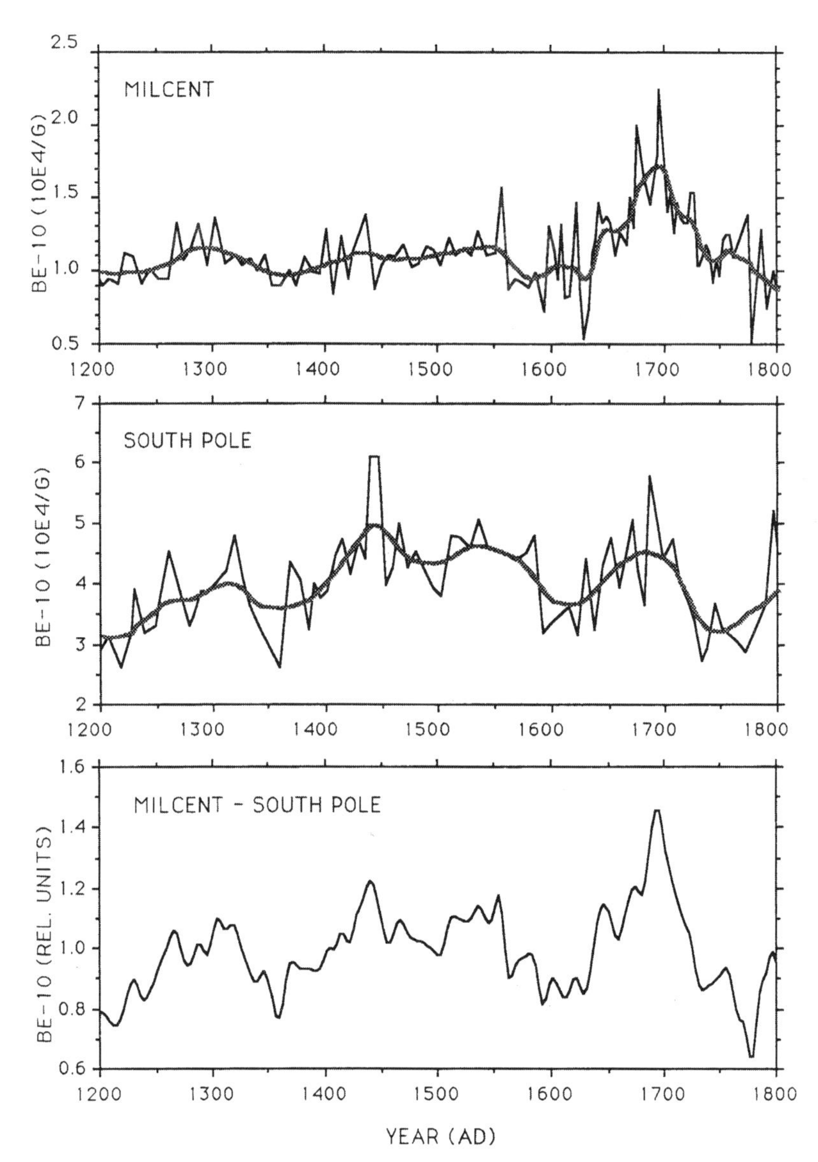

Fig. 40 Two polar samples from the North (Milcent, Greenland) and South Pole, showing enhanced ^{10}Be concentrations in ice over three solar minima (Wolf, c. 1280–1350, Spörer, c. 1420–1530, and Maunder, c. 1645–1715). Their composite result (constructed to reduce meteorological noise) is shown as the bottom panel. (Note that dating between peaks could be as uncertain as seventy years owing to uncertain ice-layer stratigraphy.) (After Beer *et al.*, 1991.)

Thus, two powerful terrestrial isotopes can serve as independent, and perhaps complementary, records of solar activity extending further back in time than the sunspot record can cover. The key conclusion drawn from these isotopic studies amassed over the last 50 years appears to be that there is large-amplitude solar-terrestrial variability. That variability operates not only on the scale of the 11-year Schwabe sunspot cycle, but also the 80–100-year one as seen in both the ^{10}Be and sunspot records. Significant changes are also noted in the forms of 2,200–2,400-year and 210-year cycles in the residual ^{14}C record, after removing the effects from the Earth's changing magnetic field strength.

Once again, even if there are large-amplitude *intrinsic changes* in the Sun's radiative and corpuscular (charged particles) emissions, the missing half of the puzzle is still about the finding of any climatic changes that may correspond to those external solar forcings.

* * *

What then can paleoclimatic study tell us? In terms of variations over millennium timescales, there is, indeed, some positive news. Large climate cycles of 1,470 years ± 500 years have now been seen in the Greenland ice core records and marine sediment records of the North Atlantic (Figure 41).

The ice core research uses the relative abundance of the stable oxygen isotope oxygen-18 (^{18}O) over oxygen-16 (^{16}O) in snow/ice to reveal large and rapid temperature fluctuations[37] over Greenland, with possible implications and connections to world-wide changes.[38] In addition, the marine sediment work

[37] The idea here is that natural water contains a fixed fraction of isotopically heavy molecules with oxygen-18 (that is a regular oxygen atom with two extra neutrons). Water molecules with oxygen-18, except for being 12% heavier, are essentially the same as the common water molecules with oxygen-16. The main difference is that heavier water molecule condenses more easily and does not evaporate as readily as the light water molecule.

When the air is warmer, the heavier water vapors containing oxygen-18 more readily condense out to form the snow that falls in a land-based ice-core season's layer than when it is colder. Thus, ice core layers with an enriched ratio of oxygen-18 to oxygen-16 indicate times of warmer air temperature and *vice versa.*

[38] Broecker, W.S., "Massive iceberg discharges as triggers for global climate change," *Nature,* Vol. 372, 1994, pp. 421–424; Mayewski, P. *et al.*, "Major features and forcing of high-latitude northern hemisphere atmospheric circulation using a 110,000-year-long glaciochemical series," *Journal of Geophysical Research*, Vol. 102, 1997, pp. 26345–26366; Steig, E.J. *et al.*, "Synchronous climate changes in Antarctica and the North Atlantic," *Science*, Vol. 282, 1998, pp. 92–95.

measures change in the amount of small rock grains[39] that settle into the sea bed of the North Atlantic. The rock grains provide estimates for the surface ocean circulation and surface ocean temperature associated with various episodes of ice-rafting events. Also to be noted is that those rock grains are either too large or too angular to be interpreted as materials from river run-off.

Additional supporting evidence for the climatic oscillation could be found from studies of the intensity and pattern of polar atmospheric circulation,[40] the atmospheric concentration of methane trapped in ice core air bubbles,[41] and the population of surface-dwelling, planktonic foraminifera.[42] During the last deep glacial period, the millennial-scale climate oscillation was characterized by a recurring rapid Earth warming, with a change of 5 to 8 °C of air temperature over Greenland in just decades to centuries. Then, a more gradual move back to cold conditions again after the brief warming occurred. In fact, such a climate cycle has been known to be operating consistently between 11,000 (or 11 kiloyears [kyr]) and 140 kyr BP. The cold portion of the cycle is thought to be connected to major discharges of icebergs into the North Atlantic from the unstable Laurentide continental ice sheet, or Barents ice shelf, or both of these. The drifting icebergs, furthermore, are thought to have potential impacts on the North Atlantic thermohaline circulation because they are a large supplier of freshwater, which significantly lowers the surface ocean salinity. Simultaneously, there is evidence for massive reorganisations of the atmospheric circulation. This makes sense, because as previously mentioned, ocean and atmosphere motions are closely related.

The research on the rock grains in North Atlantic sediments by Gerard Bond and colleagues have been able to identify how the intriguing millennial cycle operates not only for the glacial period between 11 kyr and 140 kyr BP. The cycle also seems to be persisting through the current, relatively warm and stable Holocene phase (that is, over the last 10 kyr) that we are living in now (see Figure 41).

---

[39] Namely, lithic glass, fresh volcanic glass and hematite-stain grains (see Bond *et al.,* 1997).

[40] Mayewski, P. *et al.* (1997).

[41] Brook, E.J., Sowers, T., Orchardo, J., "Rapid variations in atmospheric methane Concentration during the past 110,000 years," *Science*, Vol. 273, 1996, pp. 1087–1091.

[42] For example, Oppo, D.W., McManus, J.F., Cullen, J.L., "Abrupt climate events 500,000 to 340,000 years ago: Evidence from sub-polar North Atlantic sediments," *Science*, Vol. 279, 1998, pp. 1335–1338.

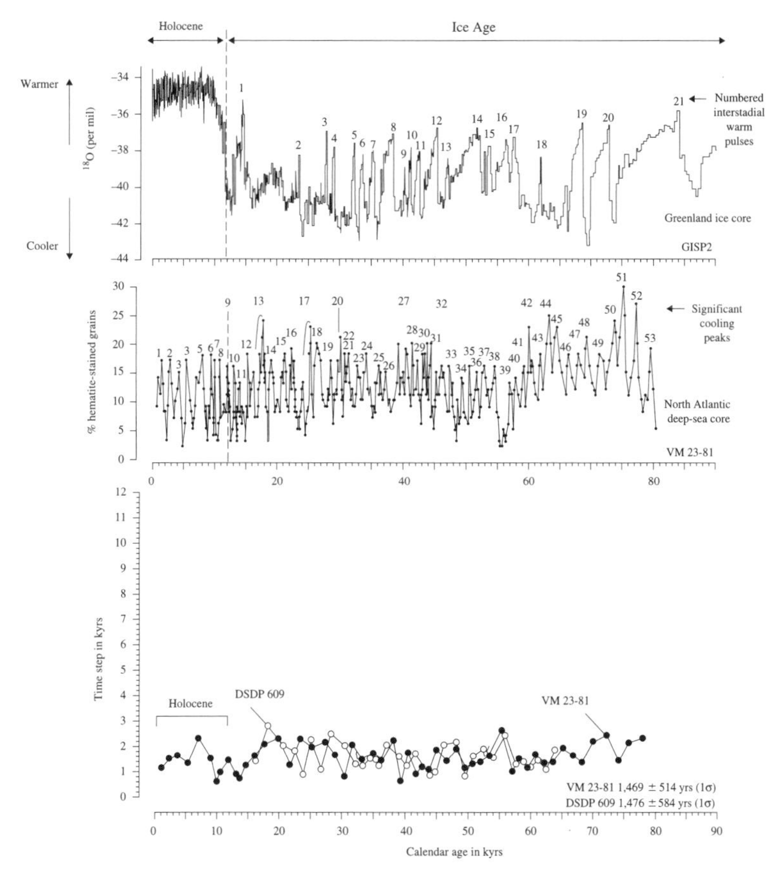

Fig. 41 Millennial climate oscillations as recorded in Greenland ice cores and North Atlantic deep sea cores. The top panel illustrates the recurring signatures of warm-cool events persisted over the last 90 kyr. Amplitudes of the millennial-scale changes during the relatively warm and stable Holocene are much smaller (but are detected by the presence of small rock grains in sediment cores shown in the middle panel) than equivalent events during the glacial period (such events are thought to be connected to ice-rafting episodes in the North Atlantic ocean). The estimated temperature change indicated by the ice core $\delta\ ^{18}O$ record is about 10–20 °C locally for Greenland during extreme excursion, e.g. between the Last Glacial Maximum near 20–25 kyr BP and the present. The middle panel shows the remarkable signature of the millennial-scale climate oscillation detected through the amount of the hematite-stain grains contained in the North Atlantic deep sea sediment core. The bottom panel illustrates the pace of each climate event and confirms the fact that there is a pervasive millennial-scale climate cycle operating in the Holocene (actual results from two independent North Atlantic deep sea sediment cores are shown). The oscillation time scales between the Holocene and the glacial interval are indistinguishable and have a mean time of 1470 years ± 500 years (the relative pacing of events using the $\delta\ ^{18}O$ indicator is more irregular so they are not indicated in the bottom panel). More importantly, G. Bond *et al.* notes that the Little Ice Age (ca. 1550–1850 or cooling peak number 0 in the middle panel) could be interpreted as the most recent cold phase of this series of millennial-scale cycles known to have persisted at least over the last 140,000 years. (Adapted from G. Bond *et al.*, 1997, 1999.)[43]

Bond *et al.*,[44] concluded that: "The millennial-scale variability in our records reflects the presence of a pervasive, at least quasi-periodic, climate cycle occurring independently of the glacial-interglacial climate state." Furthermore, "[s]urface winds and surface ocean hydrography in the subpolar North Atlantic appear to have been influenced by variations in solar output through the entire Holocene."[45] In other words there are large and abrupt 1,000–2,000-year changes in the climatic record that may be forced by *intrinsic changes* on the Sun and that is *independent* of the forced oscillations imposed by the geometrical Sun-Earth orbital changes on longer timescales.

If the chronology of Bond and colleagues was precise enough, then the main phase of the cold, unsettled weather dynamic of about (with modifications to its dating) 1300–1900 A.D. known previously as the "Little Ice Age," or "LIA," is about 1,100 years from the previous cold phase of the millennial cycle. A continuing effort by Bond and colleagues in 1999 was able to confirm this earlier (1995–1997) suspicion. In fact, a conclusion was reached that "[i]ndependent of any anthropogenic forcing, the North Atlantic's climate eventually will shift (or in fact may be shifting now) toward the warm phase of the cycle."[46] This new and important perspective clearly points to the "LIA" as not being an isolated event. Instead, it tells the story of something like the unevenly cold and wet events of what was known as "Little Ice Ages" to be part of a natural cycle that is known to have been operating as far back as 140,000 years ago. Caution, naturally, must be exercised here, because the detailed factors for both internal climate and external forcing mechanisms involved in this 1,000–2,000 years' oscillation are still poorly understood.

---

[43] Bond, G. et al., "A pervasive millennial-scale cycle in North Atlantic Holocene and glacial climates," Science, Vol. 278, 1997, pp. 1257–1266 and Bond, G. et al., "The North Atlantic's 1–2 kyr climate rhythm: Relation to Heinrich events, Dansgaard/Oeschger cycles and the Little Ice Age," in *Mechanisms of Global Climate Change at Millennial Time Scales* (American Geophysical Union: Washington DC), eds. R.S. Webb, P.U. Clark and L.D. Keigwin, 1999, pp. 35–58.

[44] Bond, G. *et al.*, "A pervasive millennial-scale cycle in North Atlantic Holocene and glacial climates," *Science*, Vol. 278, 1997, p. 1263.

[45] Bond, G. *et al.*, "Persistent solar influence on North Atlantic climate during the Holocene," *Science*, Vol. 294, December 7, 2001, 2130–2136.

[46] Bond, G. *et al.*, "The North Atlantic's 1–2 kyr climate rhythm: Relation to Heinrich events, Dansgaard/Oeschger cycles and the Little Ice Age," in *Mechanisms of Global Climate Change at Millennial Time Scales* (American Geophysical Union: Washington DC), eds. R.S. Webb, P.U. Clark and L.D. Keigwin, 1999, p. 55.

The "Little Ice Age" and the Maunder Minimum Reassessed

How the Spörer, Maunder or other earlier minima (say, the Wolf Minimum) could figure into the extended period of non-linear cold from roughly the fourteenth through nineteenth centuries, as some experts now place it,[47] is still a difficult puzzle. Besides the intrinsic change in the Sun outputs from varying solar magnetism, the effects from both the latitudinal distribution of the seasonal pattern of sunlight and its gradual changes—related to variations in the precession of the equinoxes along the Earth's orbit, the tilt of the Earth's rotation axis, and the eccentricity of Earth-Sun orbit[48]—must also be considered. Most of these later aspects of change are not the focus of this book, but the fact is that a cold period existed within the fourteenth and nineteenth centuries.

This period—still called the Little Ice Age, or "LIA"—should not be confused with the last massive glaciers of thousands of years ago such as the Wisconsin Ice Sheet. As Robert Marks put it, this time

> should not be read to mean uniformly colder temperatures, nor should it be interpreted to mean that each year during this period was colder than those preceding or following. Indeed, even during the Little Ice Age, some years and even decades were rather warm. But on balance, though, global temperatures cooled enough for glaciers to keep growing, rather than receding[49]

which puts this "age" into the focus of a realistic dynamic instead of a mere semantic label.

Additionally, Jean Grove takes the dating of the LIA apart (one common sequence being AD 1550–1880), by stating that "it [1550–1880] could be called a general cold period, linked to [the LIA] but not the whole phenomenon closely linked to glaciers ... since glaciers involve more than temperature ... they also involve the hydrologic cycle ..."[50]

---

[47] The expert in this case is the late Jean Grove.

[48] This train of thought implies Milankovitch cycles, and trends counted in tens of thousand of years (that is, cyclical in times of 19, 21, 40 and 100 thousand years). See Imbrie, J., Imbrie, K.P., *Ice Ages—Solving The Mystery* (Enslow Publishers, 1979) and Berger, A., "Milankovitch theory and climate," *Reviews of Geophysics*, Vol. 26, 1988, pp. 624–657 for a thorough review.

[49] Ibid, Marks, *Tigers, Rice, Silk, and Silt,* p. 138. Admittedly, Marks is not a physical scientist: but often, a non-physicist can describe physical processes with a preferred general—if not 100% technically perfect—clarity.

[50] Personal communication with Jean Grove, September, 2000.

Grove further remarks that it is not useful to consider either the LIA or the medieval warm period as continuous or unbroken global cooling and warming periods, respectively. The trends for either are sporadically warmer or cooler weather than usual, but overall, lower (LIA) or higher (medieval warm period) than the normal mean. Additionally, she remarked that the non-linear cooling dynamic called the LIA could be dated far earlier than 1550,[51] that is as far back as 1300 AD or so based on glacial evidence from the Canadian Rockies.[52]

Detractors of what the "Little Ice Age" was sometimes trace the origin to when the "Little Ice Age" label was first applied (1939) to describe the recurrence of moderate glaciation, and casts doubt on what we know of it. Jones and Bradley remarked:

> the term "Little Ice Age" has been used frequently ... without any discussion of what the term means. The term originated with Matthes (1939) who stated, "we are living in a epoch of renewed but moderate glaciation—a "little ice age" that already has lasted about 4,000 years."[53]

That is not wholly accurate.

The actual American Geophysical Union (AGU) report by the committee on glaciers from 1939–1940 pointed out that the Sierra Nevada snowfields (cirque glaciers and glacierets) are only 4,000 years old. Whereas, those in northern Washington (state) and Canada, along the same cordillera, "did not melt away entirely during the warm middle third of the post-Pleistocene interval but were greatly reduced in size."[54] The author of this 1940 report, Francois Matthes, notes that, in some "specific cases" there had lately appeared to be a "mild recrudescence"

---

[51] For the LIA covering 1300–1900 or so (Ibid, Grove, September, 2000).

[52] Most notably from the works of Brian Luckman and colleagues.

[53] Jones, P.D. and Bradley, R.S., *Climatic Variations Over the Last 500 Years,* p. 658. The authors emphasize the confusion due to this term, and call for more research. The authors point out the association of F. Matthes with the "Little Ice Age" (LIA), but that it is not accurate. The Matthes's quote Jones and Bradley used: "we are living in a epoch of renewed but moderate glaciation … a "little ice age" that already has lasted about 4,000 years" came from Matthes, 1939, p. 520 ("Report of Committee on Glaciers," *Transactions, American Geophysical Union,* Vol. 20, pp. 518–523).

[54] Matthes, F.E., "Committee on Glaciers, 1939–40," *Transactions, American Geophysical Union,* 1940, pp. 396–406. Matthes's words on the activity of post-Pleistocene glaciers at that time were: "They have re-expanded since then to the limits from which they are even now receding, and as their re-expansion has been of considerable magnitude, to judge from certain specific cases, there appears to be warrant for the assertion that the present age is witnessing a mild recrudescence of glacial conditions—that it is, as a clever journalist has suggested, a separate "little ice age." (p. 398).

of at least these post-Pleistocene glaciers, and that *some journalist of the time gave this mild re-emergence of ice the name "little ice age."*[55] [Author's italics]

In this report, Matthes further stated that:

> Whether of not any close synchronism actually exists between individual glacier-maximum of short duration (small peaks on a curve of long-time swings) in the Old World (say, the last glacial maximum of about 21,000 years ago) this much can be asserted with some confidence on the strength of the data now at hand, that throughout the last three centuries the glaciers in Europe and in the western United States were appreciably larger than during the preceding centuries, and their maxima in the seventeenth, eighteenth, and nineteenth centuries were without a doubt the greatest ice-extensions that have occurred since the Pleistocene ice-age.

He added:

> Lest there remain any misapprehensions on the score, that the rapid decline of the glaciers that is now (1930s) being witnessed in the United States and Europe does not necessarily mark the end of this recent period of glacier-expansion. It may be only a temporary recession—one more of the many temporary recessions that have occurred during the last 300 years and that are at least partially recorded by the intervals between the successive moraine ridges ... it is significant in any event that in spite of their recent recession the Swiss and French glaciers today are still much longer than they were during the Middle Ages [the "medieval warm period"]. The sites of the villages that were overwhelmed by them during the catastrophic advance of 1643–44 [that is, the early to mid Maunder Minimum] are still under the ice.[56]

However, let us take this non-linear cool weather dynamic beyond terminology, and give it concrete justification: "widespread indications of rapid twentieth

[55] Matthes, F.E., "Committee on Glaciers, 1939–40," *Transactions, American Geophysical Union*, 1940, pp. 396–406, pp. 397–398. From the direct quote in the footnote above, it seems that Matthes was not the coiner of the term "little ice age." It is interesting to note that many climate history experts had also inaccurately attributed the term "little ice age" to Matthes. For example, on p. 222 of the translated Ladurie, L.R. (1971), Matthes was called the "inventor" of the phrase. However, to his credit, on pp. 222–226 Ladurie provides more descriptive contexts of this time, as well as different usages of the term by other authors. For example, the advance of 1570–1850 was called the "Fernau stage" by Germans, not the little ice age.

[56] Ibid, Matthes (1940), p. 400.

century glacial retreat and the striking similarity of data coming from widely separated areas seem to indicate a coherence which justifies a single name."[57]

More Proxy Reconstruction of Past Climate

Did the solar activity cycle completely stop during the Maunder Minimum? Studies of observed aurora, at least in central Europe between 1645–1705,[58] indicate—cautiously—that the solar cycle *did* exist during at least the 1645–1705 leg of the Maunder Minimum. But the lower activity of aurora and the rare appearance of sunspots are also true. The fact that the Maunder Minimum interval was a relatively cold time can also be shown for Europe and China (with the evidence being less clear for North America) from the composite summer temperature anomaly in Figure 42.

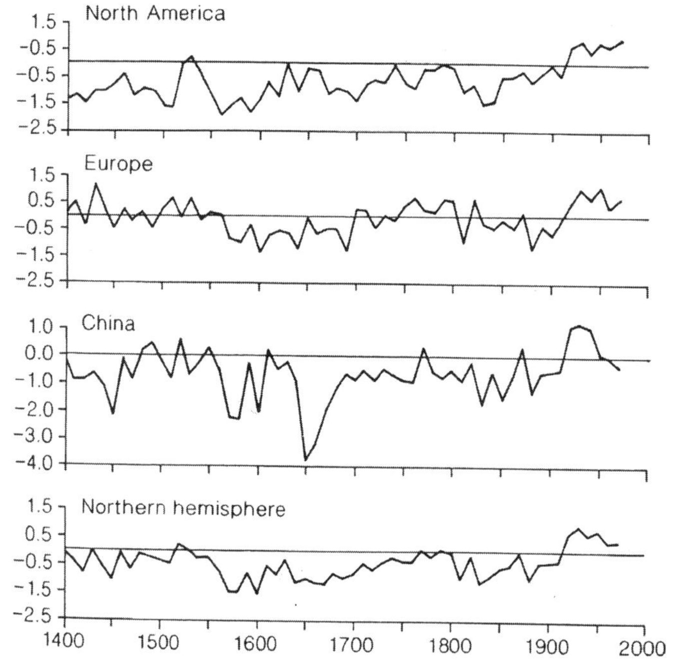

Fig. 42 Composite Summer temperature anomaly series for Europe, North America, East Asia, and the Northern Hemisphere (derived from normalized decadally averaged anomalies, with reference to the period 1860–1959). (From Bradley and Jones, after Grove 1996.)

---

[57] Grove, J., *The Little Ice Age* (Routledge, 1988), p. 5. She defends this dynamic throughout this and other treatises.

[58] Schröder, W., "Aurorae During the Maunder Minimum," *Meteorology and Atmospheric Physics* (Springer Verlag, 1988), p. 247.

Another point to take up which may put the data in Tables 1 and 1a (Chapter 5) in a better light regarding weather, at this time, are data collected by Christian Pfister[59] in the 1980s and fairly recently in 1992. Very detailed diaries were kept by, for example, the Swiss in the 1500s, 1600s, and later, detailing the spread and contraction of glaciers, lending veracity to the 1628 report from Switzerland of cold and wet summers and snowy winters. Pfister collected 80,000 diary entries, 33,000 plant germination and foliation (phenological) observations, and hundreds of observations of snow cover at different altitudes, and made thermal and wetness indices for this part of Central Europe over all of the Maunder Minimum interval (see Figure 43).

Glaciation was strongly evident in Switzerland as well, at least as it was recorded in the Grindelwald Glacier (see Figure 44). A very long period of snow accumulation is noted in the 1690–1720 timeframe—coinciding with the sharp drops in temperature recorded in Switzerland at this time. To quote Jean Grove, "the 1690s [in Switzerland] stand out as having been cold at all seasons," with heavy precipitation being the norm.[60]

Grove remarks that past climate in many other parts of the world is becoming possible to discern, due to many new collecting efforts using documentary sources, tree ring records, coral, stalagmite, lake and seafloor sediment records, etc. Or at least a hemispheric-scale picture of climate change over the last few centuries[61] is starting to emerge. For example, in Central Argentina (28–36°S, 61–67°W) Claudio Carignano and Marcela Cioccale, adopting the multi-proxy approach, documented that:

> [D]uring the Little Ice Age the [central Argentina's Córdoba] plains had temperate, semi-arid to arid climates, and Andean glaciers advanced. … [There was] the occurrence of two cold pulses separated by an intermediate period of more benign conditions. The first cold pulse extended from the beginning of the XV century to the end of the XVI century; the second cold pulse (the main one) [sic.] began at the beginning of the XVIII century and lasted until the beginning of the XIX century. Both cold pulses can be related to the Spörer and Maunder Minimums

[59] Ibid, Grove, J., "The Century Time-Scale," pp. 48–54.

[60] Ibid, Grove, J., "The Century Time-Scale," p. 51.

[61] Ibid, Grove, J., "The Century Time-Scale," p. 51.

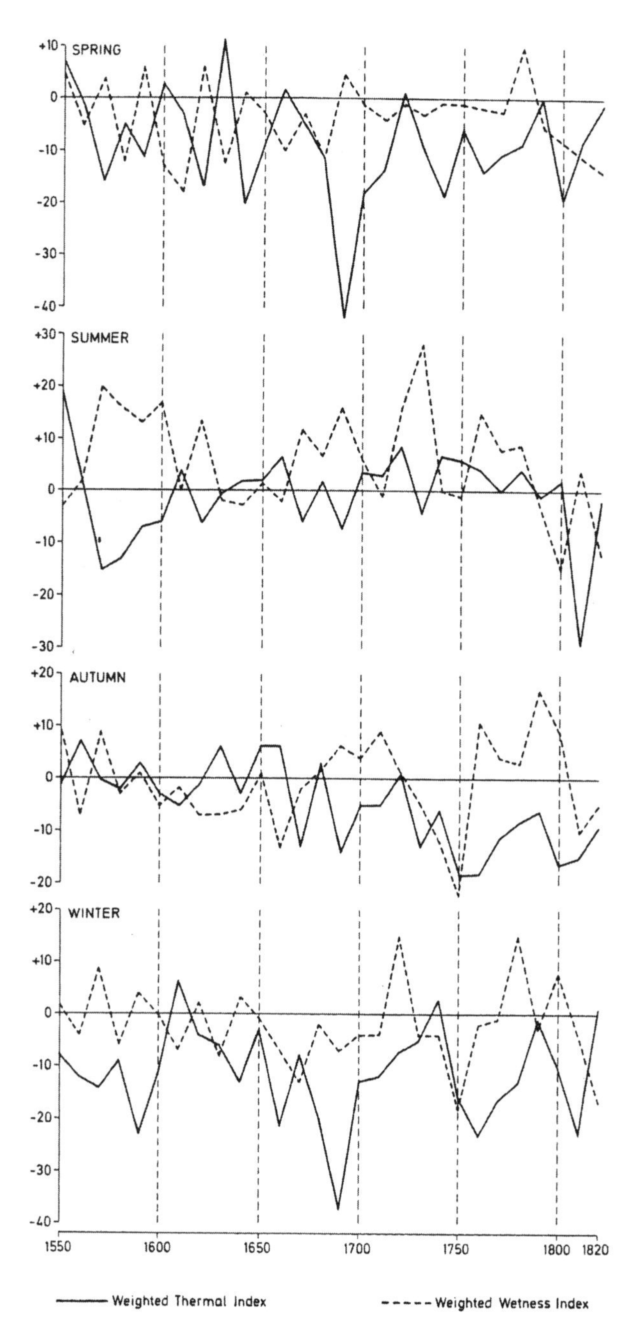

Fig. 43 Seasonal weighted thermal (solid curves) and wetness (dotted curves) indices for Switzerland, 1550–1820 (based on data in Pfister, after Grove, 1988). Note the 1690s show extremely cold springs and winters.

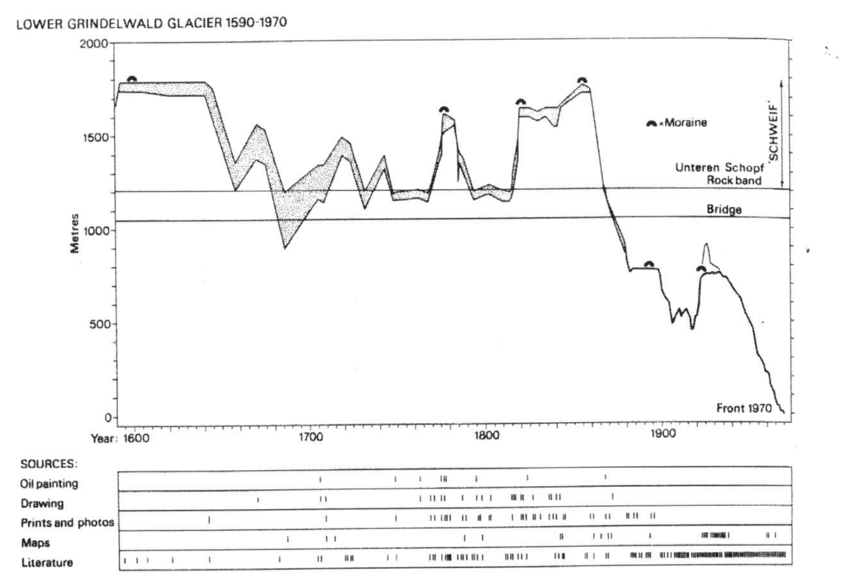

Fig. 44 Frontal positions of the lower Grindelwald Glacier, 1590–1970. The changing appearance and positions of the front of the glacier attracted so much attention from contemporaries that it is possible to trace its detailed history (Messerli, after Grove, 1988).

> respectively. ... The chronicles of the time indicated frequent snowfalls where snow does not presently occur.[62]

Still, in all fairness, the record is far from complete, and a general despair of collecting enough—not only tree ring data—from all regions runs deep. For the most part, proxy data in all forms is held in great suspicion.[63]

---

[62] Carignano, C.A., "Late Pleistocene to recent climate change in Córdoba Province, Argentina: Geomorphological evidence," *Quaternary International*, Vol. 57/58, 1999, pp. 117–134 and Cioccale, M.A., "Climate fluctuations in the Central region of Argentina in the last 1000 years," *Quaternary International*, Vol. 62, 1999, pp. 35–47. Carignano further detailed that "[t]he existence of the Little Ice Age is known throughout the region. ... Historical sources are particularly rich in data, and document a notable aridification of Córdoba. Geomorphological activity replicated patterns associated with the previous semiarid cycles, although on a lesser scale. Iriondo (1993, 1994) pointed out that similar conditions had occurred during the Last Glacial Maximum. The vegetation suffered a notable deterioration. The Eastern Chaco forest was replaced by drier climate assemblages, such as the Western Chaco forest and scrub ... The northwest lakes disappeared completely, and their beds are currently occupied by salinas. Mar Chiquita lake suffered a notable recession, ... with levels lower than during the twentieth century. In many locations, aeolian erosive processes resumed. Small barchan dunes, lunettes, and deflation hollows developed in the eastern and southern areas of the plain ... Parabolic dunes formed in the Salinas Grandes basin were 150–200 m long, 60–80 m wide, and 2–3 m high." (p. 130).

[63] Ibid, Jones and Bradley, Section E, Summary (*Climatic variations over the last 500 years*).

* * *

If the warming in the recent few decades is a reason for concern, one can look back at palynological records (fossil pollen/spores) found along the 100 degree West meridian in Central Canada to have a fossil-look at the tree limits of 5,000 and 6,000 years ago in North America. This is not very long after the end of the Late Wisconsin Ice Age's end—and tree limits of those times are very far north and away from the current tree line limits we see today.

Pollen and spores are useful environmental markers found across the planet in geological records. The record pollen and spores left in the soil and in other substrate is a clear indication that what is permafrost tundra, now, was forest, then: and what is northern pine forest, now, was once temperate (see Figure 45). Additionally, where temperate climates now exist it was much warmer and possibly even drier, once, as the burials of native Lamoka people in Upper State New York near this time (roughly 5,000 years BP) indicate.[64]

As an illustration of climate change and glacial activity from living organisms' perspectives, cold tundra plants, for example those normally found on the ground in Greenland, still grow far south in "sub-alpine" conditions. Two examples are Mount Mansfield in Vermont and Roan Mountain on the Tennessee-North Carolina border (Bald Mountains). Both have tundra growth on them that may have been left there possibly as far back as 18,000 years ago during the last (deep) ice age. The best word to use to describe these incidents is "stranded," as it were, at great height. These same biota survived periods of climate warming like the Holocene or Medieval Maximum, and now face possible extinction as climate and human intervention (ironically, as people try to "protect" them—at least in the Roan Mountain case) naturally conspire to change their habitat.[65] These are just two examples of this type of climate-steered global redistribution of living things, with more being likely.

[64] The climate in Upper New York State then may have been like lower Atlantic states such as Missouri, today. Ritchie, William, A. *The Archaeology of New York State*, Revised edn. (Purple Mountain, New York, 1994), pp. 42–57. "When the Lamoka folk lived in this territory, almost five thousand years prior to white settlement, according to radiocarbon determinations, the climate may have been warmer than at present and perhaps somewhat drier." Palynological results of lake bottom samples (Lamoka Lake) proved inconclusive in a 1958 study. However, the forest at the time was a mast forest, presumably consisting of southern variations in species of trees like hemlock and oak, and so on.

[65] Knowles, T.W., "Twilight on Bald Mountain," *The Sciences*, March/April, 1999, pp. 25–29.

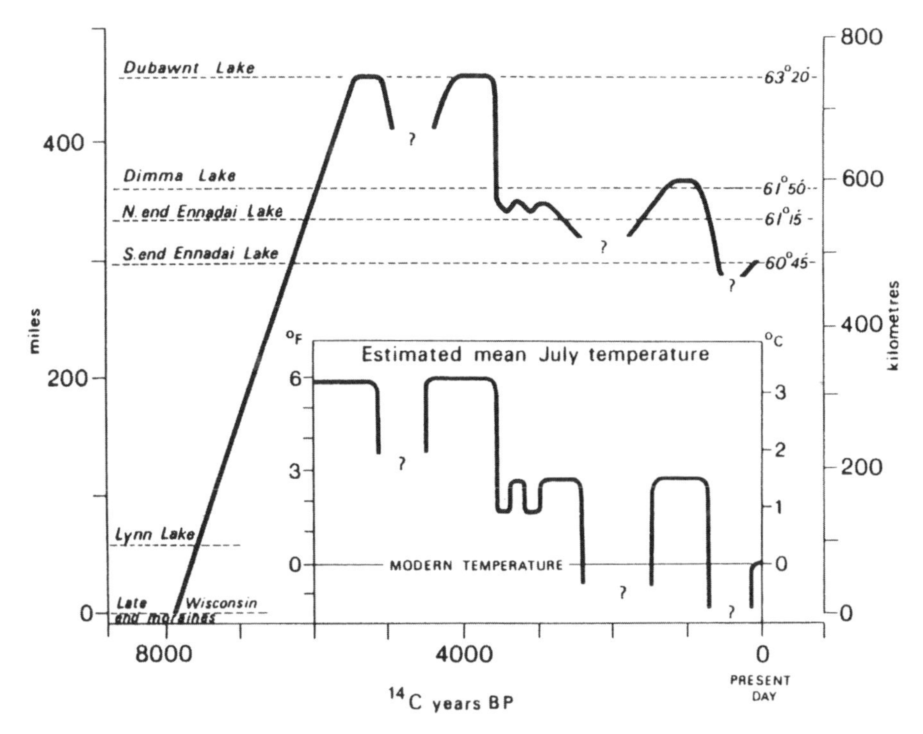

Fig. 45 A reconstruction of the position of the northern limit of continuous forest along the 100 degree W meridian in central Canada during the Holocene [epoch, which is what we live in, now] based mainly on radiocarbon-dated pollen diagrams from the Ennadai (Keewatin, Nunavut, Canada) and Lynn (Manitoba, Canada) Lake areas. (The departures from the modern mean July temperatures were calculated from the varying distance of the forest limit from Ennadai.) (Nichols, 1974, after Grove, 1988.)

Taken together, the wealth of solar and terrestrial information from ^{14}C, ^{10}Be, $\delta^{18}O$, ice-borne rock grain and several indirect climate proxy studies contribute to expanding the already enormous range of Sun-climate-relation studies. But before any true progress can be made, yet another trip back to the Sun seems important in order to understand what could be responsible for the intrinsic solar magnetic variability.

14

The Maunder Minimum and Modern Theories of the Sun's Cyclical Machinery

Go, wondrous creature! Mount where Science guides,
Go, measure earth, weigh air, and state the tides;
Instruct the planets in what orbs to run,
Correct old Time, and regulate the sun.[1]

Alexander Pope, Essay on Man (Epistle II)

Nothing could be worse than to work on a riddle, knowing that in one's lifetime, one will not know the whole answer. As a senior statesman of solar science, Walter Maunder said nothing would give him more pleasure than to understand sunspots' "real causes, origins and effects."[2]

As discussed in many parts of this book, Maunder, perhaps alone or with help, created the butterfly diagram. Some of his techniques are still used. Today[3] scientists use butterfly diagrams to study and understand the Sun's behavior—sometimes even to time the "beginnings" of new solar cycles. From Maunder's 1904 work one sees a "full" butterfly diagram filled with spots during the 1876–1902 period (Figure 47). The modern extension of Maunder's legacy since his time at the Royal Observatory is summarized in Figure 46. It is remarkable to see how each butterfly wing differs from one to the other as time progresses from 1870 on.

It is also obvious that the influence of the Maunders' sunspot studies is not limited to professional scientists alone. For example, one can also find the

---

[1] From *The Poems, Epistles and Satires of Alexander Pope* (Dutton, 1931), p. 190. In this extract, a footnote from Warburton describes that this section "alludes to Sir Isaac Newton's Grecian Chronology, which he reformed on those two sublime conceptions: the differences between the reign of kings and the generations of men—and the position of the colures of the equinoxes at the time of the Argonautic expeditions."

[2] Spoken on November 28, 1906 by the Victoria Embankment during a meeting of the B.A.A. (*JBAA*, Vol. 17, 1907, p. 66).

[3] Notably, the late Elisabeth Nesme-Ribes of the Meudon Observatory, France.

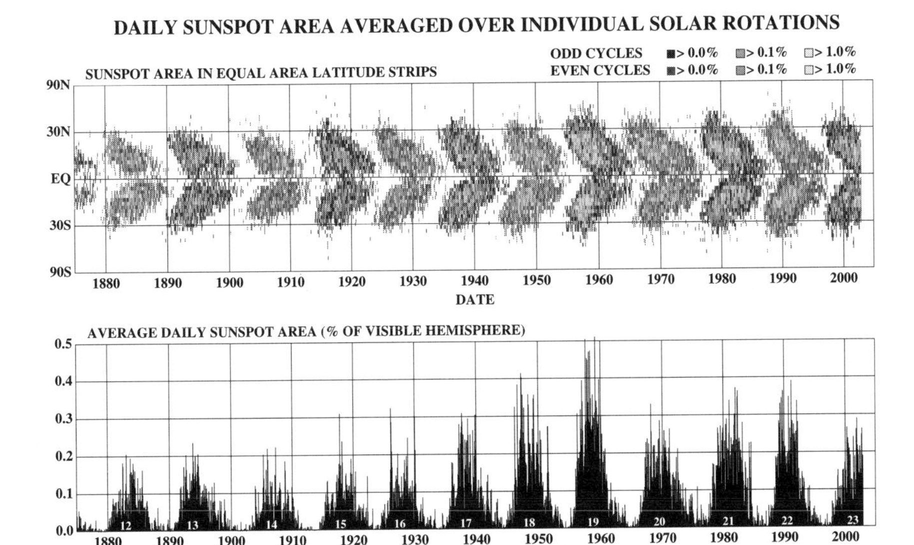

Fig. 46 Daily sunspot area average over individual solar rotations from the year 1876 to the present [circa 1999]. (Courtesy David Hathaway.) [A color version of this plot is available at http://science.nasa.gov/ssl/pad/solar/images/bfly_new.ps]

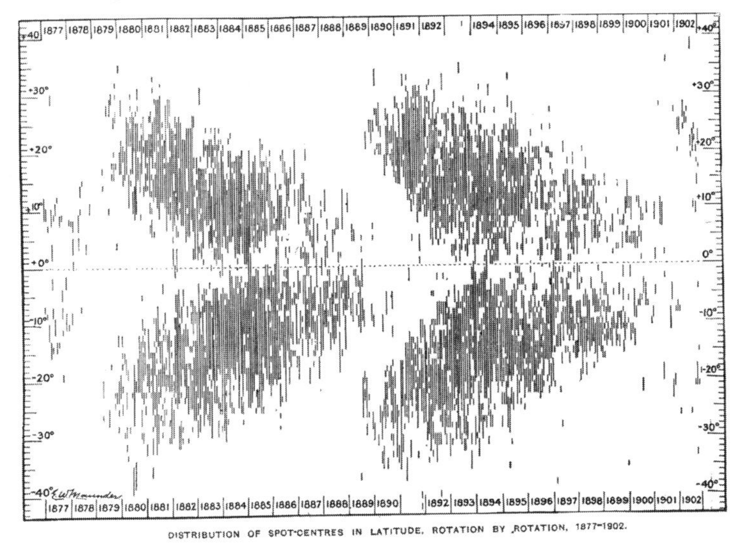

Fig. 47 Maunder's original "butterfly diagram" (1904) for the distribution of sunspots between 1876–1902 (North, above; South, below).[4] The idea thus created by Maunder was used to reconstruct the Maunder Minimum sunspot arrangement from 1620–1670, and 1670–1719 by Elizabeth Nesme-Ribes and others (see Figure 48).

---

[4] Ibid, Maunder's 1904 paper. Maunder, E.W., *Monthly Notices of the Royal Astronomical Society*, Vol. 64, pp. 747–761, "Note on the Distribution of Sun-spots in Heliographic Latitude" (1904).

extraordinary butterfly diagram summarizing the 6,701 individual observations (covering January 17, 1976 and December 24, 1986) by the amateur astronomer Sieglinde Hammerschmidt (a housewife from Solms, near Wetzlar, Germany) in Rudolf Kippenhahn's *Discovering the Secrets of the Sun.*[5] Constructing a butterfly diagram is not a trivial task, but it can certainly be done with sufficient dedication and the aid of a small telescope. This is a real tribute to Walter Maunder's early vision of the joy of astronomical studies in general.

But what about the actual spots on the Sun's face during the Maunder Minimum? What can be said about this, in addition to all the incidental notes given in earlier chapters? Research from solar astronomers like the late Elisabeth Nesme-Ribes, Jean-Claude Ribes and colleagues is able to shed light on this somewhat enigmatic riddle.[6]

Figure 48 shows a modern attempt to construct a butterfly diagram for the past—indeed, over the Maunder Minimum—to obtain results that can be compared to the large set of butterfly wings seen in Figures 46 and 47. What Nesme-Ribes and colleagues did was to reconstruct data over the period 1620 to 1670 and 1670 to 1719. This is based mainly on sunspot observations recorded at the "Sun King's"—Louis XIV's—creation: the Paris Observatory. Royal vanity aside, there is perhaps a *reason* for this king's sub-appellation for himself.[7] Indeed, he strengthened this image of solar magnificence by ordering the creation of hundreds of portraits and other images of himself, as such.[8]

In the reconstruction by Nesme-Ribes and others, the appearance of sunspots during the Maunder Minimum seemed to be more narrowly confined around the solar equator. Furthermore, nearly all the spots appeared in the Sun's *Southern Hemisphere.* Only three appeared in the Sun's Northern Hemisphere for the period measured (see the bottom half of Figure 48). Indeed, such a spot distribution is quite unusual compared to modern day "butterflies." It appears that those butterfly wings have been *broken.*[9] And, as we remember from earlier on in this book, the observation of sunspots (or perhaps, the careful notation of sunspots' *absence*)

---

[5] See Figure 2.10 (on p. 28) of Kippenhahn, R., *Discovering the Secrets of the Sun* (John Wiley, 1994) for this very special illustration of the Hammerschmidt's butterfly.

[6] Ibid, Nesme-Ribes, Sokoloff *et al.*, 1994 and J.-C. Ribes and E. Nesme-Ribes, *Astronomy and Astrophysics*, Vol. 276, 1993, pp. 549–563.

[7] "The reign of Louis XIV appears to have been a time of real anomaly in the behavior of the Sun." John A. Eddy (*Science*, Vol. 192, 1976, p. 1189).

[8] Ibid, Verdet, p. 67.

[9] A term used by Elisabeth Nesme-Ribes; the first author, Willie Soon, heard it during a lecture by her in a June 1993 conference on solar physics.

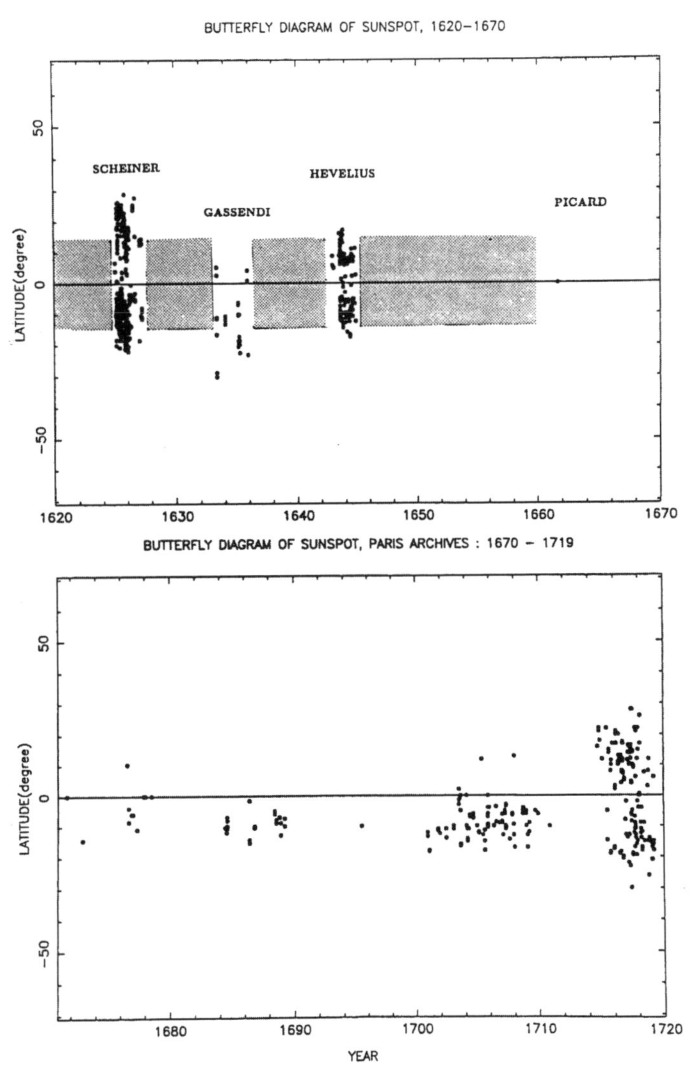

Fig. 48 Sunspots seen by selected recorders between 1624 and 1719—as covered in the first parts of this book. Although early archival data was not perfect—as amply shown—the lack of spots in the Northern Hemisphere during 1670–1719 was real.[10] The "butterfly" (bottom, covering the span c. 1680–1710) was not much of one, compared to the 1904 one recorded by E. Walter Maunder) and shows a period of "inactive" (that is, minimal or weakened) solar activity. (Courtesy of the late Elizabeth Nesme-Ribes and the Paris Observatory archives.)[11]

---

[10] See Hoyt, D., Schatten, K., *Solar Physics*, Vol. 165, 1996, pp. 181–192.

[11] For example, Nesme-Ribes, E., Sokoloff, D., Ribes, J. C., Kremliovsky, M., in *The Solar Engine and Its Influence on Terrestrial Atmosphere and Climate*, ed. E. Nesme-Ribes (Springer-Verlag, Berlin-Heidelberg, 1994), pp. 71–97.

was quite well covered for the time measured in the top half of Figure 48 (about 1620 to 1645). The quality of measurements and observations of sunspots (or their lack, thereof) in proportion to those of their modern counterparts is good enough to give us a fairly firm picture of what the Sun was undergoing during the Maunder Minimum period.

The Royal Observatory's sunspot work given to Maunder had more to tell him than he knew. With the elemental knowledge within the Sun's spectra light,[12] Maunder's diagrams[13] and other supporting data, George Ellery Hale and Seth Nicholson subsequently discovered sunspots' magnetic qualities (see Figure 49, and compare with Figure 3 in Chapter 2), and their true time-variability patterns.

It was found that most spots are "bipolar." That is, they have a leading and a following polarity having opposite "sign" (that is, direction). All spots in an activity

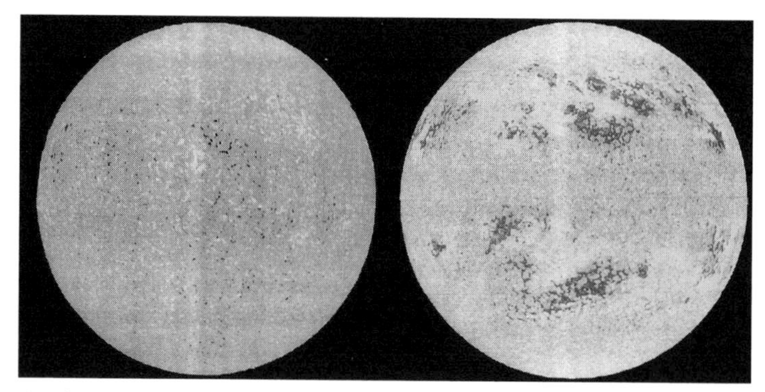

April 7, 1994 (near minimum) *February 12, 1989 (near maximum)*

Fig. 49 Full disk magnetograms of the Sun on April 7, 1994 (near activity minimum) and on February 12, 1989 (near activity maximum). Compare these with the white-light images, Figure 3 in Chapter 2, for the same days. In each magnetogram, North is at the top, East is to the left and the sun rotates from left to right. Color is used to contrast both the direction (polarity) and strength of the magnetic field. Bright color (shading) shows, for example, a north-pointing field with increasing strength from light yellow, orange to bright red. Dark color (shading) shows the reverse pointing field with increasing bluing for increased strength. Near activity minimum, the magnetic pattern composed essentially of the "salt and pepper" background of weak, small-scale magnetic flux. Near activity maximum, larger magnetic structures, like the large-scale bipolar toroidal field (that is, solar-spots), dominate. (Images courtesy of William C. Livingston, Kitt Peak Solar Observatory. Color images are available from the authors.)

---

[12] Especially the Zeeman splitting of certain spectral features due to the presence of a magnetic field. Under the inspiration of Michael Faraday (1791–1867), Pieter Zeeman (1865–1943) had originally studied this effect using the yellow spectral line produced by sodium (the sodium D-line); hence, the "Zeeman Effect."

[13] Which revealed a "new spot appearance" phenomenon happening at higher latitudes on the Sun at a sunspot cycle's start.

belt tend to have the same leading polarity—such as a north magnetic polarity—as they rush to the Sun's edge (known as the "limb.")[14] If a spot or spot group's leading polarity is north pointing for that solar hemisphere, then the leading polarity will be south pointing in the *opposite* hemisphere. With each new 11-year cycle, the spots' polarity[15] in both hemispheres is switched (see Figures 50 and 52 below). This reversal occurs roughly every 22 years, or, double the 11-year Schwabe cycle, and is called the solar magnetic polarity cycle (or even the Hale Cycle, in Hale's honor).

The Driving Solar Dynamo

What drives the giant cyclical "machinery" in the Sun that could in fact be the cause of sunspots as well as other phenomena? In this regard, the Maunders expressed a sense of frustration and perhaps pessimism:

> What causes the spots we do not yet know. We may never know, for their cause seems to lie deep down in the Sun itself ... We cannot tell if it is the same cause that gives rise to the outbreak of spots in each zone, eleven years after eleven years. If the spots are the same, cycle after cycle, we have no means of recognizing them. We can recognize again a star by its place in the pattern of stars; we can do the same with a planet by its color and its movements; ... we can even know again a whale if it has a harpoon driven to it; but how are we to know a Sun-spot when it emerges again from the solar depths?[16]

From Hale, it began to be understood that the Sun's magnetic field was the key for the character of sunspots. Magnetic field and its variability is also the key in the Sun drenching Earth in ultraviolet light, X-rays,[17] electrons, and atomic nuclei and then wrapping it in its *own* magnetic field, but it is still not understood how these intense fields are created inside the Sun or how they make solar flares,[18, 19]

---

[14] In the sense of the solar rotation: this feature explained that most bipolar spots are associated with the underlying toroidal field.

[15] Or the Hale polarity cycle.

[16] Ibid, *The Heavens and Their Story*, p. 142.

[17] Information taken from the NASA education page (Internet).

[18] As noted earlier in some detail, a white light flare was noted by R. Carrington and R. Hodgson on September 1, 1859. Both of them reported their observations in *Monthly Notices of the Royal Astronomical Society*, Vol. 20, 1860, pp. 13–15 and pp. 15–16.

[19] There are, as previously mentioned, other less well-known, earlier accounts of solar flaring phenomena. For example, Stephen Gray saw a white-flare (which is rare) on December 17, 1705 in Cambridge, England, but never published—or managed to have published—his observation.

heat the Sun's corona, or power solar winds around Earth. The still-challenging question concerning the solar magnetic field's origin remains theoretical, with only gradual progress being made.

However, a theory relates that a regenerative mechanism such as a dynamo can fulfill observations about the Sun's magnetic and 11-year sunspot cycles. This could also explain Maunder's butterfly, and potentially all butterfly diagrams.

To see how the dynamo works, one divides what is the Sun's velocity field into two components. The first component of this velocity field is orderly in motion. For example, the differential rotations, with gas near the solar equator rotating several days faster than that near solar poles.[20] These facts had been noted on the Sun's surface early on by acute observers like Spörer and Carrington. This second component of the velocity field—the "chaotic" or disorderly part—was not realized and formalized until the 1950s by Eugene Parker[21] concerning the "cyclonic" or "spiral motion" (helicity[22]) within the rotating and convecting solar plasma layer's confines. This chaotic part could be described as "whirlpool like," often having preferred rotations in clockwise or counter-clockwise directions, but which have opposite directions (signs) in either the Sun's Northern or Southern hemisphere. The strength of helicity—also referred to as the "source of generation"—is not constant with latitude but is typically assumed to have peak strength near low-to-middle latitudes.[23]

The convecting solar plasma is mostly ionized because of the high temperature, leaving the charged, ionized plasma with electrical conductivity. (The level of conductivity of solar plasma is similar to weakly-conducting metals like mercury or nickel-chromium alloy at room temperature here on Earth.[24]) It is the plasma

---

[20] A similar pattern is observed to persist throughout the whole convective zone, proper (the outer 30% of the Sun), until the base, by helioseismological study. The difference in the equator-to-pole rotation period ranges from roughly 5–20 days or so depending on the markers of rotation used.

[21] Parker, E.N., *The Astrophysical Journal*, Vol. 122, 1955, p. 293. Parker's breakthrough is the realization (without formal proof, until Steenbeck, Krause and Radler in the 1960s) of the fact that even when turbulence is isotropic, its statistical properties need not be mirror-symmetric, but that it may be cyclonic; that is, it possesses helicity.

[22] Thus, the choice of words. A printed phrase by the dynamo doyen, Dmitry Sokoloff, emphasizes the point of the elusive tensorial properties of the helicity: "A special feature of the dynamo theory is connected with the fact that helicity is a property that is very difficult to extract from observations or experimental measurements. Until now, there has been no direct observation of helicity. This is a challenge. Helicity observation is in some senses more attractive than, say, problems of observing quarks or gravitational waves, because it is based on quite classical physics."

[23] See, for example, Kunzanyan, K.M., Sokoloff, D., *Solar Physics*, Vol. 173, 1997, pp. 1–14.

[24] See, for example, Ruzmaikin, A., *Quantum*, September/October, 1995, pp. 12–17.

conductivity and the relatively strong gas pressure which allows realizations of "frozen-in" magnetic field lines. To picture what these are, recall the well-known iron filings pattern that trace out the force lines around[25] magnets. These "field lines" are then tied to solar gas motions. Subsequent relative charged-particle motions of an orderly/turbulent nature then make up the two key ingredients for the Sun's electromagnetic dynamo: differential rotation and turbulent helicity. Together, differential rotation and turbulent helicity make maintaining the cyclic solar magnetic field possible.

Here is how. Picture what is termed a "poloidal field." This is an elastic pole extending in a north-south direction, parallel to the Sun's rotation axis. Picture also a "toroidal field," which is a doughnut-shaped ring running perpendicular to the rotation axis. Essentially, differential rotation turns an initial poloidal field into a torodial one, while the helicity allows for lifting the torodial field to twist it into a regenerated poloidal field[26] with an opposite polarity. That is, if the initial poloidal field runs from south to north, the regenerated poloidal field now runs north to south, or *vice versa*—thus making the proposed large-scale dynamo a self-excited feedback process.[27] This process occurs over and over, with each recurrence having slight variations (see Figure 50).

All of these geometrical magnetic field line transformations occur *beneath* the Sun's visible surface.[28] The strongly strengthened toroidal magnetic fields' subsequent rise, helped by additional physical mechanisms, are ultimately revealed as sunspots that are bipolar in nature as the toroids rise to the surface. In such a picture, it would have pleased Maunder (and of course, Galileo and others) to know that Maunder's butterfly is most likely a cyclical dynamo wave manifestation that can be revealed by the sunspot distribution across the Sun's surface and the time

---

[25] The distinction here is "permanent" magnets: the reality of frozen magnetic field lines for the Sun is not usually discussed.

[26] A poloidal field supported by differential rotation alone will decay due to dissipative forces. For full mathematical and physical formulation of the dynamo theory, see, for example Roberts, 1967; Moffat, 1978; Parker, 1979; Krause and Radler, 1980; Zeldovich, Ruzmaikin and Sokoloff, 1983; Belvedere, 1985; Stix, 1989; Hoyng, 1992, and so on. (Detailed references are available upon request to the first author—W. Soon.)

[27] This manifestation of an alpha-omega dynamo due to the motions of charged particles is only possible, of course, with the energy of the motion supported mainly by the enormous heat generated within the core of the Sun: nuclear fusion of protons into helium. (So this is not to be fancied as that old dream about a perpetual motion machine within the Sun.)

[28] A precise location which is likely to be near the strongly-sheared region at the base of the convective zone (a constraint confirmed by helioseismological observations).

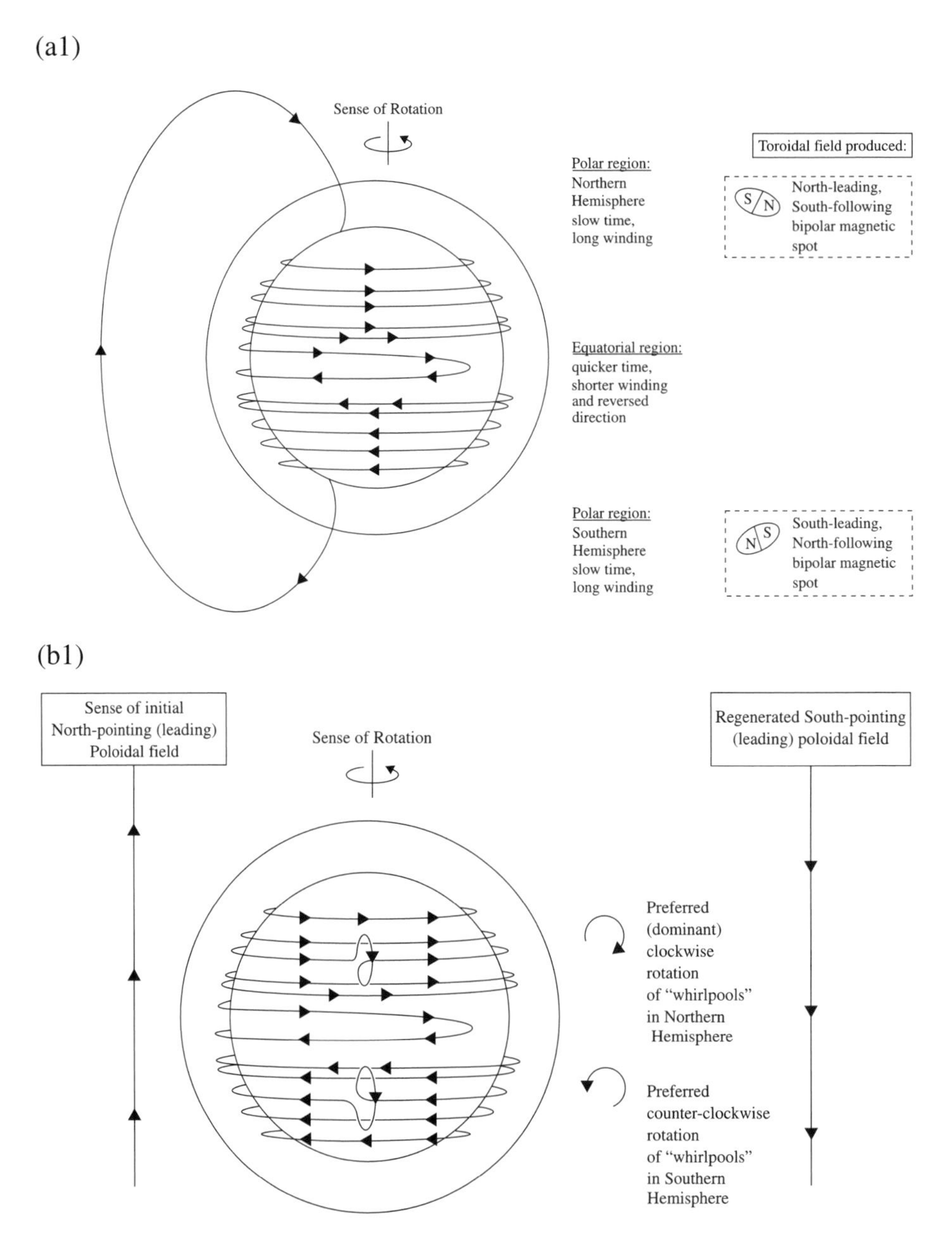

Fig. 50 (a1) Differential rotation effects: winding of an initial North-leading poloidal field into oppositely oriented toroidal fields in the Northern and Southern Hemispheres. (b1) Cyclonic (spiral) motion effects: lifting and twisting of oppositely oriented toroidal fields to produce South-leading poloidal fields.

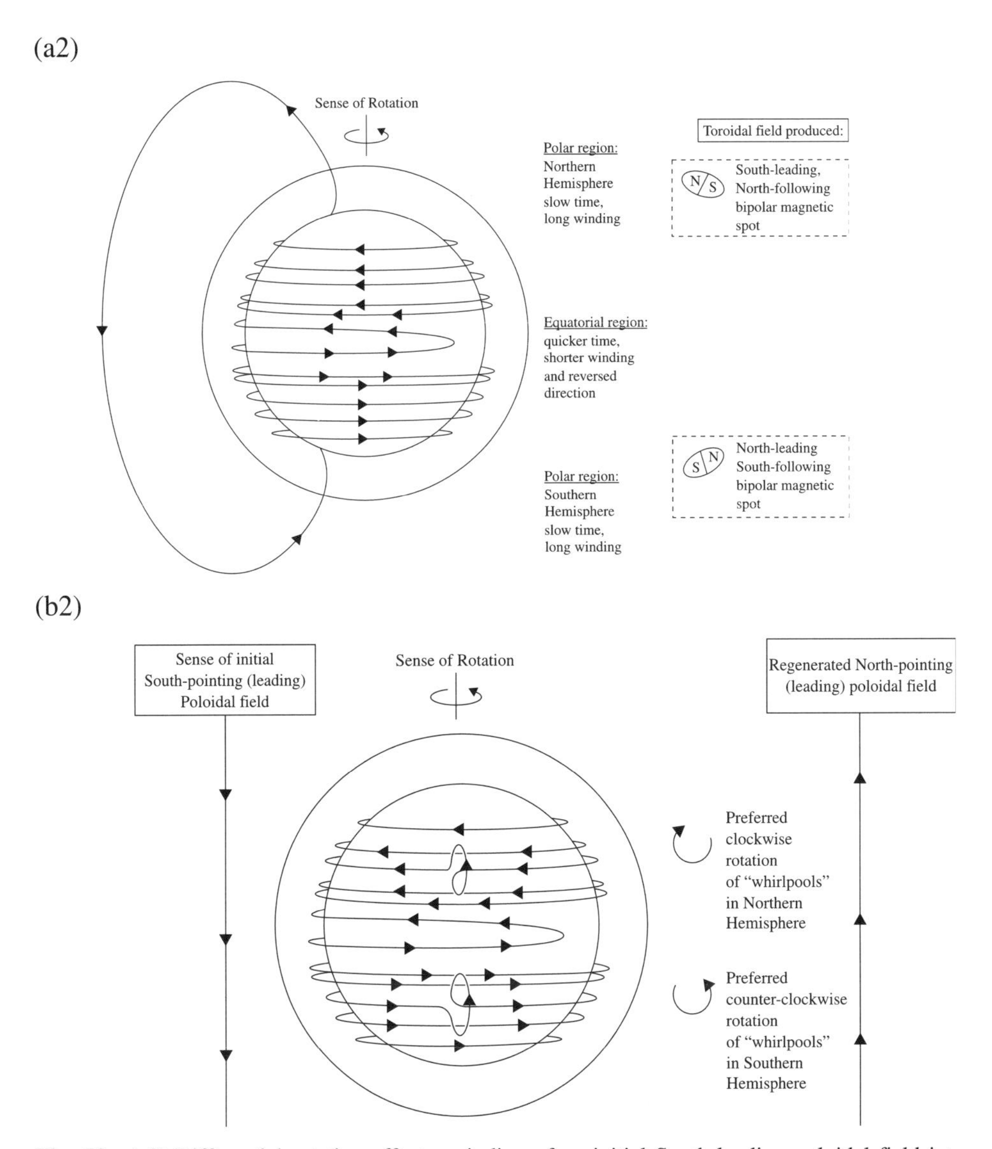

Fig. 50 (a2) Differential rotation effects: winding of an initial South-leading poloidal field into oppositely oriented toroidal field in the Northern and Southern Hemispheres. (b2) Cyclonic (spiral) motion effects: lifting and twisting of oppositely oriented toroidal fields to produce North-leading poloidal fields. The effects of differential rotation and cyclonic motion on the toroidal and poloidal fields over a full magnetic polarity cycle. Panels (a1) and (b1) indicate conditions during the first half of the magnetic cycle (or one Schwabe 11-year sunspot cycle) while panels (a2) and (b2) show the second half. The Hale magnetic polarity rules for the sunspots (that is, strong toroidal fields) across the solar hemispheres and the solar magnetic cycle, are illustrated in panels (a1) and (a2).[29]

---

[29] This figure was re-adapted with inspiration from David Hathaway's web page (http://science.nasa.gov/ssl/pad/solar/dynamo.htm).

spectrum[30] (see Figure 52). Thus, an explanation to the seemingly simple question as to why magnetic spots simply do not just appear *randomly across all latitudes* on the Sun's surface[31] may now be in sight.

The Maunder team, in *The Heavens and Their Story*,[32] used a metaphor to illustrate how sunspots probably move, relative to the sunspot zones, and what they resemble in more familiar terms. The metaphor invokes pods of whales:

> It is as if our solar whales belonged to different species in the different zones, and sought the solar depths for periods of time that varied with their distance from the solar equator; and then brought the young schools to the surface to sport for lengths of time that again varied with their distance from the solar equator. Third [point], the whole time from the beginning or end of the next does not greatly differ for the different zones. If a school of solar whales stays long upon the surface, it must stay the shorter time below; and the two times together make up about eleven years. Fourth, the outbreak of spots in the zones far from the solar equator begins earlier, and therefore ends much earlier, than those outbreaks that are near the equator. This is so marked, that we have a new outbreak starting in the highest latitudes before the old outbreak in the lowest latitudes has subsided. During a period, then, of about eleven years we seem to have an outbreak of spots on the sun beginning in the

---

[30] Near a cycle's minimum, sunspots appear near the solar equator. As a new cycle starts again, sunspots appear in higher latitudes. It was this that gave the "butterfly pattern" shown in Maunder's diagram. The physical origin for this sunspot migration pattern is not completely known, and understanding it could tell scientists something about the Sun's internal magnetic field; something that Eugene Parker, Paul Roberts, Nigel Weiss, Michael Proctor, Fritz Krause, Alexander Ruzmaikin, John Thomas, Axel Brandenburg, Gunther Rudiger, Leonid Kitchatinov, Steve Tobias, Aad van Ballegooijen, Elisabeth Nesme-Ribes, Sallie Baliunas, Paul Charbonneau, Peter Gilman, Arnab Rai Choudhuri, Mausumi Dikpati, Karel Schrijver, Manfred Schussler, Dmitry Sokoloff, and many colleagues have since begun to investigate.

[31] "It is very rarely that any spots are found outside the parallels of the Sun's latitude that are 35° north or south of the equator. A very few have been found in a latitude so far from the equator as 40°; one spot has been seen at 50°; but all of these are very small, and lasted for no longer than two or three days. The seas in which we find the solar whales are 'tropical' or 'sub-tropical'. ... [T]here is quite a long interval of time between two outbreaks of spots in any of the zones [referring to the 7 zones divided equally between 0 to 35 degrees]; always a year or two, perhaps many years; so that there is no doubt as to when an epidemic of spots begins and ends in any particular zone. Next, the time during which the outbreaks last is shorter, and the interval between the outbreaks is longer for the zones which are farther from the equator, than for those nearer to it." (pp. 135–137 of *The Heavens and Their Story*).

[32] Ibid, *The Heavens and Their Story*, pp. 137–138.

> regions farthest from his equator, and spreading to the lower zones, increasing for a while as it spreads. Finally it dies out at the equator, when a new outbreak of spots has already started in the high latitudes. This period of change is called the sun-spot cycle.

Aside from sunspots, there are other magnetically active regions appearing on the Sun. For example, there are the large bright areas called faculae, and plages, active magnetic networks and intra-network magnetic elements, grouped to manage the Sun's total energy output. Sunspots and plages, taken together, actually control most of the Sun's radiated energy flow on 11-year activity-cycle time scales. The more subtle changes in active magnetic networks or intra-network magnetic elements—which cover larger surface areas on the Sun—are able to influence the outputs over longer timescales, such as several decades to centuries. When solar magnetic activity is at its maximum, increased plage areas and intensities enhance for example ultraviolet light (UV) outputs from the Sun.

Still being researched are the Sun's effects on Earth's magnetosphere, as covered in previous chapters: these are literally the effects on Earth's magnetic field, with radiation belts, Earth's plasma ionosphere, and Earth's upper atmosphere. Solar plasma streams around the Earth's magnetic shield could possibly create micro-scale dynamo effects. These dynamo effects accelerate charged particle entry into Earth's atmosphere creating for example, northern and southern "lights"—aurora borealis and aurora australis—the context of which led Clerke and Maunder to ponder their absence to mean "profound [solar] magnetic calm" so long ago.

The Dynamo's Dynamics: The Prolonged Activity Minima of the Sun

Fraunhofer's famous solar spectral lines (and what Kirchhoff and Bunsen found out about them), called simply "spectra," can be partially used as longer-term indicators to explore variation in the Sun's magnetic activity peaks and dips versus Earth's weather—and the influence these peaks and dips have on Earth's climate.

Various known elements make up the Sun's spectra. A ubiquitous spectra signature is ionized Calcium H and K emission lines. These lines tell scientists much about the changing magnetic activity near the Sun's, as well as any solar star's,[33] surface. Satellite observations devoted to the Sun over the last 16–20

---

[33] One of Walter Maunder's expressions seen in his writing as early as 1893, also variously used in this book to denote stars with similar mass, age and magnetic variability like the current Sun (that is, "Sun-like" stars).

years[34] have yielded adequate data that tells a good tale of contemporary solar variability. Especially important is the hard-to-measure[35] total (summed over all wavelengths of light) solar irradiance. But now there is a dilemma that Maunder or Wolf did not have: this database is too short to make any long-term surmises on how the Sun worked in the past.[36]

To get a better picture of solar chronology, then, stars similar to our Sun in type have been monitored to see what they could yield regarding "Sun-like behavior" as in studies pioneered by Olin C. Wilson in the 1960s. To date, Sallie Baliunas, Willie Soon and colleagues at the Mount Wilson Observatory have done a restricted study on stars similar to the Sun in mass and age, revealing some quick drops and sudden spurts in magnetic activity (see Figure 51[37]). That is, some solar stars show cyclic or phasal variations much like our own Sun. There are indications[38] from observations that some stars like our own go through "Maunder Minimum-like" activity phases, too.[39] These observations may allow us to further understand this particular solar minimum's—the Maunder Minimum's—root cause. The solar star examples proffer additional Maunder Minimum-like physical conditions that are capable of yielding insights into the causes of minima.

In addition, the total light outputs using broad-band photometry in solar stars have also been carefully monitored by stellar astronomers: for example, Lou Boyd (Fairborn Observatory), Gregory Henry (Tennessee State University), Wes Lockwood and Brian Skiff (Lowell Observatory) and Richard Radick (Sacramento Peak Solar Observatory). Since the variation in brightness of the Sun and Sun-like

---

[34] For a fuller historical account of the study of the Sun in the space age, including the interesting development of accurate measurement of the solar total irradiance (or the "solar constant" according to meteorologists/climate scientists), read Karl Hufbauer's *Exploring the Sun* (John Hopkins University Press: 1991). This research continues with SOHO, a spacecraft that monitors the Sun on a 24-hour basis. See Kenneth R. Lang, "SOHO Reveals the Secrets of the Sun," *Scientific American*, March, 1997.

[35] Hard to measure because changes in the total solar irradiance are very small, of the order of a few tenths of a percent over a one-decade long activity cycle. See also K. Hufbauer (1991) for a historical perspective in this quest of solar research.

[36] Nesme-Ribes, E., Baliunas, S.L., Sokoloff, D. "The Stellar Dynamo," *Scientific American,* Vol. 275, No. 2, August, 1996, pp. 31–36, and Baliunas, S., Soon, W. "The Sun-Climate Connection," *Sky & Telescope*, December, 1996; the problem of getting around the lack of data and the acknowledgment of the shallow data base, respectively.

[37] The bright (V magnitude = 5.9) star HD 3651 (54 Piscium) in the constellation of Pisces is one such example.

[38] Ibid, "The Stellar Dynamo."

[39] See e.g., Ibid, "Eighteen Hundred and Froze to Death" for discussion on the Dalton Minimum of ca. 1795–1823.

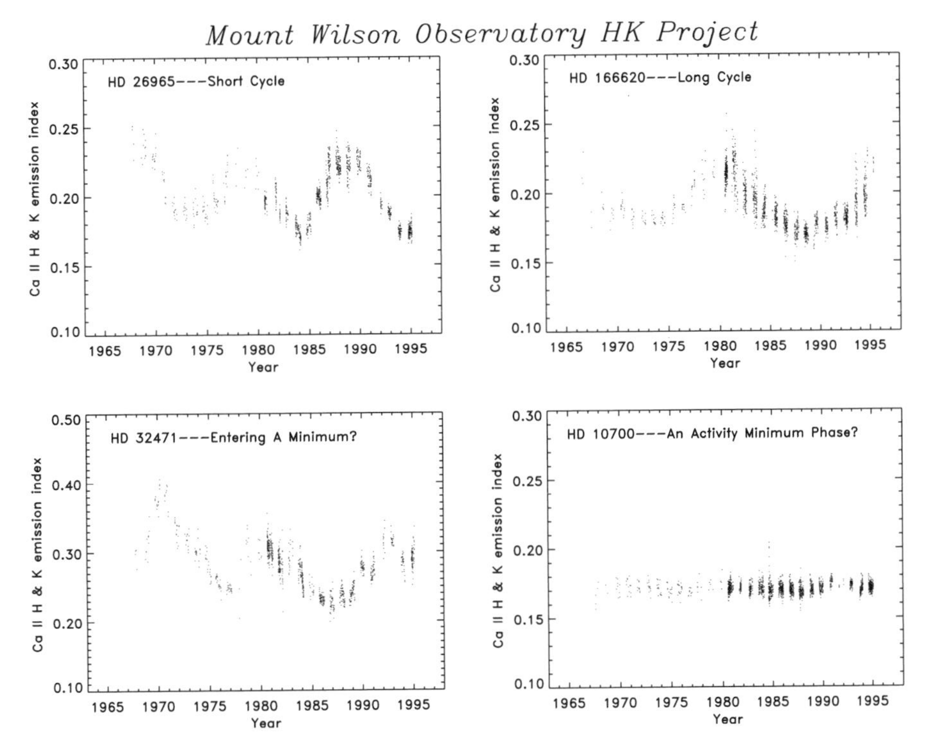

Fig. 51 Four examples of activity cycle morphology in solar stars: HD 26965, showing short, roughly 10-year activity cycles. HD 166620 shows a long, roughly 16-year activity cycle. HD 3651 shows a possible cycle entering into a Maunder-like phase. HD 10700 shows a relatively weak inactivity phase perhaps not unlike the Maunder Minimum. (Courtesy of Sallie Baliunas and Robert Donahue of The Mount Wilson Observatory HK Project.)

stars is related to changes in magnetic activity (as determined by the Calcium H and K fluxes) it has now been shown how these solar-star studies could perhaps explain how our own Sun's brightness varies in connection to the long and short cycles of solar magnetic activity.[40]

As previously discussed, some other elements useful for tracing the Sun's magnetism are done with the help of isotopes like Carbon-14 (^{14}C) and Beryllium-10 (^{10}Be). Records from the trace changes to these terrestrial isotopes can provide views on how solar activity cycles vary on much longer time scales than possible by sunspots alone. ^{14}C and ^{10}Be are more favorably produced at the upper Earth atmosphere when solar activity weakens via Earth's interaction with Galaxy-originating, high-energy charged particles. The isotopes are then "brought down"

[40] Baliunas, S., Soon, W., *The Astrophysical Journal*, Vol. 450, 1995, p. 896.

to Earth through systematic weathering patterns into proxy recorders like trees,[41] pollen, zooplankton, compacted ice, and so on. Another powerful and independent tracer that has been used to examine the reality of prolonged solar activity minima over the last several centuries are the production of intermediately long-lived isotopes like Titanium-44—^{44}Ti (with a mean life of 96 years) and Argon-39—^{39}Ar (with a mean life of 388 years) in meteorites. These isotopes are a more direct and sensitive tracer of heliospheric magnetic field conditions—and hence, the solar wind and solar activity—because they avoid many complicating terrestrial factors like geomagnetic field strength and atmospheric transports associated with the production of ^{14}C and ^{10}Be isotopes in Earth's atmosphere.

However, measurements of ^{44}Ti and ^{39}Ar in various stony and iron meteorites remain difficult because of their very small amounts. Thus far, enhancements of ^{44}Ti and ^{39}Ar concentration[42] have indeed been found in Earth-bound meteorites with known dates of entry, supporting the idea of a much weakened solar activity during many of the previously-mentioned solar minima periods. These extended minimas include the Spörer (c. 1420–1530), Maunder (c. 1620–1720) and Dalton (c. 1795–1820) minimas.

Regarding the Sun's dynamo, Dmitry Sokoloff and the late Elisabeth Nesme-Ribes have offered an explanation on what might actually have occurred in order to produce the lull in activity that was the Maunder.[43] This first suggestion involves pure energy exchanges between the Sun's dipole and quadrupole magnetic field components.[44] Under normal cyclic conditions, it is proposed that the quadrupolar field oscillates at a lower intensity than the dipolar field, known

---

[41] Many things can absorb such particles: all carbon-based life, including plants, can take on ^{14}C. Wines and even beers can, too. It should be remembered that this isotope can date "carbon-based substances" such as fossil bones or fossilized wood to obtain age approximations.

[42] There are of course many detailed technical issues associated with measurements of the ^{44}Ti and ^{39}Ar isotope production in meteorites, including calibration, inhomogeneous samples, exposure age, shielding depth, etc. See Bonino, G. *et al.*, *Science*, Vol. 270, 1995, p. 1648, and Forman, M.A., Schaeffer, O.A., in *The Ancient Sun,* eds. R.O. Pepin *et al.* (Pergamon Press, Houston, 1980), pp. 279–292.

[43] Sokoloff, D.D. and Nesme-Ribes, E., *Astronomy and Astrophysics*, Vol. 288, 1994, p. 293. See also earlier works by Brandenburg *et al.*, *Astronomy and Astrophysics,* Vol. 213, 1989, p. 411, Brandenburg, A., Krause, F., Tuominen, I., in *Turbulent and Nonlinear Dynamics in MHD Flows*, eds. M. Meneguzzi, A. Pouquet and P. L. Sulem (Elsevier Science Publ., North-Holland, 1989), pp. 35–40 and Jennings, R.L., *Geophysical and Astrophysical Fluid Dynamics*, Vol. 57, 1991, pp. 147–185.

[44] Nesme-Ribes, E., Baliunas, S.L., Sokoloff, D., "The Stellar Dynamo," *Scientific American,* Vol. 275, No. 2, August, 1996, pp. 31–36.

as the "odd-parity"[45] dominant solution (see Figure 52, below). Because the dipole field is dominant and anti-symmetric with respect to the solar equator, one is expected to see about equal dominance of magnetic field of opposite polarity for the Sun's northern and southern hemispheres, which is what is being observed today.[46]

However, during the Maunder Minimum, the Sun's dipole was somehow less intense and the solar magnetic field's quadrupolar component began to reach similar intensity: this is recognized as a solution with "even-parity" dominance. Because the quadrupolar field is symmetrical with respect to the equator, while the dipole is anti-symmetrical, one can expect large magnetic flux cancellation in one solar hemisphere, leaving the surface manifestation of the dynamo's magnetic field visible only for the other solar hemisphere.[47] The sunspots were found—as stated and shown—to be mostly confined to the southern-half of the Sun's face during the Maunder Minimum interval (Figure 48).

Steve Tobias, Edgar Knobloch and Nigel Weiss have clarified the situation and point to a second, more realistic mechanism. This mechanism invokes non-linear exchanges between magnetic energy and kinetic energy that have a direct consequence on the large-scale velocity field.[48] Fritz Krause[49] has further remarked that the weakened dipolar solar field during the Maunder Minimum can explain the increased cosmic ray intensity, traceable via increased ^{10}Be and ^{14}C production in ice cores and in tree ring records on Earth, respectively. Summarizing magnetic

---

[45] The most easily excited mode from the dynamo equation is that of the "odd parity" (such as that seen dominating the regular cyclic sunspot activity phase) meaning the mode which has mainly dipolar poloidal field and toroidal field which consists of an even number of rings (or toroids) which are arranged symmetrically to the equatorial plane, but in the opposite direction. "Even parity" can also be excited, but it requires stronger inductive actions. The "even parity" solution gives magnetic fields which have poloidal field in quadrupolar type, while the toroidal field consists of an odd number of rings (toroids).

[46] Note that each pair of the butterfly wings is not exactly symmetrical in the North-South extension, and so indicates slightly de-coupled oscillations of magnetic field in the Northern and Southern Hemispheres. This could be a hint of imbalances due to the presence of a weaker quadrupolar component during the regular cyclic phase.

[47] This is what was observed by the Paris Observatory researchers.

[48] In the language of the practitioners, they have included an additional, necessary equation for the velocity field that accounts for the back-reaction of the Lorenz force generated by magnetic field on the fluid flow (the mechanism of 'Malkus-Proctor'). See Tobias, S. M., *Astronomy and Astrophysics*, Vol. 322, 1997, pp. 1007–1017; Knobloch, E., Tobias, S.M., Weiss, N.O., *Monthly Notices of the Royal Astronomical Society*, Vol. 297, 1998, pp. 1123–1138.

[49] For example, Krause, F., "Oscillatory dynamos showing change of parity on a large time scale," in *The Solar Engine and Its Influence on Terrestrial Atmosphere and Climate*, ed. E. Nesme-Ribes (Springer-Verlag, Berlin-Heidelberg, 1994), pp. 49–56.

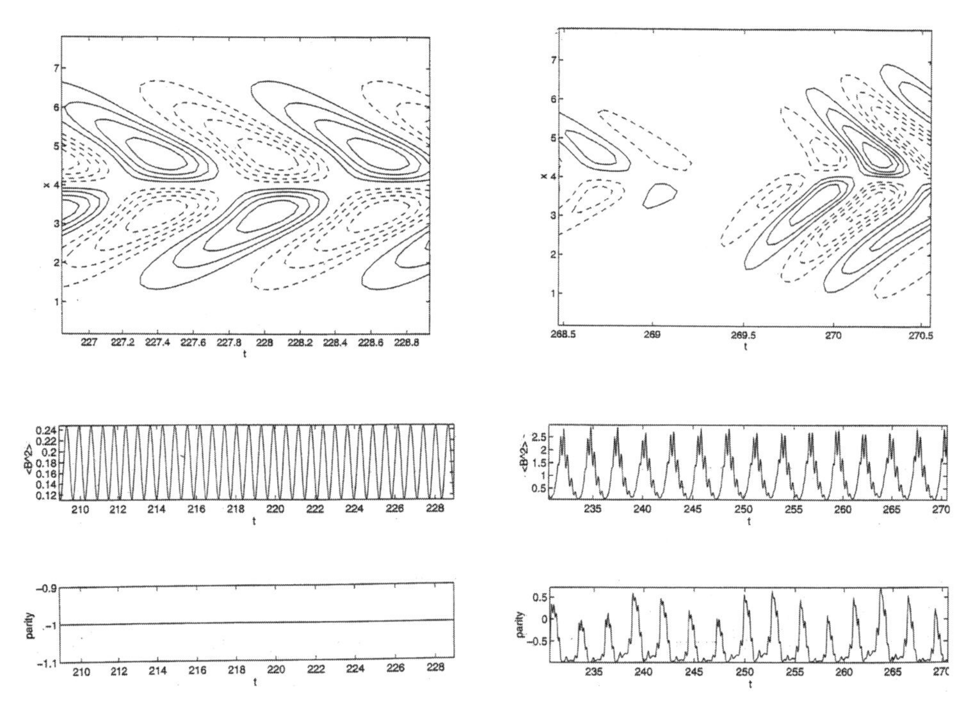

Fig. 52 The time evolution of magnetic field properties contrasting the periodic dipole solution (left column—revealing the regular, periodic toroidal field phase) versus the chaotically-modulated solution (right column—revealing the strongly suppressed toroidal field phase, a grand minima-type). Solid and dotted lines in the calculated butterfly diagram represent the positive and negative polarity of the magnetic field, respectively. (Time is presented in non-dimensional units scaled to the parameterization of simulation; a low value of B-square means lower magnetic energy, parity value of -1 means total dominance of the odd parity solution and larger positive p means increasing dominance of the even parity solution. The solution for the left column is for a low dynamo number case, while the right column is for a higher dynamo number.) The right column emphasizes the dominance of the even parity solution with the high degree of north-south asymmetry in the distribution of the toroidal magnetic field that is not unlike the interval of the Maunder Minimum in 1645–1715. Also note the large changes in the magnitude of the toroidal field as the Sun enters a period of weak minimum, complete disappearance of the toroidal field, and the highly asymmetric field in the Southern Hemisphere as the Sun emerges from the minimum between $t = 268.5$ and 270.0. The dipole pattern is restored at later times as the field grows stronger. (Courtesy Steve Tobias [1997].)[50]

field properties both during the cyclic and Maunder Minimum phases can be detailed in recent computer simulations made by Tobias and colleagues, shown below. Thus, a "Maunder Minimum" (right in Figure 52) effect is illustrated versus "normal" conditions (left in Figure 52).

[50] Tobias, S.M., *Astronomy and Astrophysics*, Vol. 322, 1997, pp. 1007–1017.

The dynamo model results may also be used to study another important question. One persistent enigma of the Maunder Minimum has always been the question about the nature of the 11-year solar activity cycle. That is, did the solar activity cycle simply vanish? There are hints from auroral records[51] and ^{10}Be isotope studies[52] that solar magnetic cycles persisted throughout the minimum, even though the overall activity was weak from about 1645 to 1715.

Once again, Tobias and colleagues are able to use the physics-motivated model to demonstrate that, indeed, the solar dynamo did not just die out. The solar cycle was still operating but it had been strongly modulated so that the produced spots (that is, for the model, it is the toroidal field) could be quite asymmetrically distributed on the solar surface, and at times, may even completely disappear— as observed during a deep activity minimum like the Maunder. This is a non-trivial answer based on modern theoretical analysis. Yet it was born of a question which—remarkably—had been anticipated years before by Walter Maunder:

> It ought not to be overlooked that, prolonged as this inactivity of the Sun certainly was, yet few stray spots noted during "the seventy years' death"—1660, 1671, 1684, 1695, 1707, 1718—correspond, as nearly as we can expect, to the theoretical dates of maximum. ... If I may repeat the simile which I used in my paper for *Knowledge* in 1894, "just as in a deeply inundated country, the loftiest objects will still raise their heads above the flood, and a spire here, a hill, a tower, a tree there, enable one to trace out the configuration of the submerged champaign," so the above-mentioned years seem to be marked out as the crests of a sunken spot-curve.[53]

There are still significant hurdles to meet in turning these non-linear dynamo results into estimates for solar wind conditions during the Maunder Minimum.[54] This in order to directly confront the terrestrial auroral and isotopic

---

[51] Gleissberg, W. and Damboldt, T., *Journal of British Astronomical Association*, Vol. 89, 1979, pp. 440–449; Schroder, W., *Journal of Geomagnetism and Geoelectricity*, Vol. 44, 1992, pp. 119–128. It is important to caution that a complete proof for the operation of the 11-year solar cycle during Maunder Minimum using auroral records may not yet be at hand. S. Silverman's analysis (*Reviews of Geophysics*, Vol. 30, 1992, pp. 333–352) showed that the cycle perhaps exists before and after the 1650–1725 interval, but not during the deep activity minimum.

[52] Beer J., Tobias S., Weiss, N., *Solar Physics*, Vol. 181, 1998, pp. 237–249.

[53] Maunder, E.W., *JBAA*, Vol. 32, 1922, p. 144.

[54] For example, Edward Smith (in *The Sun in Time*, eds. C. Sonett *et al.*, University of Arizona Press, Tucson, 1991, pp. 175–201) discussed even the conditions for the solar wind to cease during the Maunder Minimum, leaving only a "solar breeze" with wind speed of about 20 km/s (to get a feel for this number, consider the 29.8 km/s of the Earth's orbital motion around the Sun). The average solar wind speed at present (over 1964–1994) is about 440 km/s. However, Edward Cliver and colleagues

records. Even if the solar cycle may still be operating, one would still wish to know the precise outputs from the 1645–1715's Sun for a better understanding of its probable effects on the Earth's magnetic and climatic conditions.[55]

The Question of Solar Activity Maxima

The story of extended solar minima has been quite well covered. But what about solar maxima? Observationally, increased solar activity is associated with abundant sunspots. Take, for example, the period right before and after the Maunder Minimum. The "weakest" solar maximum—so far as is known—was noted in and around 1814–1820 during the Dalton Minimum's height. (The Dalton Minimum was literally a weak maximum during an overall solar minimum.) But this time around, we may be in a strong maximum. Indeed, the last several decades[56] may be among the most solar-active since instrument-based record keeping began.[57]

What Earth might now be going through is a period like the "Medieval" or "Grand Maximum" of the tenth to twelfth centuries (see Figure 5 in Chapter 2). In other words, a possible "Modern Maxima" may be ongoing. The Medieval Maximum caused the cosmogenic isotope production minimum during the tenth to twelfth Centuries as reflected by now-traceable ^{14}C and ^{10}Be records lasting over a period of some 150 years.[58] These records suggest solar activity had returned to Medieval Maximum highs after a prolonged, reduced solar-activity

---

(*continued*)

(*Geophysical Research Letters*, Vol. 25, 1998, pp. 897–900) estimated a somewhat higher solar wind speed of about 340 km/s. In addition, they have also deduced that the interplanetary magnetic field strength during the Maunder Minimum may have been about four times weaker than under present conditions. Cliver *et al.* (1998)'s result seems consistent with the aurora and ^{10}Be isotope studies.

[55] Echoing the 1801 call of William Herschel (*Phil. Trans. Roy. Soc. London*, Vol. 91, p. 316). "The result of this review of the foregoing five periods [from 1650 to 1713] is, that, from the price of wheat, it seems probable that some temporary scarcity or defect of vegetation has generally taken place, when the sun has been without those appearances which we surmise to be symptoms of a copious emission of light and heat. In order, however, to make this an argument in favour of our hypothesis, even if the reality of a defective vegetation of grain were sufficiently established by its enhanced price, it would still be necessary to show that a deficiency of the solar beams had been the occasion of it. " (Note, however, that Herschel's solar beam is connected to his suspicion of the relevance of transmission of some "sun-beams" for the growth of vegetation.)

[56] That is, since the 1930s–1940s.

[57] We are not relying only on measures like sunspot number in order to make this statement.

[58] See for example Jirikowic, J.L., Damon, P.E., *Climatic Change,* Vol. 26, 1994, p. 309. In the abstract, Jirikowic and Damon say: "Paleoclimatic studies of the Medieval Solar Maximum (c. A.D. 1100–1250, corresponding with the span of the Medieval Warm Epochs) may prove useful because it provides a closer analogue to the present solar forcing than the intervening era. The Medieval Solar Maximum caused the cosmogenic isotope production minimum during the 12th and

period. Perhaps this reduced period was the Oort Minimum, timed to about the years A.D. 1010–1050—and roughly coinciding with Viking exploration records for "more arable land,". south.[59] A reported climate cooling began in Greenland in the thirteenth century[60] which culminated in an abandoned/destroyed colony, and which may be connected to the earlier-noted, long-duration Spörer Minimum of circa A.D. 1420–1530. A "Modern Maximum" could be characterized by a solar magnetic "strong maxima" which may be sustained for decades, on Earth—or even for a century—as the Medieval Maximum apparently did.[61] Additional evidence supporting this notion may be noted from the fact that magnetic activity levels in the past few and present decades on the Sun are higher than average compared to some solar stars' levels.[62]

(continued)

13th centuries A.D. reflected by delta ^{14}C and ^{10}Be records stored in natural archives. These records suggest solar activity has returned to Medieval Solar Maximum highs after a prolonged period of reduced solar activity (possibly referring to the Oort Minima timed to about 1010–1050). Climate forcing by increased solar activity may explain some of this century's temperature rise without violating paleoclimatic constraints."

[59] Vikings sailed from Scandinavia to Iceland and Greenland from A.D. 985 to circa A.D. 1011, and subsequently founded colonies in Iceland, Greenland, and even North America. Founding the Nova Scotia colony may have been an attempt to gain more arable land in a cooler climate during the Oort Minimum. Try reading "The King Mirror" (written ca. 1250 by Einar Gunnarson, as recommended in Brekke's and Egeland's *The Northern Light*). Schaefer, B.E., *Sky & Telescope*, April 1997, pp. 34–38 gave some historical accounts of Leif Ericson's voyage around A.D. 1000, when he purportedly landed in America.

[60] In 982 Erik Röde and his son discovered Greenland. They had heard rumours about new land only a couple of days' sailing from Iceland. They arrived at Kap Farvel, turned north and discovered valleys covered with grass. Here they stayed the winter. The next spring they returned to Iceland to try to bring some more people with them to this new country. Erik got over 500 new settlers who followed him to the new country. The colony on Greenland came under Norwegian rule during the 13th century, but the contact with Norway slowly faded away. At the same time the climate changed for the worse. Still they clung to the colony for over 500 hundred years before they had to leave it because of the worsening living conditions. In the 15th century the colony died and left only ruins of some lonely houses. The last colonists are believed to have been killed by the plague, but no one really knows what happened. Perhaps they were killed by natives. (Högskolan I Luleå's Hemsida, Internet, Luleå Universitet) (Note: many historical accounts really need serious re-examination in order to filter out the climatic information.)

[61] Hence, a long duration of sustained high solar activity is also a part of the reality about our Sun.

[62] See for example, Baliunas, S. *et al.*, *The Astrophysical Journal*, Vol. 438, 1995, pp. 269–287. See also observational results on brightness changes of many solar stars by Wes Lockwood, Brian Skiff, Greg Henry and Richard Radick. Certainly, more progress in these terms is expected.

15

Summary: Cycles of the Sun and Their Tie to Earth

He had been Eight Years upon a Project for extracting Sun Beams out of Cucumbers,which were to be put into Vials hermetically sealed, and let out to warm the Air in the raw inclement Summers.

Jonathan Swift (1726, A Voyage to Laputa, Balnibarbi, Luggnagg, Glubbdubdrib and Japan)[1]

What was related in this book regarding the incidents in the Maunder Minimum was a tale told from the viewpoint of the witnesses who lived through the interval. These pioneering observers recorded and in some cases, made syntheses of the data they collected and even witnessed the mechanisms of solar behavior in action, commenting on them and describing them when and where they could. But for the most part the fact of a "weakening" solar activity was either not known or slow to be revealed.

As we and others get the impression from art in the late sixteenth through eighteenth centuries, people across most of the Northern Hemisphere had been aware of milder, warmer times in a not-so-distant past that was so unlike the colder, more variable times they were living in. At times, they recorded the effects of heightening freezing and precipitation in paintings, diaries and even in literature. One could not help but also note the care they took to ward off the cold or sudden storms (from architecture or inventions). People from the late nineteenth century well into the twentieth, at times admitted that the weather of their forebears' time was often more brutal than they knew it to be in their regions in their own lifetimes.

In glancing over the end of the Spörer Minimum to the putative start of the Maunder Minimum (in the 1620s or 1630s), there are in fact no sharp lines demarcating the end of the one (Spörer) and the beginning of the other minimum. That

---

[1] Swift (1726) *Gulliver's Travels*, Part III, Chapter 5.

is, from actual climate and sunspot records/aurora sightings, this end is not clear, even if there is good evidence between the minima that regular sunspot activity occurred (i.e., from sunspot studies by Scheiner and Galileo and others). That the oddly-punctuated poor weather began lifting in the second decade of the 18th century, along with a marked and well-recorded increase in sunspot and aurora activity versus "deep Maunder" (1645–1695) is, however, much clearer. This lends further empirical weight to the extended minima somehow shifting back into more regular cyclical behavior at or around the first decade of the eighteenth century. The mid-eighteenth century aurora record, for example, is very well represented. This is a sign of very much increased solar activity and the end of the Maunder Minimum.

Though likely responsible in itself for reduced global warming due to solar inactivity, the Maunder Minimum was at the same time locked into an already-existent, unsettled, unusually cold weather dynamic having probable roots in the mid-1400s if not earlier (the LIA interval) and which ended in the late 1800s. The Maunder Minimum found itself within the LIA and it served perhaps as a platform on which even deeper unsettled weather/cold dips of the LIA ran its course. This dynamic, as J. Grove found, was further stretched out than most believed, being longer than the A.D. 1550–1880 period claimed. For the trends in the LIA (or the Medieval Maximum—the opposite tendency of LIA) were not continuous or unbroken periods of cold or warmth, dryness or wetness; storminess, etc. Nor was it tied only to temperature—but to hydrologic cycles and other cycles involving ice, vegetation and other environmental variables as well. In such occurrences, we are perhaps seeing one hint of the forces that may have caused the great ice ages to spread and contract on Earth in the distant past, as the Lamoka Lake (Central New York) and Ennadai and Lynn Lake (Central Canada) isotope reconstructions illustrate. Additionally, we also see the quick changes—to warm, to cold, with odd storm effects, severe drought in unpredicted places, etc.—all happening on shorter time scales. Here, we formulate only an idea rather than any testable form of a hypothesis since there are still so much data to collect and synthesize.

In graphs, we presented the evidence of significant ups and downs of solar variability. One extended maximum, for instance, was the "Medieval Maximum" of approximately the eleventh through twelfth centuries. This is shown in the low ^{14}C production in the atmosphere at that time, and also by the lack of ^{10}Be in snow core samples. (The weather in Europe at this time was warm and often sunny, according to understandably incomplete evidence.) Possibly, we are now undergoing a similar period of extended maximum; literally, this could be called a "Modern Maximum." Yet, there is nothing to say that the current warm "spike" will not lead to a period such as 6,000–7,000 years ago (roughly the Holocene Maximum) witnessed by the Brewerton (Lamoka) people of what is now Upper New York state. According to pollen samples in undisturbed layers of soil deep in their burial places, they lived in a New York where the climate possibly resembled that of modern-day Missouri's. The pollen tells us that at least the trees indicated a more southern-only spread (perhaps Carolina Hemlock [*Tsuga caroliniana*] to name one) and to which the archaeologist, W.A. Ritchie, stated that "radiocarbon dating indicated that the climate may have been warmer [after a "thermal maximum"] and perhaps drier" c. 6,500 years ago. This and other evidence reinforces the fact that the tree-line was much higher north at times many thousands of years ago. That meant widespread climate change (that is, the climate in the current U.S. Northeast was approximately as warm as the lower-to-mid-Atlantic states; the mid-Canadian climate resembled approximately that of the current the U.S. Northeast, and so on) occurring naturally, and often abruptly, across history.

Human and Natural Global Warming

Though this matter is not the focus of our book the issue of natural global warming by solar and purely geothermic means must meet with the concern of CO_2-global warming produced by humans. This natural global warming, among other issues, is about how the Sun can affect Earth by linking to a subset of planetary warming/cooling processes and as such, it cannot and should not be ignored. That men and nature are contributing to the recent climate and ecosystem change is no secret, but the fear of global warming engendered by the unknown is also great.

The available generations of thermometers suggest that Earth's average surface temperature has warmed. But it has also been noted that this mean world-wide temperature has increased since 1850, before coal and fossil fuels became heavy factors for the attendant increases in carbon dioxide (CO_2) of about 25% in Earth's atmosphere. Accounting for other minor greenhouse gases,[2] the numerical estimate of the total human contribution or forcing on climate now rises to a "radiative equivalent" of a 50% increase in CO_2 alone or about a global radiative forcing of 2.5 W/m^2 over the last 100–150 years or so. This enhanced human factor is riding on top of our Sun's powerful radiance which puts out about 1370 W/m^2 at Earth's orbital distance.

---

[2] Additional minor trace gases that are suspected to be able to contribute to the greenhouse-effect (aside from the dominant greenhouse gas: water) are Methane (CH_4), Nitrous Oxide (N_2O), Ozone (O_3) and Chlorofluorocarbons (such as CFC-11 and CFC-12). The major greenhouse gas in the atmosphere is water vapor.

That world-wide mean temperature, however, did not rise steadily. Statistical temperature analyses since 1850 reveal year-to-year and decade-to-decade temperature change patterns.[3] Longer-term climate records (although with increasingly less area coverage as the records lengthen) hint that Earth's surface temperatures could have been on the overall *upswing* since the late 1600s. And this is well before world-wide industrialization began in volume.[4] To emphasize the very serious lack of records for a truly global representation of climatic entities like, for example, temperature over any extended time, we quote Malcolm Hughes and Henry Diaz:[5]

> Much work remains to be done to portray in greater detail the climatic essence of the ninth through fourteenth centuries. In particular, the simplified representations of the course of global temperature variation over the last thousand years reproduced in various technical and popular publications [for example, Eddy *et al.*, 1991; Frior, 1990; Houghton *et al.*, 1990[6]; Mayewski *et al.*, 1993] should be disregarded, since they are based on inadequate data that have, in many cases, been superseded. Equally obviously is the fact that temporal changes displayed by nearly all of the long-term paleotemperature records examined [here] indicate

[3] Allen, M.R. and Smith L.A., *Geophysical Research Letters*, Vol. 21, 1994, pp. 883–886.

[4] Bradley, R.S., Jones, P.D., *Holocene*, Vol. 3, 1993, pp. 367–376; Mann, M.E. *et al.*, *Nature*, Vol. 378, 1995, pp. 266–270. As noted in the previous chapter, the Little Ice Age and the recent 100-year rise of temperature could be interpreted as part of the natural millennial climatic oscillations that have been known to have persisted since at least 140,000 years ago, based on works from ice cores and marine sediment cores (for example, Keigwin, L.D., "The Little Ice Age and Medieval Warm Period in the Sargasso Sea," *Science*, Vol. 274, 1996, pp. 1504–1508; Bond, G. *et al.*, "A pervasive millennial-scale cycle in North Atlantic Holocene and glacial climates," *Science*, Vol. 278, 1997, pp. 1257–1266; Mayewski, P. *et al.*, "Major features and forcing of high-latitude northern hemisphere atmospheric circulation using a 110,000-year-long glaciochemical series," *Journal of Geophysical Research*, Vol. 102, 1997, pp. 26345–26366). For a summary of available evidence of climatic variation and change worldwide, see e.g., Soon, W., Baliunas, S., "Proxy climatic and environmental changes of the past 1000 years," *Climate Research*, Vol. 23, 2003, pp. 89–110 and Soon, W., Baliunas, S., Idso, C., Idso, S., Legates, D.R., "Reconstructing climatic and environmental changes of the past 1000 years," *Energy & Environment*, Vol. 14, 2003, pp. 233–296.

[5] "Was there a Medieval Warm Period, and if so, where and when?" Hughes, M., Diaz, H., *Climatic Change*, Vol. 26, 1994, pp. 109–142, see p.136. For an updated summary of available evidence of climatic variation and change world-wide, see e.g., Soon, W., Baliunas, S., "Proxy climatic and environmental changes of the past 1000 years," *Climate Research*, Vol. 23, 2003, pp. 89–110 and Soon, W., Baliunas, S., Idso, C., Idso, S., Legates, D.R., "Reconstructing climatic and environmental changes of the past 1000 years," *Energy & Environment*, Vol. 14, 2003, pp. 233–296.

[6] Houghton *et al.*, United Nations IPCC, on the state of climate change (1990).

that substantial decadal to multidecadal scale variability is present in regional temperature over the last millennium. This is, in all likelihood, a characteristic of most regional climates for the last several thousand years."

Yet new research suggested that the LIA could indeed be seen as part of the natural millennial climatic oscillations known to have persisted for over the last 140,000 years. This latest work is based on accumulating research knowledge from marine sediment cores and polar ice cores.[7] With the increase in temperature, we must not bypass the fact that solar activity has also been increasing over the last 100 years. This fact is witnessed by more than mere sunspot abundance.

What then is one supposed to make of the human-made greenhouse gases and their putative, long-term impacts on global climate system?

One useful question to start is: Do greenhouse gases directly heat the air? It has been pointed out that the "Greenhouse Effect" due to gaseous heating is not the correct operating procedure involved in the actual physical effect. This is simply because it is not the trapping of thermal radiation that brings increased warmth expected for the lower troposphere. The added warm anomaly from increasing concentrations of gases like water vapor, carbon dioxide, and nitrous oxide, and presumably, for the other trace elements in "greenhouse gas warming," such as ozone and CFC, in the troposphere is mainly caused by reduced flow of wind and sea current as well as enhanced latent heat from condensation of water vapor. So it should be said that the greenhouse works by preventing advective and convective cooling or latent heat warming: not by blocking escaping infrared radiation as the popular paradigm would have it.

There are many warming and cooling agents to consider in global warming overall rather than just man-made CO_2 and all the other minor greenhouse gases. There are changes to Earth's surface reflectivity (albedo) caused by shifting growth patterns in trees, green plants (rain forest removal/revival activities) and changing distribution of snow and ice. There are even complex urban development and large and small glacial fields to ponder; either can account for surface

[7] See Bond, G. *et al.*, "A pervasive millennial-scale cycle in North Atlantic Holocene and glacial climates," *Science*, Vol. 278, 1997, pp. 1257–1266. Bond, G. *et al.*, "The North Atlantic's 1–2 kyr climate rhythm: Relation to Heinrich events, Dansgaard/Oeschger cycles and the Little Ice Age," in *Mechanisms of Global Climate Change at Millennial Time Scales* (American Geophysical Union, Washington DC, 1999), eds. Webb, R.S., Clark, P.U., Keigwin, L.D., pp. 35–58. Please see also Keigwin, L.D., "The Little Ice Age and medieval warm period in the Sargasso sea," *Science*, Vol. 274, 1996, pp. 1504–1508.

reflectivity of sunlight. There are also oceanic current shifts and imbalances with intimate links to salinity and the atmosphere's temperature and precipitation fields. A role is even suggested for charged particles from the Sun and indeed the Galaxy which may act as mediators for nucleating water, ice, and aerosol particles that in turn convert into condensation nuclei for cloud formation[8]—literally, the making of clouds by cosmic means which contribute to Earth's warming and cooling. (As we are reminded by Eugene Parker in the Preface: "The worse mistake a scientist can make is to assert prematurely that some exotic new effect "cannot be" because our present limited knowledge does not cover the effect." Another reminder would be the incidence of Lord Kelvin's calculation incorrectly claiming to rule out the link between geomagnetic variations and the changing Sun outputs.) Changes in solar UV radiation modulate the ozone concentration in the stratosphere. This subsequently influences the troposphere's circulation patterns—including mid-latitude storm tracks.[9] Thus greenhouse gases are but one heating factor in a whole range of counterbalancing cooling/heating factors.

That human-produced greenhouse gases are added to the atmosphere is true: but whether or not CO_2 is the *determining* factor in large-scale global warming that could further lead to major, future global disaster for all is the hard question. The popular statement[10] that increases in CO_2 in the atmosphere resulting from increased industrial activity have not only caused global temperatures to rise over the past century, but if unchecked, will also cause catastrophic warming, is likely more emotional than correct.[11] In order to confirm the dominant role of human-made greenhouses in any climatic change scenario, there are the enormous tasks of rejecting the contribution by the Sun's variable outputs as well as various naturally changing patterns of air and sea circulation.

---

[8] See for example, Tinsley, B.A., *Eos, Transaction, American Geophysical Union*, Vol. 78, No. 33, 1997, p. 341.

[9] Haigh, J.D., "The impact of solar variability on climate," *Science*, Vol. 272, 1996, p. 981.

[10] The description is from a press release issued by the Fraser Institute, Centre for Studies in Risk and Regulation based on the publication *Global Warming: A Guide to the Science*, by W. Soon *et al.*, Fraser Institute, November 2001. See http://www.fraserinstitute.ca/admin/books/files/GlobalWarmingGuide.pdf or forward a request to W. Soon at wsoon@cfa.harvard.edu

[11] Other than the models not showing the results predicted by the CO_2 greenhouse gas "hypothesis," it has been pointed out that the "Greenhouse Effect" does not work as popularly believed—since it is not trapping thermal radiation that brings increased warmth expected for the lower troposphere. The added heat from increasing concentrations of gases like water vapor, carbon dioxide, and nitrous oxide (N_2O) in the troposphere is mainly caused by reduced wind and evaporation. So it should be said that the "greenhouse effect" works by preventing convection cooling: not blocking escaping infrared radiation. See Lee, R., "The 'Greenhouse Effect'," *Journal of Applied*

Here are some more relevant details. Although the temperature at the surface of the Earth, measured by thermometers placed on land and sea at locations around the globe has risen by about 0.5°C to 0.6°C over the last one hundred years or so, the main-stage of the warming occurred before most of the greenhouse gases were added to the air by human activities. The record shows that world-wide average temperatures peaked by around 1940, then actually cooled until the 1970s. Since then, there has been a warming of the Earth's surface. But as approximately 70–80 percent of the rise in levels of CO_2 during the twentieth century occurred after the initial major rise in temperature in the 1940s, the increase in CO_2 cannot have caused the bulk of the past century's rise in temperature. Thus, the bulk of the 20th century warming must have been natural.

What could be even more unsettling is the fact that the primary impact of the greenhouse effect of added CO_2 is in the lower atmosphere rather than on Earth's surface, but accurate measurements of that layer of air by U.S. National Oceanic and Atmospheric Administration (NOAA) satellites over the last 23 years have not shown any clear sign of global warming.

Concern that the continued increase in the concentration of CO_2 in the air will lead to a disastrous rise in the global temperature stems mainly from computer models of the climate system that have been making forecasts through the next century. The common tool for a computer simulation of the climate is the General Circulation Model (GCM). The climate models are an integral part, not only of the science of climate change, but also of the global warming policy issue so much at the forefront of current world politics.

However, present models are not sufficiently accurate in forecasting future climate change.[12] At present, it is impossible to isolate the effect of an increased concentration of atmospheric CO_2 on climate. Yet it is not surprising that it is still

---

(continued)

Meteorology, Vol 12, 1973, pp. 556–557; Essex, C., "What do climate models tell us about global warming?" *Pageoph* (*Pure and Applied Geophysics*), Vol. 135, 1991, pp. 125–133; Goody, R.M., *Principles of Atmospheric Physics and Chemistry* (OUP:1995), pp. 129–130; and Soon, W. *et al.*, "Modeling climatic effects of anthropogenic carbon dioxide emissions: Unknowns and uncertainties," *Climate Research*, Vol. 18, 2001, pp. 259–275. For more on convection cooling rather than infrared radiation trapping, see Held, I.M., Soden, B.J., "Water vapor feedback and global warming," *Annual Reviews of Energy and the Environment*, Vol. 25, 2001.

[12] See for example the scientific discussion outlined in Soon, W. *et al.*, "Environmental effects of increased atmospheric carbon dioxide," *Climate Research*, Vol. 13, 1999, pp. 149–164, and Soon, W. *et al.*, "Modeling climatic effects of anthropogenic carbon dioxide emissions: Unknowns and uncertainties," *Climate Research*, Vol. 18, 2001, pp. 259–275.

impossible to reliably calculate the climatic impact of increases in the concentration of atmospheric CO_2 and to distinguish the CO_2 imprints from other naturally varying changes. At any moment, around five million different variables have to be followed in a computer mock-up of the climate.[13] All their important impacts and interactions must be known, yet it is certain that they are not sufficiently well understood.

The Uneven Course of Solar and Geophysical Science in Sun-Earth Connection Study: From the Past into the Future

Conceiving the Maunder Minimum even in a rudimentary form took many years to accomplish, as this book has tried to show, just one aspect in the broad range of solar science. Much data describing this minimum was left in storage, as it were: much was forgotten or never properly collected in the first place—solar and terrestrial. A smaller, less politically and scientifically sophisticated world found in the seventeenth and eighteenth centuries can be alternately blamed and forgiven for this transgression. Dating the Maunder's Minimum occurrence is still not straightforward. Mild debates for a longer (1620–1720, 1645–1710) and shorter (1675–1715) duration persist and have not at all been settled. The insistence that such extended solar minima (alternately, maxima) are mere *hypothetical* (hence with no plausible terrestrial impacts) concepts still captures the attention of many people, influential or otherwise.

Thankfully, the sunspot record—such as it was—coinciding with a time of energetic and accurate use of early telescopes was left behind in a somewhat

[13] In order to convey a sense of *immaturity* in our ability to model the climate of the world, here is an extensive quote (Kerr, R.A., *Science*, Vol. 276, 1997, p. 1040): "The effort to simulate climate in a computer faces two kinds of obstacles: lack of computer power and a still very incomplete picture of how real-world climate works. The climate forecasters' basic strategy is to build a mathematical model that recreates global climate processes as closely as possible, let the model run, and then test it by comparing results to the historical climate record. But even with today's powerful supercomputers, that is a daunting challenge, says modeller Michael Schlesinger of the University of Illinois, Urbana-Champaign: 'In the climate system, there are 14 (to 16) orders of magnitude of scale, from the planetary scale, which is 40 million meters, down to the scale of one of the little aerosol particles on which water vapor can change phase to a liquid cloud particle—which is a fraction of a million of a millimetre.' Of these 14 (to 16) orders of magnitude, notes Schlesinger, researchers are able to include in their models only two of the largest: the planetary scale and the scale of weather disturbances: 'To go to the third scale—which is that of thunderstorms down around 50-kilometer's resolution—we need a computer a thousand times faster, a teraflops machine that maybe we'll have in five years.' And including the smallest scales, he says, would require 10^{36} to 10^{37} more computer

better condition than were the climate records. Still we despair that these historical records may not be fully recovered and restored. An age passed before this great dip in solar activity was noticed and connected by Walter Maunder and others to changing weather patterns, such as increased cold, abundant violent storms and heavy precipitation; fluctuating Earth magnetic activity, and other phenomena. For all this, even his and his contemporaries' observations were tenuous and incomplete. After Walter Maunder's death in the 1920s, it took another fifty years before the low dip in solar activity was finally pinned down to specific data. Solar astronomer J.A. Eddy attached Maunder's name to these findings and connections, so relevant to solar astronomical research, today, and rightfully so. Beyond Eddy's restoration of Maunder's place in these proceedings, Maunder and his wife's pioneering work stands on its own merits:

- Butterfly diagrams
- Sun-Earth geomagnetic phenomena observations and discoveries, including witnessing solar bursts affecting Earth's largest magnetic storms, recognizing the role of fast solar wind (definite rays or beams) from M regions or coronal holes
- Other important insights like a brighter, rather than a dimmer, Sun when solar magnetism is strong

It is to E. Walter Maunder and his wife Annie and to their solar work that solar research is very much indebted. Even relatively modern theorists such as Sydney Chapman and astronomers like G.E. Hale found much grist for the mill of solar physics research due to the Maunders' painstaking work and basic scientific

---

(*continued*)

power. 'So we're kind of stuck.' " Indeed we suspect that the latter of the two major problems will be the ultimate limit to known present observations, let alone predicting future climates.

The distinguished dynamic meteorologist Edward Lorenz somewhat humorously reminded us how one early twentieth century meteorologist, Napier Shaw described what we called a "model" today as simply a "fairy tale" (*The Essence of Chaos*, University of Washington Press, Seattle, 1993, p. 87). In describing the current state-of-the-art, five million variables European Centre for Medium Range Weather Forecast, we hear that: "Lest a system of 5,000,000 simultaneous equations in as many variables appear extravagant, let us note that, with a horizontal grid of less than 50,000 points, each point must account for more than 10,000 square kilometres. Such an area is large enough to hide a thunderstorm in its interior. I have heard speculations at the Centre that another enlargement of the model is unlikely to occur soon ..." (pp. 101–102).

Philosophically, and in a similar context to our statement about models not being reality, the renowned mathematician and computing pioneer, Peter D. Lax in 1987 is noted to have said: "... there is a danger that the computation becomes a substitute for a non-existent theory." (p. 132 of Essex, C., "What do climate models tell us about global warming?" *Pageoph [Pure and Applied Geophysics]*, Vol. 135, 1991, pp. 125–133).

constructs surrounding their 30-plus year's worth of solar observation and study. These were solar observations and research not in one particular form, but in many forms (solar spots' definition and behavior, faculae; rudimentary understanding of solar magnetism and its behavior; eruptive loops, the corona and so on). The Maunders in particular interconnected the phenomena where and when science allowed. Their practical astronomical studies and methods (use of photography and illustration) should also not be ignored. Their work placed into sharp relief the then-ongoing preliminary work on physical interactions between solar and terrestrial charged particles as done by Birkeland, Paulsen, and Arrhenius, subsuming it to the greater—and eventual—understanding of the magnetosphere, formulated to a great extent by Chapman, Ferraro, Asakofu and refined by many others—such as Parker, who helped delineate the solar wind.

For all this, whether in making the fundamental observations and piecing together discoveries in the Sun-Earth connection, all those mentioned above depended heavily on their forebears' contributions. Maunder appreciated and valued the findings of Wolf, Spörer, Schwabe, and Sabine, as Chapman and even Bartels valued the work of Maunder, Birkeland, Paulsen and others. Without the useful scepticism of Lord Kelvin and F. Lindemann at crucial times, the gain in solar science would have been less. All would have been at a loss without the work of Scheiner, Galileo, Flamsteed, de Mairan, and others still. Additionally, how would it now be possible to get a factual view into the past without the tools found and shaped in part by W. Libby and H. Suess? The curious questioning by Douglass—working on trees and not stars—led not only to a concrete connection between the two, but to an entirely new branch of science—dendrochronology.

By choosing the way we related these linkages between thinkers and discoverers, grasping at times at straws of truth, and having to reject others whilst looking for tools and processes, we have tried to illustrate science's sometimes tenuous and very often difficult, oft-repeated, and crooked paths. In particular, we have tried to show the difficulties and tangibility of Sun-Earth physics in terms of obtaining concrete, useable, and conclusive findings.

* * *

Modern research, where it has been built on the Maunders' work, has been positive as, for example, non-linear dynamo theoretical interpretations of solar cycles now suggest. There was, in the distant past—as modern evidence has gleaned from isotopic records left on Earth by the Sun in the tenth, and twelfth to thirteenth centuries—times when some cycles operated at a higher intensity. Additionally, as the Maunder Minimum illustrates, there were indeed times (so the isotopes indicate) when the cycles were more sluggish. With solar proxy evidence in abundance and better Earth climate reconstructions ongoing—including the pointing out of a better understanding of what the LIA was, and what it meant—the hope for a more positive regimen in solar physics and geophysics is certainly apparent. But as the latter chapters of this book show, we have only theory to buttress most claims as to why solar cycles, alone, are as they are. How they operate is still, by and large, a mystery. How they affect Earth's climate, apart from or with Earth's own geophysical frame, is also as yet out of reach.

Knowledge gained as to how the Sun works in tandem with Earth has been slow to accumulate, and at times, hard to keep in focus over longer periods of time. As we have tried to show, the database for exploration is rich and continuously growing—however hard it was to create and, at times, reconstruct. Due to neglect or avoidance, this database can grow cold or be lost. Yet it is always ready to be mined by those willing to investigate the riddles posed by our Sun's and Earth climate's operating faculties. The question is, who will be the next solar physicists and geophysicists who will rely heavily on the work of their forebears to establish even greater understanding of the cycles of the Sun as they affect Earth? This, like much that is unforeseeable, lies in the future.

16

The Maunders and Their Final Story

Science fiction is replete, leitmotif upon motif, in relationships men and women have in imagined worlds, deep in space, far into time—back into the past or into the future. We speak in terms of an interdependence of thought and body: a fluid, protean world of man and woman, where form and function intertwine, taking and forming new shapes to fit new functions. The often equally protean nature of ancient Greek or Roman myth comes to mind. Even though all this is usually associated with the non-factual, seldom does one realize how often, among men and women astronomers—famous or otherwise—a similar bonding, interdependence of thought, if not of frame, must occur.

As this book has hopefully shown, we have seen a few famous pairings, some like the Maunders more than others. In science fiction, whole bodies are transformed into androids, or cyborgs, or other such phantasmagoria. Such a description is to underline the fluidity of form in the face of function, perhaps: deep space still a mystery to us, as Earth's oceans once were, and somewhat still are. But the marriage bond, or the love bond, if one will, creates with it a kind of strange but often wonderful blend in individuals that sees fit to compensate for the weaknesses of either, such as they are, to make for a better, more complete whole.

It is not the intention here to draw an overly romantic picture of what A.S.D. and Walter Maunder found together as husband and wife. It would be overly-cynical, by the same token, to assert that the better-educated (and, by association, the more skilled) Annie was making her "will known" through a well-placed and respected adjunct to the Royal Observatory by a marriage of convenience. That Walter Maunder may have taken an avuncular approach to his relationship with a woman young enough to be a daughter is not unlikely, serving at times as a mentor. Little is known of the personal relationship here and we wish to go no further. In the absence of other evidence, what drew these two together was, in Mary Bruck's words, their evangelical Christian spirit.

Married first at age twenty-three to Edith Bustin—with whom he had all his children—Maunder was a widower by the time he was 36, Edith having died of acute tuberculosis just short of her thirty-sixth birthday. She had been ill for over a year.[1]

It seems quite possible that Annie did, indeed, bring a rounder, more professional view to much of Walter's, and her own, work, albeit in the professional manner of those trained beyond undergraduate rank, and it clearly shows in much of her own work, and perhaps, had empowered and rounded out more fully some of the observations, annotations, and mathematics of her husband.

She was a skilled solar observer, and photographer. Both astronomers kept up a world-wide correspondence with astronomers and others, right up to their deaths. Walter was a good public speaker, and a fascinating one, being not in the least shy in this regard. Annie, on the other hand, was reluctant to speak, so she claimed, due to her voice, and refused posts in the B.A.A., for example, that would cause her to be a moderator or speaker before public gatherings. As such, the picture of this man-and-woman team is complete. It need only be noted in the obituary for Mrs. Maunder of 1948,[2] written by a friend of nearly fifty years' standing—astronomer Mary Orr Evershed—who stated that "love on both sides was deep and true," besides the obvious references to their mutual scientific interests.

Annie outlived Walter by nineteen years. E. Walter Maunder, after an extended period of "great disability," died on March 21, 1928,[3] to Annie's "great grief," as another obituary by Mrs. Orr Evershed related.[4] For some years previous to this, he had been "afflicted with painful internal trouble."[5] It was a month before his 77th birthday. He had taken an active part, as we could well think, in the affairs of the Church of England.

---

[1] Edith Hannah Bustin was born in Shoreditch, Sub District Kennington, October 29, 1852 at 11 Claremont Place. Her father was Antony Bustin, "Gentleman," and mother, Matilda Sarah Bustin. E. Walter Maunder, listed as an "Assistant at the Royal Observatory," was married in the Wesleyan Chapel, High Street, Clapham, by his own father on 11 September 1875. E. Walter's address at this time was 4 Vansillart Terrace, Greenwich. Edith's address was 17 Althorpe Road, Wandsworth Common. Her father being deceased, their wedding was witnessed by Thomas Frid Maunder, Jane Eliza Martindale, Ann Eliza Wheeler (E. Walter Maunder's sister), and probably Hannah's brother, Antony Bustin. Her death was recorded at Greenwich East for 26 October 1888, Hyde House, Ulundi Road. Her death was certified by A. Forsyth (MD) "Cause: Phthisis Pulmonalis (1 year, 3 months), Acute Tuberculosis (21 days)." (Personal communication with Alan Maunder, Maunder's direct descendant.)

[2] A.S.D. Maunder Obituary, Notices, *MNRAS,* Vol. 108, 1948, pp. 48–49.

[3] Ibid, Maunder Obituary, *JBAA,* Vol. 38, No. 8, Session 1927–1928.

[4] A.S.D. Maunder Obituary, Notes, *JBAA,* Vol. 57, 1947, p. 238.

[5] Ibid, "Report to the Council to the ...," *JBAA*, LXXXIX, February 4, 1929.

In the *Nature* obituary, for April 7, 1928[6] two points stand out, in particular, that emphasize Maunder's "unrecognized recognition": that his being brought into the Royal Observatory represented one of the first and most important thrusts into making the observatory a "physical observatory" (that is, Maunder's embodiment of the essence of a pioneering astrophysicist) and second, his "ready pen," which was not merely this, but of course the skilled tool of an acute mind.

The laborious nature of the first film developing process at Greenwich is described, since Maunder was appointed the photographer of the Sun before the "gelatin dry-plate" technique was fully adopted. Maunder is shown to be a pioneer in this revealing obituary. Only later, after Maunder's determined dedication to the studies of the Sun are shown, is he given an assistant who can take some of the burden off of him. Clearly, his mind is running ahead of limitations in technology and technique, and is busy devising new ones. It is without doubt that he transferred a great deal of knowledge on photography to Annie, who became more adept at it than him (as her work bore out). It is also said that Maunder was "freed up" to do more important work, like the focus on sunspots and other phenomena, and that the staff grew in the 1880s—whether at his instigation or due to his zeal, or both, or other factors, is not well known. He returned to this work after retirement, in the First World War, taking up his old role of working with sunspots, being put "in charge" of this in 1916.[7]

What had perhaps been glossed over is his active role in founding the B.A.A., which by 1928 was recognized as a national achievement and as a society of scientists, relatively independent of class concerns, free to think in whatever way they pleased, be they a Stephen Gray or a Lord Kelvin. Annie helped him form the B.A.A. journal's shape and structure,[8] and freed Walter of being editor after a year, when he was president. Noticeable, too, is his rather-modern respect and understanding for not only other religions, but even the opposite sex—insofar as his helping found the B.A.A. had as one of its goals, the providing of an open forum where all could pursue astronomical study.

Finally, it is Walter's unpretentious and interesting writing style that takes center stage. His "command of language" is made much of, and here, we see his

---

[6] Ibid, Maunder obituary in *Nature*, No. 3049, Vol. 121, 7 April, 1928, pp. 455–456.

[7] Ibid, Bruck, "Alice Everett and Annie Russell Maunder torch bearing women astronomers," *Irish Astronomical Journal,* p. 289.

[8] Ibid, A.S.D. Maunder Obituary in J.B.A.A. According to Bruck, the only income A.S.D. Maunder personally made (that is, by herself) for written work after was winning the Gibson Prize from Girton College in 1923. It was an essay on a Biblical subject and earned her £123.

support of Annie in some of her work. As she writes, we can see much of his assistance, seeing as he had a flair for words, and had been an editor and writer for years in journals like *The Observatory*, the J.B.A.A., the organ for the RAS, the journal *Nature*, and others, still. He was also the author of numerous books, one of them being *Astronomy Without a Telescope*—a very popular item. It is found in observatory libraries (Harvard College Observatory for example) to this day.

His eldest brother, Thomas, who died at the age of 92, outlived him by seven years. An obituary statement remarks that he and E. Walter were the "founding fathers" of the society, both acting as "honourable secretaries." Thomas eventually became the legs and arms of the B.A.A., becoming Assistant Secretary, and held this position for the next 38 years.

Annie lived until September 15, 1947, dying at the age of 79. She had served as the J.B.A.A.'s editor for two years in the middle 1890s, and for thirteen years between 1917 and 1930. Like her husband, she contributed notes and articles on ancient and current astronomy and her papers were even published at the Victoria Institute, where her husband was a pious associate of many years' standing.[9] She was known "ever as a steadfast friend and cheerful colleague whose loyalty was the outstanding feature of her character."[10]

[9] Ibid, Bruck, "Alice Everett and Annie Russell Maunder torch bearing women astronomers," *Irish Astronomical Journal,* p. 284.

[10] Obituary of Mrs. E. Walter Maunder, December, 1947, The Observatory, Vol. 67, pp. 231–232.

Index of Sources

Historical Journals (Defunct)

Child, A.L., "Annual growth of tree rings," *Popular Science Monthly*, Vol. XXII, December, 1882.

Child, A.L., "Concentric rings of trees," *Popular Science Monthly*, Vol. XXIV, December, 1883.

Clerke, A.M., "Letters," *Knowledge*, Vol. 17, September 1, 1894.

Douglass, A.E., "Climatic records in the trunks of trees," *American Forestry*, Vol. XXIII, December, 1917.

Manley, G., "A preliminary note on early meteorological observations in the London region, 1680–1717, with estimates of the monthly mean temperatures, 1680–1706," *Meteorological Magazine*, Vol. 90, 1961, pp. 303–310.

Maunder, E.W., "A prolonged sunspot minimum," *Knowledge*, Vol. 17, August 1, 1894.

Maunder, E.W., "The solar origin of terrestrial magnetic disturbances," *Popular Astronomy*, Vol. XIII, No. 2, February, 1905 (cf. *Astrophysical Journal*, in 1905).

Mitchell, W.M., "The history of the discovery of the solar spots," *Popular Astronomy*, Vol. 24, 1916 (series of papers extending from p. 22 to p. 562).

Monck, W.H.S., *English Mechanic*, July 18, 1890.

Paulsen, A., "The radiation theory of northern light," *Nyt – Tidskrift for Fysik og Kemi*, 1896.

Popular Journals (Existing)

Baliunas, S., Soon, W., "The Sun-Climate Connection," *S&T*, Vol. 92, No. 6, December, 1996.

Billington, M., "A grand design: Kepler and renaissance science in China," *21st Century Magazine*, Summer, 1996.

Burke, J., "And now the weather," *Scientific American*, Vol. 280, No. 1, January, 1999.

Burke, J., "Lend me your ear," *Scientific American*, Vol. 280, No. 3, March, 1999.

Fleischman, J., "Tropics on ice," *Earth*, Vol. 6, No. 5, October, 1997.

Lang, K.R., "SOHO reveals secrets of the sun," *Scientific American*, March, 1997.

Lemmonick, M.D., "The end of evolution?" *Time*, August 23, 1999.

Maneker, R., "Fit for a king," *Art & Antiques*, November, 1998.

Nesme-Ribes, E., Baliunas, S., Sokoloff, D., "The stellar dynamo," *Scientific American*, Vol. 275, No. 2, August, 1996.

Sandel, J., "Observing the Sun by projection," *S&T*, Vol. 94, No. 4, October, 1997.

Soon, W., Yaskell, S.H., "Cycles of the Sun," *Astronomy Now*, Vol. 12, No. 10, October, 1998.

Soon, W., Yaskell, S.H., "Year without a summer—Eighteen hundred and froze to death," *The Mercury*, Vol. 32, No. 3, May/June, 2003.

Stephenson, F.R., "Early Chinese observations and modern astronomy," *S&T*, Vol. 97, No. 2, February, 1999.
Taylor, K., "Rapid climate change," *American Scientist*, Vol. 87, No. 4, July–August, 1999.

Scientific Journals, Supplements, Conference Notes, Contributed Chapters, Letters, Diaries, Memoirs, Addresses, etc.

Adams, W.G., "IV. Comparison of simultaneous magnetic disturbances at several observatories," *Phil. Trans. Roy. Soc. of London*, Vol. 183, 1891.
Akasofu, S.-I., "A note on the Chapman–Ferraro theory," *Physics of the Magnetopause*, eds. Song, P. *et al.* (American Geophysical Union, 1995).
Allen, M.R., Smith, L.A., "Investigating the origins and significance of low-frequency modes of climate variability," *Geophysical Research Letters*, Vol. 21, 1994.
Anonymous, "The Monck Plaque," News and Comments, *Irish Astronomical Journal*, Vol. 18, 1989.
Arrhenius, S., "On the electric equilibrium of the Sun," *Proceedings Royal Society London*, Vol. 73, 1904, pp. 496–499.
Baldwin, M., "Magnetism and the anti-Copernican polemic," *JHA*, Vol. XVI, 1985.
Baliunas, S.L. *et al.*, "Chromospheric Activity Variation II.," *Astrophysical Journal*, Vol. 438, 1995.
Baliunas, S., Soon, W., "Are variations in the length of the activity cycle related to changes in brightness in solar-type stars?" *Astrophysical Journal*, Vol. 450, 1995.
Baliunas, S., Frick, P., Sokoloff, D., Soon, W., "Timescales and trends in the central England temperature data (1659–1990): A wavelet analysis," *Geophysical Research Letters*, Vol. 24, 1997.
Bartels, J., "Terrestrial-magnetic activity and its relation to solar phenomena," *Terrestrial Magnetism and Atmospheric Electricity*, Vol. 37, March, 1932.
Baumgartner, F.J., "Sunspots or Sun's planets: Jean Tarde and the sunspot controversy of the early seventeenth century," *JHA*, Vol. XVIII, 1987.
Beer, J., Raisbeck, G.M., Yiou, F., "Time Variations of ^{10}Be and Solar Activity," in *The Sun in Time*, eds. C. P. Sonett *et al.* (University of Arizona, 1991), p. 343.
Beer, J., Tobias, S., Weiss, N., "An active Sun throughout the Maunder minimum," *Solar Physics*, Vol. 181, 1998.
Bennett, J.A., "Cosmology and the magnetical philosophy," *JHA*, Vol. XII, 1981.
Berger, A., "Milankovitch theory and climate," *Reviews of Geophysics*, Vol. 26, November, 1988.
Bond, G. *et al.*, "A pervasive millenial-scale cycle in North Atlantic Holocene and glacial climates," *Science*, Vol. 278, 1997.
Bond, G. *et al.*, "The North Atlantic's 1–2 kyr climate rhythm: Relation to Heinrich events, Dansgaard/Oeschger cycles and the Little Ice Age," in *Mechanisms of Global Climate Change at Millennial Time Scales* (American Geophysical Union: Washington DC), eds. R.S. Webb, P.U. Clark and L.D. Keigwin, 35–58, 1999.
Bond, G. *et al.*, "Persistent solar influence on North Atlantic climate during the Holocene," *Science*, Vol. 294, December 7, 2001, pp. 2130–2136.
Bonino, G., *et al.*, "Behavior of the heliosphere over prolonged solar quiet periods by ^{44}Ti measurements in meteorites," *Science*, Vol. 270, 1995.
Boyle, R., "An observation of a spot in the Sun," *Philosophical Transactions of the Royal Society*, Vol. 74, April 27, 1671.

Bozhong, L., "Changes in Climate, Land and Human Efforts" (from R.B. Marks' *Tigers, Rice, Silk and Silt*, CUP, 1998).

Bradley, R.S., Jones, P.D., "Little Ice Age summer temperature variations: their nature and relevance to recent global warming trends," *Holocene*, Vol. 3, 1993.

Brandenburg, *et al., Turbulent and Nonlinear Dynamics in MHD Flows*, eds. Menuzzi *et al.* (North-Holland, 1989).

Broecker, W.S., "Massive iceberg discharges as triggers for global climate change," *Nature*, Vol. 372, December 1, 1994.

Brook, E.J., Sowers, T., Orchardo, J., "Rapid variations in atmospheric methane concentration during the past 110,000 years," *Science*, Vol. 273, August 23, 1996.

Brooks, R.C., "The development of micrometers in the seventeenth, eighteenth, and nineteenth centuries," *JHA*, Vol. XXII, 1991.

Brooks, R.C., "Errors in measurements of the solar diameter in the 17th and 18th centuries," *JHA*, Vol. XIX, 1988.

Bruck, M.T., "Agnes Mary Clerke, chronicler of astronomy," *Quarterly Journal of the Royal Astronomical Society*, Vol. 35, 59–79, 1994.

Bruck, M.T., "Alice Everett and Annie Russell Maunder: Torch bearing women astronomers," *Irish Astronomical Journal*, Vol. 21, 281–291, 1994.

Bruck, M.T., "Lady computers at Greenwich in the early 1890s," *Quarterly Journal of the Royal Astronomical Society*, Vol. 36, 83–95, 1995.

Bruck, M.T., Grew, S., "The family background of Annie S.D. Maunder (née Russell)," *Irish Astronomical Journal*, Vol. 23, 1996.

Bruck, M.T., "Mary Ackworth Evershed Née Orr (1867–1949) solar physicist and Dante scholar," *Journal of Astronomical History and Heritage*, Vol. 1, 1998.

Camuffo, D., Enzi, S., "Reconstructing the climate in Northern Italy from archive sources," in *Climate Since 1500* A.D., eds. R.S. Bradley, P.D. Jones (Routledge, New York, 1995), pp. 143–154.

Carignano, C.A., "Late Pleistocene to recent climate change in Córdoba Province, Argentina: Geomorphological evidence," *Quaternary International*, Vol. 57/58, 1999.

Carrington, R., "Description of a singular appearance seen in the Sun on September 1, 1859," *MNRAS*, Vol. 20, 1860.

Casanovas, J., "Early observations of sunspots: Scheiner and Galileo," *1st Advances in Solar Physics Euroconference/Advances in the Physics of Sunspots* (ASP Conference Series, Vol. 118, 1997).

Chapman, S., "The energy of magnetic storms," *MNRAS*, Vol. 79, 1918.

Chapman, S., "Historical introduction to aurora and magnetic storms," *Annales Geophysique*, Vol. 24, 1968.

Chapman, S., Ferraro, V.C.A., "A new theory of magnetic storms," *Nature*, Vol. 126, 1930.

Cioccale, M.A., "Climate fluctuations in the central region of Argentina in the last 1000 years," *Quaternary International*, Vol. 62, 1999.

Clark, D.H., Murdin, L., "The enigma of Stephen Gray, astronomer and scientist (1666–1736)" *Vistas in Astronomy*, Vol. 23, 1979.

Cliver, E.W., "Solar activity and geomagnetic storms: The first 40 years," *Eos*, Vol. 75, No. 49, December 6, 1994.

Cliver, E.W., "Solar activity and geomagentic storms: The corpuscular hypothesis," *Eos*, Vol. 75, No. 52, December 27, 1994.

Cliver, E.W., "Solar activity and geomagnetic storms: From M-regions and flares to coronal holes and CMEs," *Eos*, Vol. 76, No. 8, February 21, 1995.

Cliver, E.W., Boriakoff, V., Bounar, K.H., "Geomagnetic activity and the solar wind during the Maunder minimum," *Geophysical Research Letters*, Vol. 25, 1998.

Cliver, E.W., Siscoe, G., "History of the discovery of the solar wind," from Section News, *Eos*, Vol. 75, No. 12, March 22, 1994.

Cowling, T., "Sydney Chapman," *Biographical memoirs of the fellows of the royal society*, Vol. 17 (Headley Brothers, 1971), pp. 52–89.

Crommelin, A.C.D., Maunder Obituary, *The Observatory*, May, 1928.

Crooker, N.U., Cliver, E.W., "Postmodern view of M-regions," *Journal of Geophysical Resea*rch, Vol. 99, No. A12, December 1, 1994.

Damon, P.E., "The natural carbon cycle," from *Radiocarbon After Four Decades*, eds. R.E. Taylor, *et al.* (Springer-Verlag, 1992).

Damon, P.E., Sonett, C.P., "Solar and terrestrial components of the atmospheric ^{14}C variation spectrum," in *The Sun in Time*, eds. C. P. Sonett *et al.* (University of Arizona, 1991), p. 360.

Daglis, I.A., Thorne, R.M. *et al.*, "The terrestrial ring current: Origins, formation, and decay," *Reviews of Geophysics*, Vol. 37, No. 4, 1999.

de Vries, H., "Variation in concentration of radiocarbon with time and location on earth," *Koninkl. Ned. Akad. Wetenschap, Proc. Ser.* B61, 1, 1958.

Dils, J., "Epidemics, Mortality and the Civil War in Berkshire, 1642–1646" from *The English Civil Wars*, ed. R.C. Richardson (Sutton, 1997).

Dixon, F.E., "Some irish meteorologists," *The Irish Astronomical Journal*, Vol. 9, No. 4, December, 1969.

Douglass, A.E., "Weather cycles in the growth of big trees," *Monthly Weather Review*, XXXVII, June, 1909.

Douglass, A.E., "*Climatic Cycles and Tree Growth: A Study in the Annual Rings of Trees in Relation to Climate and Solar Activity*" (Carnegie Institution, 1919).

Drake, S., "Galileo's First Telescopic Observations," *JHA*, Vol. VII, 1976.

Dyson, F., Christie, W., "IX. Drawings of the corona from photographs at total eclipses from 1896 to 1922," *Memoirs of the RAS*, Vol. 64, 1925–1929.

Dyson, F. W., Maunder, E. W., "The position of the Sun's axis as determined from photographs of the Sun from 1874 to 1912, measured at the Royal Observatory, Greenwich," *MNRAS*, Vol. LXXIII, 1913.

Eddy, J.A., "The Maunder Minimum," *Science*, Vol. 192, 1976.

Eddy, J.A., "Historical and Arboreal Evidence for a Changing Sun," in *AAAS Selected Symposia 17: The New Solar Physics*, 1978, pp. 10–33.

Eddy, J.A., "Climate and the role of the Sun," *Journal of Interdisciplinary History*, Vol. X, No. 4, Spring, 1980.

Ellis, W., "Sun-spots and magnetic disturbance," *JBAA*, Vol. LXI, June, 1901.

Ellis, W., "The aurorae and magnetic disturbance," *MNRAS*, Vol. LXIV, January, 1904.

Engardt, M.L., Rhode, H., "A comparison between patterns of temperature trends and sulfate aerosol pollution," *Geophysical Research Letters*, Vol. 20, 1993.

Essex, C., "What do climate models tell us about global warming?" *Pageoph (Pure and Applied Geophysics)*, Vol. 135, 1991.

Evans, D.S., "Historical note on astronomy in South Africa," *Vista in Astronomy*, Vol. 9, pp. 265–282, 1967.

Fairbridge, R.W., Shirley, J.H., "Prolonged minima and the 179-yr cycle of the solar inertial motion," *Solar Physics*, Vol. 110, 1987.

Fara, P., "Lord Derwenter's lights: Predictions and the Aurora Polaris," *JHA*, Vol. XXVII, 1996.

Fernow, B.E., "Age of trees and time of blazing determined by annual rings," *USDA, Division of Forestry, Circular No. 16* (Washington, D.C., 1897) (E.G. Webb citation).

Flamsteed, J., Historical note in "Notes," *JBAA*, Vol. 51, No. 3, 1941.

Fleming, J.R., "The great climate debate in early America," from *Historical Perspectives on Climate Change* (OUP, 1998) (Holyoke quote).

Forman, M.A., Schaeffer, O.A., "^{37}Ar and ^{39}Ar in meteorites and solar modulation of cosmic rays during the last thousands years" in *The Ancient Sun*, eds. R.O. Pepin *et al.* (Pergamon, 1980), pp. 279–292.

Foster, K.R., Jenkins, M.E., Toogood, A.C., "The Philadelphia Yellow Fever Epidemic of 1793," *Scientific American*, Vol. 279, No. 2, August, 1998.

Gleissberg, W., Damboldt, T., "Reflections on the Maunder minimum of sunspots," *JBAA*, Vol. 89, 1979.

Gould, S.J., "Our Natural Place," from *Hen's Teeth and Horse's Toes* (Norton, 1983).

Grove, J., "The Century Time Scale," in *Time Scales and Environmental Change*, eds. T.S. Driver, G.P. Chapman (Routledge, 1996), pp. 39–87.

Haigh, J.D., "The impact of solar variability on climate," *Science*, Vol. 272, 1996.

Hale, G.E., "The spectroheliograph and its work, Part III: Solar eruptions and their apparent terrestrial effects," *Astrophysical Journal*, Vol. 73, 1931.

Hansen, J. *et al.*, "Common sense climate index: Is climate changing noticeable?"
Proceedings of the National Academy of Science (*USA*), Vol. 95, April 14, 1998.

Held, I.M., Soden, B.J., "Water vapor feedback and global warming," *Annual Reviews of Energy and the Environment*, Vol. 25, 2000.

Henry, J., "Observations on the relative radiation of the solar spots," *Proceedings of the American Philosophical Society*, Vol. IV, June 20, 1845.

Herschel, W., "Observations tending to investigate the nature of the Sun, in order to find the causes and symptoms of its variable emission of light and heat ...," *Philosophical Transactions of the Royal Society of London*, Vol. 91, 1801.

Hiorter, O.P., "Om-magnet-nålens Åskillige ändringar," *Kongl. Svensk Vetenskaps Akad. Handlingar*, Vol. 8, 1747, pp. 27–43 (reference made by Meadows and Kennedy).

Hobbs, P. V., "Clouds: Their beauty and challenge," *Eos*, Vol. 75, No. 13, 1994, p. 145, 150.

Hodgson, R., "On a curious appearance seen in the Sun," *MNRAS*, Vol. 20, 1860.

Hoyt, D.V., Schatten, K.H., "How well was the Sun observed during the Maunder Minimum?" *Solar Physics*, Vol. 165, 1996.

Hoyt, D.V., Schatten, K.H., "Group sunspot numbers: A new solar activity reconstruction," *Solar Physics*, Vol. 181, 1998.

Hughes, M., Diaz, H., "Was there a medieval warm period, and if so, where and when?" *Climatic Change*, Vol. 26, 1994.

Hunt, G.E., Kandel, R., Mecherikunnel, A.T., "A history of presatellite investigations of the Earth's radiation budget," *Reviews of Geophysics*, Vol. 24, 1986.

Ingram, M.J. *et al.*, "Historical climatology," *Nature*, Vol. 276, 23 November, 1978.

Jirikowic, J.L., Damon, P.E., "The medieval solar activity maximum," *Climatic Change*, Vol. 26, 1994.

Jokipii, J.R., Kota, J., "The polar heliospheric magnetic field," *Geophysical Research Letters*, Vol. 16, 1989.

Jones, P.D., Bradley, R.S., "Climatic variations over the last 500 years," in *Climate since A. D. 1500*, eds. R. S. Bradley and P. D. Jones (Routledge, 1995), p. 649.

Keigwin, L. D., "The Little Ice Age and medieval warm period in the Sargasso sea," *Science*, Vol. 274, 1996.

Kelvin, W.T., "Astro-Physical Notes—The Sun's effect on terrestrial magnetism," *Astronomy and Astrophysics*, Vol. 12, 1893.

Kelvin, W.T., "Presidential address to the Royal Society," *Nature*, Vol. 47, December 1, 1892.

King-Hele, D.G., "A view of Earth and air," (Bakerian Lecture, 1974) *Philosophical Transactions of the Royal Society of London,* 278A, 1975.

King-Hele, D.G., "The Earth's atmosphere: Ideas old and new," (Milne Lecture, 1984) *Quarterly Journal of the Royal Society*, 26, 237–261, 1985.

Knobloch, E., Tobias, S. M., Weiss, N. O., "Modulation and symmetry changes in stellar dynamos," *MNRAS*, Vol.. 297, 1998.

Knowles, T.W., "Twilight on bald mountain," *The Sciences*, March/April, 1999.

Kollerstrom, N., "The hollow world of Edmond Halley," *JHA*, Vol. XXIII, 1992.

Kunzanyan, K.M., Sokoloff, D., "Half-width of a solar dynamo wave in Parker's migratory dynamo," *Solar Physics*, Vol. 173, 1997.

Lamb, H.H., "Climatic variation and changes in wind and ocean circulation: The Little Ice Age in the North Atlantic," *Quaternary Research*, II, 20, 1979.

Lamb, H.H., "An approach to the study of the development of climate and its impact in human affairs," in *Climate and History*, eds. T. M. L. Wigley, M.J. Ingram and G. Farmer (CUP, 1981), p. 291.

Lamb, H.H., "Volcanic dust in the atmosphere: With a chronology and assessment of its meteorological significance," *Philosophical Transactions of the Royal Society of London* (*Series A*), Vol. 266, July, 1970.

Lee, R., "The 'Greenhouse Effect'," *Journal of Applied Meteorology*, Vol. 12, 1973.

Lassen, K., Friis-Christiansen, E., "Variability of the solar cycle length during the last five centuries and the apparent association with terrestrial climate," *Journal of Atmospheric and Terrestrial Physics*, Vol. 59, 1995.

Lewis, J.M., "Clarifying the dynamics of the general circulation: Philip's 1956 experiment," *Bulletin of the American Meteorological Society*, Vol. 79, Vol. 1, January, 1998.

Libby, L. M. *et al.*, "Isotopic tree themometers," *Nature*, Vol. 261, May 27, 1976.

Lin, J., Soon, W., Baliunas, S., "Theories of solar eruptions: A review," *New Astronomy Reviews*, Vol. 47, Iss. 2, April, 2003.

Lindzen, R.S., "Can increasing carbon dioxide cause climate change?" *Proceedings of the National Academy of Science* (*USA*), Vol. 94, 1997.

Locke, J., "An Essay Concerning the True Original Extent and End of Civil Government," from *Two Treatises of Government* (London, 1764) reprinted in *The Clash of Political Ideals*, 3rd edn., A.R. Chandler (Appleton-Century-Crofts, 1957).

Lockyer, J.N., "Simultaneous solar and terrestrial changes," *Nature*, No. 1789, Vol. 69, 1904.
Mann, M.E. *et al.*, "Global interdecadal and century-scale climate oscillations during the past five centuries," *Nature*, Vol. 378, 1995.
Marriott, R.A., "Norway 1896: The BAA's first organised eclipse expedition," *JBAA*, Vol. 101, 1991.
Matthes, F.E., "Report of committee on glaciers," *Transactions of American Geophysical Union*, Vol. 20, 1939.
Matthes, F.E., "Committee on glaciers, 1939–1940," *Transactions of American Geophysical Union*, Vol. 21, 1940.
Mrs. E. Walter Maunder's Obituary, *The Observatory*, December, 1947.
Maunder, A.S.D., Review of A.S.D Maunder's book, "Astronomical Publications," *JBAA*, Vol. XII, No. 4, 1902.
Maunder, A.S.D., "Suggested connection between Sun-spot activity and the secular change in magnetic declination," *MNRAS*, Vol. LXIV, January, 1904.
Maunder, A.S.D., "An apparent influence of the Earth on the numbers and areas of Sun-spots in the cycle 1889–1901," *MNRAS*, Vol. LXVII, May, 1907.
Maunder, A.S.D., "The date of the Bundahis," *The Observatory*, Vol. 35, October, 1912.
Maunder, A.S.D., "The Zoroastrian star-champions," *The Observatory*, Vol. 36, March, 1913.
Maunder, A.S.D., "Arthur Matthew Weld Downing," *JBAA*, Vol. 28, December, 1917.
Maunder, A.S.D., "The date and place of writing of the Slavonic book of Enoch," *The Observatory*, Vol. 41, August, 1918.
Maunder, A.S.D., "Review on the astronomical observatories of Jai Singh by G. R. Kaye," *The Observatory*, Vol. 42, September, 1919.
Maunder, A.S.D., "The origin of the constellations," *The Observatory*, Vol. 59, December, 1936.
Maunder, A.S.D., Maunder, E.W., "The origin of the planetary symbols," *JBAA*, Vol. XXX, April, 1920.
Maunder, E.W., "Professor Spoerer's researches on Sun-spots," *MNRAS*, Vol. 50, 1890.
Maunder, E.W., "The Zodiac explained," *The Observatory*, Vol. 21, December, 1898.
Maunder, E.W., "The oldest astronomy. II.," *JBAA*, Vol. IX, 1899.
Maunder book review, The Royal Observatory, Greenwich: A Glance at its History and Work, "New Books and Memoirs," *Proceedings, BAA Journal*, Vol. XI, No. 1, 1900.
Maunder, E.W., "Spöerer's Law of Zones," *The Observatory*, Vol. 26, August, 1903.
Maunder, E.W., Book review in "Papers Communicated to the Association," "The Earth's Place in the Universe," *JBAA*, Vol. XIII, No. 6, 1903.
Maunder, E.W., "Note on an early astronomical observation recorded in the book of Joshua," *The Observatory*, Vol. 27, January, 1904.
Maunder, E.W., "The great magnetic storms, 1875 to 1903, and their association with sunspots as recorded at the Royal Observatory, Greenwich," *MNRAS*, Vol. 64, January, 1904.
Maunder, E.W., "Note on the distribution of Sun-spots in heliographic latitude, 1874 to 1902," *MNRAS*, Vol. 64, June, 1904.
Maunder, E.W., "Magnetic disturbances, 1882 to 1903, as recorded at the Royal Observatory, Greenwich, and their association with Sun-spots," *MNRAS*, Vol. 65, November 1904.
Maunder, E.W., "The solar origin of terrestrial magnetic disturbances," *Astrophysical Journal*, Vol. 21, 1905.
Maunder, E.W., "An Indian mode of indicating time," *The Observa*tory, Vol. 28, December, 1905.

Maunder, E.W., "The solar origin of terrestrial magnetic disturbances," *JBAA*, Vol. 16, 1906.

Maunder, E.W., in BAA's November 26, 1906 Meeting, *JBAA*, Vol. 17, 1907.

Maunder, E.W., "The dates of Genesis," *The Observatory*, Vol. 32, October, 1909.

Maunder, E.W., "Ancient Chronology," *The Observatory*, Vol. 32, November, 1909.

Maunder, E.W., "Notes on the cyclones of the Indian Ocean, 1856–1867 and their association with the solar rotation," *MNRAS*, Vol. 70, November, 1909.

Maunder, E.W., "A strange celestial visitor," *The Observatory*, Vol. 39 (a special note for the publication of No. 500), May, 1916.

Maunder, E.W., Letter to A.E. Douglass, February 18, 1922, Box 64, *Douglass Papers*, UAL (Douglass' reply is probably March 23, 1922).

Maunder, E.W., "The prolonged Sun-spot minimum," *JBAA*, Vol. 32, January 1922.

Maunder, E.W., Meeting of the BAA, April 26, 1922 (where Douglass' comments and observations were read aloud) "The prolonged Sun-spot minimum," *JBAA*, Vol. 32, April 1922.

Maunder, E.W., "The Sun and sunspots, 1820–1920," Lecture of May 29, 1922, on the Centenary of the Royal Astronomical Society, *MNRAS*, Vol. 82, 1922.

Maunder, E.W., Maunder A.S.D., "Some experiments on the limits of vision for lines and spots as applicable to the question of the actuality of the canals of Mars," *JBAA*, Vol. XIII, 1903.

Maunder, E.W., Maunder A.S.D., "The solar rotation period from Greenwich Sun-spot measures, 1879–1901," *MNRAS*, Vol. LXV, June, 1905.

Maunder, E.W., Maunder A.S.D., "The rotation period of the Sun as derived from magnetic storms," *MNRAS*, Vol. LXXXIV, 1924.

Mayewski, P. *et al.*, "Major features and forcing of high-latitude northern hemisphere atmospheric circulation using a 110,000-year-long glaciochemical series," *Journal of Geophysical Research*, Vol. 102, 1997.

McCain, D.R., "Life in early America: The worse winters" (Internet reference).

McHargue, L.R., Damon, P.E., "The global Beryllium 10 cycle," *Reviews of Geophysics*, Vol. 29, 2, May, 1991.

Meadows, A.J., Kennedy, J.E., "The origin of solar terrestrial studies," *Vistas in As*tronomy. Vol. 25, 1982.

Morton, N., New England's Memorial (1699 edn. Courtesy of M.S. Humphreys, special collections librarian, Redwood Library and Athenaeum, Newport, RI).

Mourt's Relation or Journal of the Plantation at Plymouth (Kimball Wiggin, 1865).

National Weather Service, Newport, N.C., "Eighteenth century hurricanes impacting North Carolina," (Internet).

Obituary of A.S.D. Maunder, *JBAA*, No. 841, 1947.

Obituary of E. W. Maunder, *JBAA*, Vol. 38, No. 6, Session 1927–28.

Obituary of E.W. Maunder, *Nature*, No. 2049, Vol. 121, April 7, 1928.

Obituary of Thomas Frid Maunder, *JBAA*, Vol. 45, Session for October, 1935.

Odenwald, S., "Solar storms: The silent menace" (private copyright, 1998).

Ogilvie, M. B., "Obligatory amateurs: Annie Maunder (1868–1947) and British women astronomers at the dawn of professional astronomy," *British Journal for the History of Science*, Vol. 33, 67–84, 2000.

Oguti, T., "The auroral zone in historic times—The Northern UK was in the auroral zone 300 years ago," *Journal of Geomagnetism and Geoelectricity*, Vol. 45, 1993.

Oppo, D.W., McManus, J.F., Cullen, J.L., "Abrupt climate events 500,000 to 340,000 years ago: Evidence from subpolar North Atlantic sediments," *Science*, Vol. 279, February 27, 1998.

Oreskes, N. *et al.*, "Verification, validation, and confirmation of numerical models in the earth sciences," *Science*, Vol. 263, 1994.

Parker, B., "The Coriolis effect: Motion on a rotating planet," *Mariners Weather Log*, Vol. 42 (No. 2), August 1998.

Parker, E.N., "Hydrodynamic dynamo models," *Astrophysical Journal*, Vol. 122, 1955.

Parker, E.N., "Dynamics of the interplanetary gas and magnetic fields," *Astrophysical Journal*, Vol. 128, 1958.

Parker, E.N., "A history of early work on the heliospheric magnetic field," *Journal of Geophysical Research*, Vol. 106, 2001.

Peratt, A.L., in *Galactic and Intergalactic Magnetic Fields*, eds. R. Beck *et al.* (Kluwer, Dordrecht, 1990), p. 149.

Persson, A., "How do we understand the Coriolis force?" *Bulletin of the American Meteorological Society*, Vol. 79, No. 7, July, 1998.

Pfister, C., "The years without a summer in Switzerland: 1628 and 1816" from *Klimageschichte der Schweiz*, 1525–1860 (band I: 140–141) Verlag Paul Haupt, Bern.

Proceedings of the BAA Hundred and Ninth Annual General Meeting, February 1929 (*BAA Journal*) "Report to the Council."

Quinn, W. H., Neal, V.T., Antunez de Mayolo, S. E., "El Nino occurrences over the past four and a half centuries," *Journal of Geophysical Research*, Vol. 92, 1987.

Reiter P., "From Shakespeare to Defoe: Malaria in England in the Little Ice Age," *Emerging Infectious Diseases*, Vol. 6 (No. 1), January-February, 2000.

Report of the Ordinary General Meeting, *BAA Journal*, October, 1942.

Ribes, J.C., Nesme-Ribes, E., "The solar sunspot cycle in the Maunder minimum A.D. 1645 to A.D. 1715," *Astronomy and Astrophysics*, Vol. 276, 1993.

Rizzo, P.V., Schove, D.J., "Early New World aurorae, 1644, 1700, 1719," *JBAA*, Vol. 72, 1962.

Roxburgh, W., "Suggestions on the introduction of such useful trees, shrubs and other plants as are deemed the most likely to yield sustenance to the poorer classes of natives of these provinces during times of scarcity," (Report to the President's Council, Tamil Nadu State, Vol. CLXXXL, 8 February, 1793, in Nature and The Orient (OUP, 1998), pp. 301–323 (R.H. Grove's quote).

Sabine, E., "On periodical laws discoverable in the mean effects of the larger magnetic distubances—No. II," *Philosophical Transaction of the Royal Society of London*, Vol. 142, May 6, 1852.

Schröder, W., "Aurorae during the Maunder Minimum," *Meteorology and Atmospheric Physics*, Vol. 38, 1988.

Schröder, W., "On the existence of the 11-year cycle in solar and auroral activity before and during the so-called Maunder minimum," *Journal of Geomagnetism and Electricity*, Vol. 44, 1992.

Scheiner, C., *Tres Epistolae de Maculis Solaribus Scriptae ad Marcum Welserum* (January 5, 1612).

Schove, D.J., "The sunspot cycle, 649 B.C. to A.D. 2000," *Journal of Geophysical Research*, Vol. 60, 127–146, 1955.

Schove, D.J., Reynolds, D., "Weather in Scotland, 1659–1660: The diary of Andrew Hay," *Annals of Science*, Vol. 30, No. 2, June, 1973.

Schuster, A., "Sun-spot and magnetic storms," *MNRAS*, Vol. 65, 1905.

Shanaka L.D., Zelinski, G.A., "Global influence of the A.D. 1600 eruption of Huaynaputina, Peru," *Nature*, Vol. 393, 4 June, 1998.
Sharma, V.N., "Astronomical efforts of Sawai Jai Singh—A review" in *History of Oriental Astronomy*, eds. G. Swarup *et al.* (CUP, 1985), pp. 233–240.
Shutts, G.J., Green, J.S.A., "Mechanisms and models of climate change," *Nature*, Vol. 276, 1978.
Silverman, S.M., "Joseph Henry and John Henry Lefroy: A common 19th century vision of auroral research," *Eos*, Vol. 70, No. 15, April 11, 1989.
Silverman, S.M., "Secular variations of the aurora," *Reviews of Geophysics*, Vol. 30, 4, 1992.
Silverman, S., "19th century auroral observations reveal solar activity patterns," *Eos*, Vol. 78, No. 14, April 8, 1997.
Siscoe, G.L., "Evidence in the auroral record for secular solar variability," *Reviews of Geophysics and Space Physics*, Vol. 18, No. 3, August, 1980.
Sluiter, E., "The telescope before Galileo," *JHA*, Vol. XXVIII, 1997.
Smith, E. J., "The Sun and interplanetary magnetic field," in *The Sun in Time*, eds. C. Sonett *et al.* (University of Arizona, 1991), pp. 175–201.
Sokoloff, D.D., Nesme-Ribes, E, "The Maunder Minimum: A mixed-parity dynamo mode?" *Astronomy and Astrophysics*, Vol. 288, 1994.
Sonett, P., Finney, S.A., "The Spectrum of Radiocarbon," *Philosophical Transactions of the Royal Society*, Vol. A330, 1990.
Soon, W., Baliunas, S., "Proxy climatic and environmental changes of the past 1000 years," *Climate Research*, Vol. 23, January, 2003.
Soon, W., Baliunas, S., Idso, C., Idso, S., Legates, D.R. "Reconstructing climatic and environmental changes of the past 1000 years," *Energy & Environment*, Vol. 14, April, 2003.
Soon, W., Baliunas, S.L., Idso, S.B., Kondratyev, K.Ya., Posmentier, E.S., "Modeling climatic effects of anthropogenic carbon dioxide emissions: Unknowns and uncertainties," *Climate Research*, Vol. 18, November 2, 2001.
Soon, W., Baliunas, S.L., Robinson, A.B., Robinson, Z.W., "Environmental effects of increased atmospheric carbon dioxide," *Climate Research*, Vol. 13, October 26, 1999.
Stephenson, F.R., Fatoohi, L.J., "Accuracy of solar eclipse observations made by jesuit astronomers in China," *JHA*, Vol. XXVI, 1995.
Stephenson, F.R., Willis, D.M., "The earliest drawing of sunspots," *Astronomy & Geophysics*, Vol. 40, 6.21, December, 1999..
Stern, D.P., "A brief history of magnetospheric physics before the spaceflight era," *Reviews of Geophysics*, Vol. 27, February 1989.
Stern, D.P., "A brief history of magnetospheric physics during the space age," *Reviews of Geophysics*, Vol. 34, February 1996.
Stewart, B., "On the great magnetic disturbance which extended from August 28 to September 7, 1859, as recorded by photograph at the KEW Observatory," *Philosophical Transactions of the Royal Society*, Vol. 151, 1862.
Stuiver, M., "Carbon-14 content of 18th and 19th century wood: Variations correlated with sunspot activity," *Science*, Vol. 145, 30 July, 1965.
Stuiver, M., Braziunas, T.F., "Sun, ocean, climate and atmospheric $^{14}CO_2$: An evaluation of causal and spectral relationships," *The Holocene*, Vol. 3, 1993.

Stuiver, M., Quay, P.D., "Changes in atmospheric carbon-14 attributed to a variable Sun," *Science*, Vol. 207, No. 4426, 4 January, 1980.

Stothers, R.B., "The Great Tambora eruption and its aftermath," *Science*, Vol. 224, 15 June, 1984.

Suess, H.E., "Secular variations of the cosmic-ray-produced carbon-14 in the atmosphere and their interpretations," *Journal for Geophysical Research*, Vol. 70, No. 23, December 1, 1965.

Suess, H.E., Linick, T.W., "The ^{14}C record in bristlecone pine wood of the past 8000 years based on dendrochronology of the late C.W. Ferguson," *Philosophical Transactions of the Royal Society of London*, A300, 402–412, 1990.

Steig, E.J. *et al.*, "Synchronous climate changes in Antarctica and the North Atlantic," *Science*, Vol. 282, October 2, 1998.

Tinsley, B.A., "Do effects of global atmospheric electricity on clouds cause climate changes?" *Eos*, No. 33, 1997.

Tobias, S.M., "The solar cycle: Parity interactions and amplitude modulation," *Astronomy and Astrophysics*, Vol. 322, 1997.

Udias, A., "Jesuit astronomers in Beijing 1601–1805," *Quarterly Journal of the Royal Astronomical Society*, Vol. 35, 1994.

Udias, A., Barreto, L.M., in *Exploring the Earth: Progress in Geophysics Since the 17th Century*, eds. W. Schröder, M. Colacino, G. Gregori (Interdivisdional commission on history of IAGA, 1992), pp. 128, 139.

Van Helden, A., "The development of compound eyeieces," *JHA*, Vol. VIII, 1977.

Wilcox, J.M., "Solar activity and the weather," *Journal of Atmospheric and Terrestrial Physics*, Vol. 37, 1975.

Willis, D.M., Stephenson, F.R., "Simultaneous auroral observations described in the historical records of China, Japan and Korea from ancient times to AD 1700," *Annales Geophysicae*, Vol. 18, 2000, pp. 1–10.

Witmann, A.D., Xu, Z.T., "A catalogue of sunspot observations from 165 BC to AD 1984," *Astronomy and Astrophysics Supplement Series*, Vol. 70, 1987.

Weiss, J.E. and Weiss, N.O., "Andrew Marvell and the Maunder Minimum," *Quarterly Journal of the Royal Astronomical Society*, Vol. 20, 1979 (Marvell quote).

Wolf, R., "Abstract of his results. By Prof. Wolf. (Translation communicated by Mr. Carrington)," *MNRAS*, Vol. 21, 1861.

Zhang, Z., "Korean auroral records of the period AD 1507–1747 and the SAR Arcs," *JBAA*, Vol. 95, No. 5, 1985.

Zheng, S., Feng, L., "Historical evidence on climate instability above normal in cool periods in China," *Scientia Sinica* (Series B), Vol. 24, 1986.

Books

Akasofu, S-I., *Exploring the Secrets of the Aurora* (Kluwer, 2002).
Armitage, A., *William Herschel* (Doubleday, 1963).
Aveni, A., *Conversing with the Planets* (Times Books, 1992).
Bacon, F., "The Great Instauration," *The English Philosophers from Bacon to Mill* (Modern Library, 1939).
Ball, R., *The Story of the Sun* (D. Appleton & Co., 1893).
Ballie, M.G.L., *A Slice Through Time* (Batsford, 1995).
Bedini, S.A., *Thinkers and Tinkers* (Scribners, 1975).
Bentley, E., *Theatre of War* (Viking, 1970).
Biagioli, M., Van Helden, A., *Galileo, Scheiner, and the Sunspot Controversy: Scientific Practice in the Patronage Context* (Rice Galileo Project, Internet).
Birkeland, K.R., "On The Cause of Magnetic Storms and the Origin of Terrestrial Magnetism," in The *Norwegian Aurora Polaris Expedition*, 1902–03, Vol. 1 (Longmans, Green & Co., 1908).
Bone, N., *The Aurora: Sun-Earth Interactions* (2nd edn.) (Wiley & Sons, 1996).
Bradford, W., *Of Plymouth Plantation 1620–1647* (Knopf, 1952).
Brekke, A., Egeland, A., *The* Northern Light (Springer Verlag, 1983).
Bronowski, J., *The Ascent of Man* (Futura, 1981).
Burke, J., (ed.) *History of England* (Guild, 1988).
Caspar, M., *Kepler* (Dover Reprint, 1993).
Chapman, A., *The Victorian Amateur Astronomer* (John Wiley, 1998).
Chapman, S., Bartels, J., *Geomagnetism* (OUP, 1940).
Clerke, A.M., *The Herschels and Modern Astronomy* (MacMillan, 1895).
Cook, A., *Edmund Halley* (OUP, 1998).
Crowe, M.J., *Modern Theories of the Universe: From Herschel to Hubble* (Dover, 1994).
Dafoe, D., *A Journal of the Plague Year* (Oenguin, 1986).
Delevoy, R.L., *Brueghel* (Rizzoli, 1990).
D'Elia, P.M., *Galileo in China* (HUP, 1960).
de Mairan, J.-J.D., *Traité Physique et Historique de l'Aurore Boreale* (L'Imprimerie Royal, 1773).
Dragàn, J.C., Airinei, S., *Geoclimate and History* (NAGARD, 1989).
Drake, S., *Discoveries and Opinions of Galileo* (Doubleday-Archer, 1957).
Drake, S., *Galileo at Work: His Scientific Biography* (Dover, 1995).
Domrös, M., Peng, G., *The Climate of China* (Springer-Verlag, 1988).
Espinasse, M., *Robert Hooke* (W. Heinemann, 1956).
Fabricius, J., *De Maculis in Sole Observatis, et Apparente earum cum Sole Conversaione, Narratio* (June 13, 1611).
Fleming, J.R., *Historical Perspectives on Climate Change* (OUP, 1998).
Forbes, E.G., *Greenwich Observatory*, Vol. 1. (Taylor and Francis, 1975).
Forbes, E.G., *The Correspondence of John Flamsteed, the First Astronomer Royal*, Vol. I, 1666–1682; Vol. II 1682–1703 (Institute of Physics Publishing, 1995, 1997).
Frunzettit, I., *Classical Chinese Painting* (Meridiane, 1976).
Gaukroger, S., *Descartes: an intellectual biography* (OUP, 1995).
Goody, R.M., *Principles of Atmospheric Physics and Chemistry* (OUP, 1995).

Grove, J., *The Little Ice Age* (Routledge, 1988).
Grove, R.H., *Green Imperialism: Colonial expansion, tropical island edens and the origins of environmentalism, 1600–1800* (CUP, 1995).
Grove, R.H., *et al.*, *Nature and the Orient: The Environmental History of South and Southeast Asia* (OUP, 1998).
Goldstein, T., *Dawn of Modern Science* (American heritage Library) (Houghton Mifflin, 1988).
Haack, S., *Manifesto of a Passionate Moderate* (University of Chicago, 1998).
Hall, A.R., *Isaac Newton: Adventurer in thought* (CUP, 1992).
Harrison, J.B., Sullivan, R.E., *A Short History of Western Civilization* (Knopf, 1975).
Harrison, E., *Darkness at Night* (HUP, 1987).
Hawthorne, N., *Twice Told Tales* (Walter Black, 1837).
Heath, T., *Greek Astronomy* (Dent, 1932).
Houghton, W.E., *The Victorian Frame of Mind* (Yale University Press, 1957).
Hoyt, D.V., Schatten, K.H., *The Role of the Sun in Climate Change* (OUP, 1997).
Hufbauer, K., *Exploring the Sun* (John Hopkins University Press, 1991).
Huff, T., *The Rise of Early Modern Science: Islam, China and the West* (CUP, 1995).
Imbrie, J., Imbrie, K. P., *Ice Ages—Solving The Mystery* (Enslow Publishers, 1979).
Jago, L., *The Northern Lights* (Knopf, 2001).
James, J., *The Music of the Spheres* (Copernicus, 1993).
Jaynes, J., *The Origin of Consciousness in the Breakdown of the Bicameral Mind* (PUP, 1990).
Johnson, P., *A History of the American People* (Weidenfeld & Nicolson, 1997).
Jones, B.Z., Boyd, L.G., *The Harvard College Observatory, 1839–1919* (HUP, 1971).
Keegan, J., *A History of Warfare* (Pimlico, 1994).
Kippenhahn, R., *Discovering the Secrets of the Sun* (John Wiley, 1994) [translated by S. Dunlop].
Krupp, E.C., *Echoes of the Ancient Skies* (OUP, 1983).
Kuhn, T., *The Structure of Scientific Revolutions*, 3rd edn. (University of Chicago, 1996).
Lamb, H.H., *Climate—Present, Past and Future, Volume 2: Climatic History and the Future* (Methuen, 1977).
Lamb, H.H., *Climate, History and the Modern World* (Methuen, 1982).
Larousse *Dictionary of Painters* (Mallard, 1981).
Littmann, M., Willcox, K., *Totality—Eclipses of the Sun* (University of Hawaii, 1991).
Lorenz, E.N., *The Essence of Chaos* (University of Washington, 1993) [quote of Napier Shaw].
Lubbock, C.A., *The Herschel Chronical* (CUP, 1933).
Ludlum, D., *Early American Winters 1604–1820* (American Meteorological Society, 1966).
Ludlum, D., *New England Weather Book* (Houghton Mifflin, 1976).
Marks, R.B., *Tigers, Rice, Silk and Silt* (CUP, 1998).
Maunder, A.S.D., Maunder, E.W., *The Heavens and Their Story* (Dana Estes and Company, 1908–1909).
Maunder, E.W. (ed.), *The Indian Eclipse 1898* (B.A.A., Hazell, Watson and Viney, LD, 1899).
Maunder, E.W. (ed.), *The Total Solar Eclipse of May 1900* (B.A.A., "Knowledge" Office-Witherby and Co., 1901).
Maunder, E.W. , *Astronomy Without A Telescope* ("Knowledge" Office, 1902).
Meadows, A.J., *Greenwich Observatory*, Vol. 2 (Taylor and Francis, 1975).
McCluskey, S., *Astronomies and Cultures in Early Medieval Europe* (CUP, 1998).

Morison, S. E. (ed.), *The Francis Parkman Reader* (Da Capo Press, 1998).

Moss, J.D., *Novelties in the Heavens:Rhetoric and Science in the Copernican Controversy* (University of Chicago, 1993).

Murdin, L., *Under Newton's Shadow* (Adam Hilger, 1985).

Needham, J., Ling, W., *Science and Civilisation in China* (CUP, 1959).

Nesme-Ribes, E., *The Solar Engine and Its Influence on Terrestrial Atmosphere and Climate* (Springer-Verlag, 1994).

Noyes, R., *The Sun, Our Star* (HUP, 1982).

Panek, R., *Seeing and Believing: The Story of the Telescope, or How We Found Our Place in the Universe* (Fourth Estate, 2000).

Pannekoek, A., *A History of Astronomy* (Dover Reprint, 1989).

Parker, F., *George Peabody: A Biography* (Vanderbilt University Press, 1995).

Rawski, T.G., Li, L.M. (eds.) *Chinese History in Economic Perspective* (University of California, 1991).

Reeves, E., *Painting the Heavens* (PUP, 1997).

Reichenbach, H., *From Copernicus to Einstein* (Dover Reprint, 1970).

Rhodes, R., *The Making of the Atomic Bomb* (Touchstone, 1988).

Riccioli, G.P., *Almagestum Novum Astronomian Veterem Novamque Complectens, Observationibus Aliorum Et Propis, Novisque Theorematibus Problematibus As Tabulis Promotam* (Bologna, Italy, 1651, 1653, Vols. I and II).

Ritchie, W. A., *The Archaeology of New York State*, Revised edn. (Purple Mountain, 1994).

Rosen, E., *Johannes Kepler's Somnium—The Dream or Posthumous Work on Lunar Astronomy* (University of Wisconsin, 1967).

Rudiger, G., *Differential Rotation and Stellar Convection: Sun and Solar-type Stars* (Academic Verlag, 1989).

Salvini, R., *Michelangelo* (Milano, 1976).

Schopenhauer, A., *Works*, 1888, Vol V.

Schove, D.J., *Sunspot Cycles* (Hutchinson Ross, 1983).

Sharratt, M., *Galileo: Decisive Innovator* (CUP, 1994).

Shirley, J. W., *Thomas Harriot: A Biography* (OUP, 1983).

Sime, J., *William Herschel and His Work* (Scribners, 1900).

Simeons, A.T.W., *Man's Presumptuous Brain: An Evolutionary Interpretation of Psychosomatic Disease* (Dutton, 1960).

Simkin, T. *et al.*, *Volcanoes of the World* (Hutchinson Ross, 1981).

Snow, D.R., *The Archaeology of New England* (Harcourt, Brace, Jovanovitch, 1980).

Sobel, D., *Galileo's Daughter: A historical Novel of Science, Faith, and Love* (Walker and Co., 1999).

Soon, W., Baliunas, S., Robinson, A. Robinson, Z., *Global Warming: A Guide to the Science* (Fraser Institute, 2001).

Swift, J., *Gulliver's Travels* (Riverside, 1960).

Taylor, R.E., Long, A., Kra, R.S. (eds.) *Radiocarbon After Four Decades* (Springer-Verlag, 1992).

The Cambridge Encyclopedia of Earth Sciences, ed. Smith, D.G. (CUP, 1981).

The Concise Cambridge History of English Literature (CUP, 1947).

Theophrastes, *De Signis Temperstatum* (Enquiry into Plants and Minor Works on Odours and Weather Signs) Trans. Hort, A. (Heinemann:MCMXVI).
Tsurutani, B.T., Gonzales, W.D., *et al.*, *Magnetic Storms* (AGU, 1997).
Verdet, J.-P., *The Sky: Order and Chaos* (Gallimard, 1987).
Walford, E., *Frost Fairs on the Thames* (London, 1887).
Webb, G.E., *Tree Rings and Telescopes: The Scientific Career of A.E. Douglass* (University of Arizona, 1983).

Poems, Plays, Stories

Dickinson, E., Poem 244 (591) (circa 1862) *Final Harvest: Poems Selected and with an Introduction*, ed. Johnson, T.H. (Little, Brown, 1964).
Donne, J., "Ignatius His Conclave" from "The Pseudo-Martyr" *Complete Poetry and Selected Prose of John Donne* (Modern Library, 1952).
Hawthorne, N., "Old News," *Twice Told Tales* (Volume IV) Walter Black.
Marvell, A., "The Last Instructions To A Painter" (1667) (See Weiss and Weiss).
Pope, A, *The Poems, Epistles and Satires of Alexander Pope* (Dutton, 1931).

Recommended Reading on Earth's Climate, Physics of the Sun and Solar-Terrestrial Relationships

Akasofu, S.-I., *Exploring the Secrets of the Aurora* (Kluwer, 2002).
Chapman, S., Bartels, J., *Geomagnetism* (OUP, 1940).
Grove, J., *The Little Ice Age* (Routledge, 1988).
Hoyt, D.V., Schatten, K.H., *The Role of the Sun in Climate Change* (OUP, 1997).
Kippenhahn, R., *Discovering the Secrets of the Sun* (John Wiley, 1994) [translated by S. Dunlop].
Lamb, H.H., *Climate, History and the Modern World* (Methuen, 1982).

Legend

CUP = Cambridge University Press
HUP = Harvard University Press
JBAA = Journal of The British Astronomical Association
JHA = Journal of Astronomical History and Heritage
MNRAS = Monthly Notices of the Royal Astronomical Society
OUP = Oxford University Press
PUP = Princeton University Press
S&T = Sky and Telescope

Index